NEUROSCIENCE

LABFAX

The LABFAX series

Series Editors:

B.D. HAMES Department of Biochemistry and Molecular Biology, University of Leeds, Leeds LS2 9JT, UK

D. RICKWOOD Department of Biology, University of Essex, Wivenhoe Park, Colchester CO4 3SQ, UK

MOLECULAR BIOLOGY LABFAX
CELL BIOLOGY LABFAX
CELL CULTURE LABFAX
BIOCHEMISTRY LABFAX
VIROLOGY LABFAX
PLANT MOLECULAR BIOLOGY LABFAX
IMMUNOCHEMISTRY LABFAX
CELLULAR IMMUNOLOGY LABFAX
ENZYMOLOGY LABFAX
PROTEINS LABFAX

Forthcoming titles

MOLECULAR BIOLOGY: GENE CLONING LABFAX
MOLECULAR BIOLOGY: GENE ANALYSIS LABFAX

NEUROSCIENCE

LABFAX

EDITED BY

M.A. LYNCH

Department of Physiology
University of Dublin
Trinity College
Dublin 2
Ireland

and

S.M. O'MARA

Department of Psychology
University of Dublin
Trinity College
Dublin 2
Ireland

ACADEMIC PRESS

SAN DIEGO LONDON BOSTON
NEW YORK SYDNEY TOKOYO TORONTO

This book is printed on acid-free paper.

Academic Press
525 B Street, Suite 1900, San Diego, California 92101-4495, USA
http://www.apnet.com

Academic Press Limited
24–28 Oval Road, London NW1 7DX, UK
http://www.hbuk.co.uk/ap/

ISBN 0-12-460490-0

Library of Congress Cataloging-in-Publication Data

Neuroscience labfax / edited by Marina A. Lynch, Shane M. O'Mara.
p. cm. — (Labfax series)
Includes index.
ISBN 0–12–460490–0
1. Neurobiology—Methodology. I. Lynch, Marina A. II. O'Mara, Shane M. III. Series.
QP357.N475 1997
612.8′028–dc21 96–47962
CIP

A catalogue record for this book is available from the British Library

Typeset by Phoenix Photosetting, Chatham
Printed in Great Britain by The University Press, Cambridge.

97 98 99 00 01 02 EB 9 8 7 6 5 4 3 2 1

Front cover: Lateral view of human brain, showing the gross anatomy of the cerebral cortex, cerebellum and spinal cord. Visible aspects of the cerebral cortex include the frontal, parietal, temporal and occipital lobes, as well as prominent sulci and gyri. From a mid-19th Century French print of unknown Provenance.

CONTENTS

CONTRIBUTORS

R. BADDELY
Department of Experimental Psychology, University of Oxford, South Parks Road, Oxford OX1 3UD, UK

Z.I. BASHIR
Department of Anatomy, School of Medical Sciences, University Walk, Bristol BS8 1TD, UK

M.P. CLEMENTS
Brain and Behaviour Research Group, Department of Biology, Open University, Milton Keynes MK7 6AA, UK

M.A. DAVIES
Brain and Behaviour Research Group, Department of Biology, The Open University, Milton Keynes, MK7 6AA, UK

S.B. DUNNETT
MRC Cambridge Centre for Brain Repair and Department of Experimental Psychology, University of Cambridge, Forvie Site, Robinson Way, Cambridge CB2 2PY, UK

M.S. FAZELI
Department of Experimental Pathology, UMDS Guys Hospital, London Bridge, London SE1 9TR, UK

D.M. FLAVELL
Division of Cardiovascular Genetics, Department of Medicine, University College London Medical School, London WC1E 6JJ, UK.

E.S. HAWLEY
Department of Physiology, SUNY Health Science Center Brooklyn, 450 Clarkson Avenue, Brooklyn, NY 11203, USA

L.W. HAYNES
School of Biological Sciences, University of Bristol, Bristol, BS8 1UG, UK

C. HÖLSCHER
Department of Pharmacology and Therapeutics, University of Dublin, Trinity College, Dublin 2, Ireland

I.C. KILPATRICK[a]
Department of Pharmacology, School of Medical Sciences, University Walk, Bristol BS8 1TD, UK

---

[a]Present address: Department of CNS Biology, Knoll Pharmaceuticals, Pennyfoot Street, Nottingham, NG1 1GF, UK.

J.L. KUBIE

Department of Anatomy and Cell Biology, SUNY Health Science Center Brooklyn, 450 Clarkson Avenue, Brooklyn, NY 11203, USA

M.A. LYNCH

Department of Physiology, University of Dublin, Trinity College, Dublin 2, Ireland

G.J. McBEAN

Department of Biochemistry, University College, Belfield, Dublin, Ireland

J. MILLAR

Department of Physiology, Queen Mary and Westfield College, Mile End Road, London E1 4NS, UK

R.U. MULLER

Department of Physiology, SUNY Health Sciencсs Center Brooklyn, 450 Clarkson Avenue, Brooklyn NY 11203, USA

K.M. O'BOYLE

Department of Pharmacology, University College Belfield, Dublin 4, Ireland

S.M. O'MARA

Department of Psychology, University of Dublin, Trinity College, Dublin 2, Ireland

G.D. PRATT

Neurosciences Unit, Gallaxo Wellcome Medicines Research Centre, Gunnels Wood Road, Stevenage SG1 2NY, UK

C. REW

Bio-Rad Laboratories, Bio-Rad House, Maylands Avenue, Hemel Hempstead HP2 7TD, UK

J.A.P. ROSTAS and **A.T.R. SIM**

The Neuroscience Group, Faculty of Medicine and Health Sciences, University of Newcastle, Callaghan, NSW 2308, Australia

M.G. STEWART

Brain and Behaviour Research Group, Department of Biology, The Open University, Milton Keynes MK7 6AA, UK

M. VIGNES

Department of Anatomy, School of Medical Sciences, University Walk, Bristol BS8 1TD, UK

V. WALSH

Department of Experimental Psychology, University of Oxford, South Parks Road, Oxford OX1 3UD, UK

F.G. WOUTERLOOD

Department of Anatomy, Vrije Universiteit, Van der Boechorststraat 7, 1081 BT Amsterdam, The Netherlands

ABBREVIATIONS

ABC	avidin–biotin–peroxidase
ABS	Artola Brocher and Singer
Ach	acetylcholine
AChE	acetylcholineesterase
ACPD	1*S*,3*R*–aminocyclopentane dicarboxylic acid
aCSF	artificial cerebrospinal fluid
AD	Alzheimers disease
ADR	adrenaline
cAMP	cyclic adenosine monophosphate
AMPA	9-amino-3-hydroxy-5-methylisoxazole-4-proprionate
APP	amyloid precursor protein
AQC	6-aminoquinolyl-*N*-hydroxysuccinimidyl carbamate
Asp	aspartate
AT	angiotensin
ATP	adenosine triphosphate
ATP	γ Adenosine trithiophosphate
BCIP	β-chloroindolyl phosphate
BCM	Bienenstock Cooper and Munro
BDA	biotinylated dextram amine
BDNF	brain-derived neurotrophic factor
BPB	bromophenol blue
BSA	bovine serum albumin
BSS	balanced salt solution
CaMPK	Calcium–calmodulin–dependent protein kinase
hCaMKII	human calcium–calmodulin kinase II
CAT	chloramphenicol acetyltransferase
CCK	cholecyctokinin
CE	capillary electrophoresis
CNQX	6-cyano-7-nitroquinoxaline-2,3-dione
CNS	central nervous system
CS	conditioned stimulus
CTB	cholera toxin subunit
CZE	capillary zone electrophoresis
DA	dopamine
DAB	diaminobenzidine
DABSYL	4-(dimethylamino) azobenzene-4′-sulfonyl chloride
DAG	diacylglycerol
DANSYL	5′-dimethylaminonaphthalene-1-sulfonyl chloride
DBH	dopamine β-hydroxylase
DEPC	diethylpyrocarbonate
DMEM	Dulbecco's modified essential medium
DMSO	dimethyl sulphoxide
DOPA	dihydroxyphenylalanine
DT	diptheria toxin A chain
DTT	dithiothreitol
EC	electrochemical detection
EEG	electroencephalogram
EDTA	ethylenediaminetetraacetic acid
EGTA	ethylene glycol-bis((-amino-ethyl ether)*N*,*N*,*N*,*N*-tetraacetic acid
EHS	Emgelbreth-Holm Swarm
ELISA	enzyme-linked immunosorbent assay
EM	electron microscopy
EPSC	excitatory postsynaptic current
EPSP	excitatory postsynaptic potential
FETs	field-effect transistors
FGF	fibroblast growth factor
FMOC	9-fluorenylmethyl-chloroformate
fMRI	functional magnetic resonance imaging
GABA	γ-aminobutyric acid
GAP-43	growth-associated protein-43

GDNF	glial-cell-line-derived neurotrophic factor
Glu	glutamate
cGMP	cyclic guanosine monophosphate
GnRH	gonadotrophin hormone releasing hormone
hpf	hypogonadal
HPLC	high performance liquid chromatography
HRP	horse radish peroxidase
HSV	herpes simplex virus
5-HT	5-hydroxytryptamine
ICC	immunocytochemistry
IL-1R	interleukin-1 receptor
IP3	inositol-4,5-trisphosphate
IPSC	inhibitory postsynaptic current
IPSP	inhibitory postsynaptic potential
LCR	locus control region
LDCV	large dense core vesicles
LED	light emitting diode
LTP	long-term potentiation
mAb	monoclonal antibody
MARCKS	myristoylated alanine-rich C-kinase substrate
MCS	multiple cloning site
MEM	Eagles modified essential medium
mGluR	metabotropic glutamate receptor
MMTV	mouse mammary tumor virus
NDA	2,3-naphthalenedialdehyde
NDA-CN	2,3-naphthalenedialdehyde/ sodium cyanide
NGF	nerve growth factor
NK1, NK2	neurokinin , neurokinin 2
NMDA	*N*-methyl-*D*-aspartate
NPY	Neuropeptide *Y*
NT4/5	neurotrophin 4/5
OPA	*o*–phthaldialdehyde
OT	oxytocin
PACAP	pituitary adenylyl cyclase – activating polypeptide
PCD	precolumn derivatization
PDGF	platelet-derived growth factor
PET	Positron Emission Tomography
PFA	paraformaldehyde
PKA	protein kinase A
PKC	protein kinase C
PKI	protein kinase A inhibitor
PLP	periodat/lysine/ paraformaldehyde
PNS	peripheral nervous system
PP1, PP2A, PP2B	protein phosphatase 1, 2A, 2B
PPF	paired pulse facilitation
PR	progesterone receptor
QX 314	*N*-(2,6-dimethyl-phenylcarbamoylmethyl)-triethylammonium bromide
RIA	radioimmunoassay
RT	room temperature
SCA1	spinal cerebellar ataxia type 1 gene
SCCP	Schaffer collateral commissural pathway
SDS	sodium dodecyl sulfate
SDS-PAGE	sodium dodecyl sulfate polyacrylamide gel electrophoresis
SSV	small synaptic vesicles
SubP	substance P
Tag SV40	large T antigen
TBS	tris buffered saline
TCA	trichloroacetic acid
TESPA	3-(triethoxysilyl-) propylamine
TMB	trimethylbenzidine
TMS	transcranial magnetic stimulation
UCS	unconditioned stimulus
UPW	ultrapure water
VIP	vasoactive intestinal peptide
YAC	yeast artificial chromosome

PREFACE

Contemporary neuroscience has become a hugely interdisciplinary affair over the past twenty years; it is no longer possible for the psychologist to conduct behavioral analyses without knowing something of what the neurochemist does and likewise the neurochemist needs to have some idea of the procedures involved in the behavioral analyses conducted by the psychologist. In planning this *Neuroscience Labfax* volume, we set out to achieve several things. Our major goal was to provide a comprehensive guide to methods, techniques and data-handling in different areas of neuroscience for the neuroscientist who is working in one area and who wishes to widen his/her understanding, collaborate with neuroscientists from other disciplines or even commence experiments using techniques from other areas. Our hope therefore is to demystify the variety of different techniques and methods used by neuroscientists from differing backgrounds and experimental perspectives.

A unique feature of the *Labfax* series, in general and this book in particular, is the compendium of data provided by the volumes in the series. This data compendium will save many hours searching for appropriate references and experimental results and will therefore be an invaluable time-saving aid for the busy neuroscientist. We also hope that the data provided in these chapters will enable the neuroscientist embarking on new experiments to quickly assess if new methods are being used appropriately and if valid results are being obtained, thereby avoiding many hours of wasted time.

A recurrent theme in contemporary neuroscience is the need to understand the brain from a variety of differing levels of analysis. We therefore have attempted to provide as comprehensive an introduction as possible to a variety of different levels of analysis ranging from the molecular and cellular, on the one hand, through to systems and behavior on the other. Our hope is to encourage cross-fertilization between different levels of analysis to everyone's mutual understanding. One good example of this kind of interaction is exemplified in the chapters by Hölscher & O'Mara, and Bashir & Vignes. The former provides an account of the possible role played by the phenomenon of long-term potentiation (LTP) during learning and the latter an account of the cellular mechanisms of LTP and how it is analyzed in the hippocampel slice using patch clamp techniques. This *Labfax* was designed to gather together, in one volume, a compilation of techniques that are widely used but whose rationale and methodology are rarely, if ever, explained in detail. The chapter by Muller and colleagues on single and multiple neuron recording in the Freely Moving Animals is unique as virtually no other available source gathers together the methods and data for this increasingly important technique in one article. The chapter by O'Mara and Walsh attempts to collate information on methods and data gathered on lesioning techniques, which judging by the volume of papers published every year, is perhaps one of the most popular techniques in behavioral neuroscience, but about which very little has specifically been written.

The range of chapters chosen then is deliberately broad, but can be categorized into those broadly dealing with molecular and cellular analyses, systems-level analyses and behavioral analyses. As will also become clear, many chapters will easily fall into several categories as information from one level of analysis is vital for understanding those at another. To take one example, information on the connectivity of brain areas (Systems Level Analysis by Wouterlood *et al.*) is vital for understanding synaptic transmission in reduced preparations (the Hippocampal Slice by Bashir & Vignes) and in intact, freely moving animals by Muller *et al.* To take another example, Techniques for electrode manufacture and for measuring electrode performance (chapter by Miller) are important for a wide variety of techniques at different levels of analysis, such as understanding brain connectivity (chapter by Wouterlood *et al.*), synaptic transmission in the brain slice (chapter by Bashir & Vignes) and of course for neuronal recordings in the freely moving animal (Muller *et al.*). Another example is the wide applicability of neural network models (by Baddeley) to many differing sorts of experimental analysis, depending on the variables of interest. Neural network modeling requires detailed data regarding brain connectivity and, in turn, can provide detailed predictions and simulations of the behavior of individual neurons (Cellular Neurophysiological Modeling) or of populations of neurons (Systems Modeling). In the former case detailed information is required for the simulations on such variables as Receptor Localization and Receptor Density (chapters by Pratt and McBean), and Information About Changes Mediated by Gene Expression (by Flavell). In the latter case information is required about activation thresholds of neurons (chapters by Bashir & Vignes, Muller *et al.*), as well as on Neuronal Connectivity (by Wouterlood *et al.*). We could provide many other such examples, but our overall point should be clear.

We hope this volume will prove equally useful for advanced undergraduate and graduate students who increasingly need to understand and acquire several different techniques. For the practicing neuroscientist, the chapters in the present volume will provide comprehensive starting points for understanding techniques and methods drawn from the broad discipline of neuroscience.

M.A. Lynch
S.M O'Mara
(March 1997)

CHAPTER 1
INTRACELLULAR AND EXTRACELLULAR RECORDING WITH MICROELECTRODES

J. Millar

1. ELECTRODES FOR INTRACELLULAR RECORDING

1.1. Physical considerations

These electrodes are designed to measure changes in the membrane potential of nerve cells. For this we need to have one electrode inside the cell and one outside. The outside electrode is usually not a problem as it can be anywhere in the tissue or organ bath that has a low-resistance electrical pathway to the cell. But since nerve cell bodies range from a maximum of about 50 microns diameter (for large motoneurons in the spinal cord) down to less than 5 microns (for some kinds of interneurons) the inside electrode must have a tip diameter small enough to impale the cell without too much damage. In practice electrodes with tip sizes of 0.1–1 microns are normally used.

It is also obviously necessary to have a method for moving the electrode in precisely defined distances forward and back so that the tip can be accurately positioned inside or outside the cell. Finally, the whole physical assembly must be sufficiently vibration free so that, once inside, the electrode tip will stay inside the cell while the necessary measurements are being made. The requirements for physical stability and immobility are such that most intracellular recording is done on isolated cells or cells in a brain slice. There is normally too much movement (pulsations due to respiratory and circulatory mechanisms) for intracellular recording to be feasible in whole animal preparations. A full discussion of the means of physical stabilization of the preparation is beyond the scope of this chapter. Fundamentally, an air-suspension table is needed with 'micromanipulators' (micropositioners specifically designed for use with microelectrodes) to hold the microelectrodes and enable them to be placed in the tissue under study. A selected list of manufacturers of air tables, manipulators and associated equipment for intracellular recording is given in *Table 1*.

Perusal of these manufacturers' catalogues will give the novice neuro-impaler a good idea of what is required. It should be stressed, however, that there is absolutely no substitute for going to visit the labs of established workers in the field and seeing how the machinery is actually set up. The golden rule is this: Before you start intracellular recording, visit as many labs that are doing it as you possibly can. (It is also a good way to make like-minded friends and ensure that you have a shoulder to cry on when, as always happens, electrode manufacture fails completely at some crucial point in the experimental series.)

Table 1. Manufacturers of micromanipulators, air tables etc.

Adams & List Associates Ltd, 1100 Shames Drive, Westbury, NY 11590, USA
Fax 313 756 9737

Burleigh Instruments Inc., Burleigh Park, Fishers, NY 14453, USA
Fax 716 924 9072

Table 1. Continued

David Kopf Instruments, PO Box 636, Tujunga, CA 91403, USA Fax 818 352 3139
Ealing Co., 22 Pleasant Street, South Natick, MA 01760, USA
Narishige Scientific Instruments, 9–28 Kasuya 4-Chome Setagaya-Ku, Tokyo, Japan Fax 813 3308 2005 (in USA fax 516 759 6138)
Newport Corporation, PO Box 8020, 18235 Mt Badley Circle, Fountain Valley, CA 9728-8020, USA Fax 714 963 2015
Technical Manufacturing Corporation (TMC), 15 Centennial Drive, Peabody, MA 01960, USA Fax 508 531 8682

1.2. Electrode manufacture

Intracellular electrodes are made from glass tubes (usually 1–2 mm outside diameter) heated and pulled in a specialized 'puller' so that they taper down to a very fine tip. Changing the puller settings changes the final tip diameter. Smaller tips produce electrodes with excessive electrical noise, which may also be difficult to fill (see below), but these small tips may be necessary to get inside particularly small or fragile cells. The key piece of apparatus that is needed to make an intracellular electrode is thus a specialized electrode puller. There are numerous manufacturers of these instruments (they may have local agents in different countries). A selection is shown in *Table 2* (some of the manufacturers of manipulators in *Table 1* are also manufacturers of pullers, particularly Kopf and Narishige).

Table 2. Manufacturers of microelectrode pullers

Campden Instruments, King Street, Silby, Loughborough LE12 7LZ, UK Fax 1509 816 097
Mecanex, Vuarpilliére 29, 1260 Nyon, Switzerland Fax 41 22 61 52
Stoelting Co., 620 Wheat Lane, Wood Vale, IL 60191, USA Fax 708 860 9775
Sutter Instruments Co., 40 Leveroni Court, Novato, CA 94949, USA Fax 415 883 0572
World Precision Instruments, International Trade Center, 175 Sarasota Boulevard, Sarasota, Florida 34240-9258, USA Fax 813 371 1003

This pipette once pulled needs to be filled with conducting fluid, and then a wire has to be inserted in the blunt (shank) end to make contact with the fluid. There are two problems here:

(a) What fluid to use?
(b) How to stop air bubbles being trapped in the very fine tip region and thus preventing electrical contact with the tip?

The fluid is traditionally potassium chloride (1–3 M). Concentrated solutions are used to reduce the electrolyte resistance. However, since the Cl^- can leak out and change the chloride concentration in the cell, chloride is not used when changes in the intracellular chloride concentration can affect the experiment. In these cases citrate or acetate anions can be used. Distilled water, high purity chemicals and filtration of solutions (a 0.2 micron millipore filter) before use is strongly recommended to prevent blockage of the electrode tip by undissolved particles.

Many methods have been used in the past to get electrodes to fill properly, i.e. to have an uninterrupted column of fluid to the tip. Nowadays two main methods are used, both of which rely on capillary forces to pull the fluid to the tip. With omega glass, there is a small glass filament incorporated into the glass tube (*Figure 1a*).

When the glass is pulled this filament is also pulled into a proportionately minute whisker. If the top (blunt) end or the tip of the freshly pulled electrode is dipped into the filling fluid, capillary forces drag fluid into the tip. This is a slow but sure method of filling. With 'theta' glass, there is a septum down the middle of the tube (*Figure 1b*). In this case the angle between the septum and the wall provides the capillary space to pull the fluid into the tip. For most purposes both sides of the theta glass can be filled with the same fluid, and the connecting wire placed in either side. There are some specialist manufacturers of omega and theta glass (*Table 3*); these companies also specialize in the supply of electrode wires and

Table 3. Manufacturers of microelectrode glass

Clark Electromedical Instruments, PO Box 8, Pangbourne, Reading, RG8 7HU, UK Fax 1734 845 374
FHC Inc., 69 Pleasant Street, PO Box 337, Brunswick, ME 04011, USA Fax 207 729 1603
Jencons Scientific Ltd, Cherrycourt Way Industrial Estate, Stanbridge Road, Leighton Buzzard, Beds IU7 8UA, UK Fax 01525 379 547
W. Krannich, 37079 Göttingen, Elliehäuser Weg 17, Germany Fax 640 966 7673

Figure 1. (A) 'Omega' tubing (end-on). (B) 'Theta' tubing (end-on).

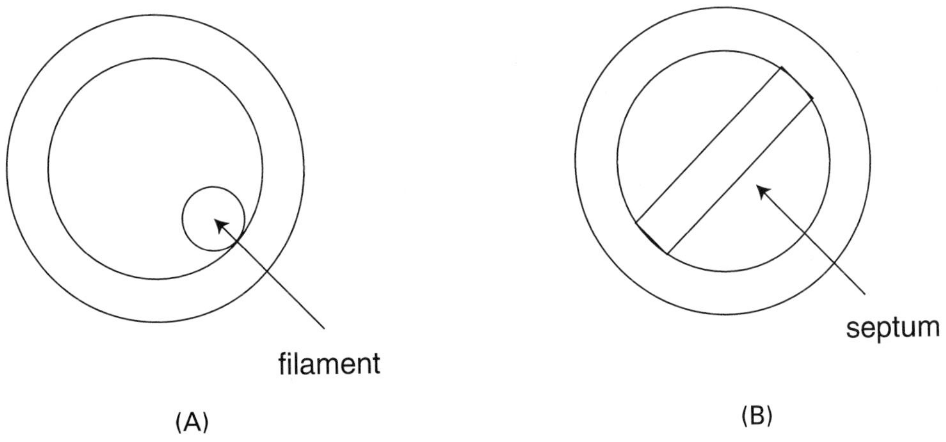

silver chloride pellets. Not surprisingly, the manufacturers of pullers tend to be agents for electrode glass and vice versa, so all the puller manufacturers listed in *Table 2* can supply or are happy to give advice on which glass to use with their make of puller. (World Precision Instruments also sell a very useful micro-syringe (MicroFil™) which can be used in conjunction with omega or theta glass to fill electrodes.)

Because the tip is so small, the resistance of intracellular micropipettes is very high. (It is related to the conductivity of the filling fluid and the cross-sectional area of the tip.) Values in the range 10–500 MΩ may be obtained, depending on the tip size and shape. Since the tip is usually too small to see with a light microscope, the electrode resistance is often used as the main criterion for electrode quality. For normal intracellular work a value of 50–100 MΩ (depending on the filling fluid) is a good starting point. Experience will soon show that more successful penetrations are made with higher or lower resistance electrodes, and the puller settings can then be adjusted accordingly.

Because of the way they are made, the tips of micropipettes may have a jagged end, which can cause extra damage to the impaled cells. A good way round this is to bevel the ends with a special device so that a smooth glass surface is produced. The glass surface can be made oblique, like a hypodermic needle end, and this is believed to reduce the trauma during impalement and produce a better seal around the electrode end. Narishige, Sutter and WPI all make good bevelling instruments.

The wire that makes contact with the fluid needs to form a stable electrochemical junction, otherwise noise and drift in the electrical signal will result. The best procedure is to use a small silver/silver chloride pellet which has been welded onto the end of a fine silver wire. The pellet is inserted into the fluid in the blunt end of the electrode. Hardy veterans in the intracellular world often regard this as wimpish behavior, and will give you recipes on how to 'chloride your own'. But do not be fooled! Badly home-chlorided wires are a source of problems that you don't need! Buy a pelleted wire from one of the puller or electrode glass vendors listed in *Tables 1, 2 or 3* above and you will have a stable and reliable connection.

1.3. Electrical considerations

The higher the electrode resistance, the greater the electrode noise. This is an unavoidable fact, due to the physics of conduction. In practice a 50 MΩ electrode will usually have a noise in the range 0.5–1 mV. Since the signals that are to be recorded are usually in the tens of millivolts range, this gives an acceptable signal-to-noise ratio of about 10:1 or better. The noise is a function of the recording bandwidth: if electronic filters are used to restrict the recording bandwidth (normally from DC to 5–10 kHz), then an improvement in signal-to-noise ratio may be obtained.

Trying to record a millivolt signal through a 100 MΩ resistance is like trying to see a small luminous fish at the bottom of a very deep well full of dirty water. The large amount of water between you and the fish restricts the amount of light that gets to your eyes. In the same way the high resistance of the electrode restricts the amount of current that the voltage source in the tissue can send through to the amplifier. One therefore needs an amplifier that needs very little current into its input to turn it on. This fortunately is not a problem nowadays. Microelectrode amplifiers that need only a few picoamperes to turn on are available. These are called high input impedance amplifiers, and are commonly specified in terms of their 'input resistance'. (They are not really resistances at all, but the terminology has stuck.) A good rule of thumb is to have an amplifier with an input resistance 100 times greater than the electrode resistance. Thus for a 100 MΩ electrode, you need a 10 000 MΩ (10 GΩ) amplifier.

The wire feeding into an amplifier with a 10 GΩ input resistance acts like an aerial, i.e. it will pick up electromagnetic signals from the atmosphere. In particular, it will pick up signals at 50 or 60 Hz, depending on the local line (mains) frequency. To minimize these unwanted signals, the length of the connecting wire must be kept to a minimum. The best solution is to mount the microelectrode actually on the first stage of the amplifier, which is possible as this first stage can be brought out on a long cable from the main amplifier to a small 'headstage' mounted next to the tissue.

There are several manufacturers of specialist amplifiers for work with microelectrodes; a selection is listed in *Table 4*. (Again, as the positioners, glass, pullers and amplifiers are used together, the manufacturers in *Tables 1–3* are often also agents or even manufacturers of amplifiers as well.)

With the electrode made and connected up, all that is needed to complete the circuit is a reference electrode (the electrode outside the cell) to put in the tissue bath. A second silver/silver chloride pellet is the best choice. Platinum or stainless steel wires are not really adequate.

Table 4. Manufacturers of intracellular microelectrode preamplifers (also see *Tables 1–3*)

Axon Instruments, 1101 Chess Drive, Foster City, CA 94404, USA Fax 415 571 9500
Bak Electronics Inc., 13220-N2 Wisteria Drive, German Town, MD 20875, USA Fax 301 601 9018
Dagan Corp., 2855 Park Avenue, Minneapolis, MN 55407, USA Fax 612 827 6535
Digitimer Ltd, 37 Hydeway, Welwyn Garden City, Herts AL7 3BE, UK Fax 1707 373 153
HEKA electronik GmbH, Wiesenstrasse 71, D-67466 Lambrecht/Pfalz, Germany Fax 63 25 80 39
List-Electronic, Pfungstaedter Strasse 18–20, 6100 Darmstadt, Germany Fax 615 156 060
Medical Systems Corp., 1 Plaza Road, Greenvale, NY 11548, USA Fax 515 621 8503
Neuro Data Instruments, 35 Waterside Plaza, New York, NY 10010, USA Fax 212 685 8508

If the glass microelectrode is placed with a micromanipulator in the same bath of saline as the reference electrode the potential that the amplifier displays will not be zero. This is because the recorded potential is a composite of several voltage-generator batteries in series. The difference in concentration between the electrolyte in the tip and that in the bath solution generates a 'tip junction potential' which may be several millivolts in size. This is due to the unequal rate of diffusion of the anions and cations out of the micropipette tip. It will depend on the tip geometry. In addition, uneven adsorption of ions by the very thin glass of the tip can lead to a second 'tip adsorption potential'. (The difference between the two is that the adsorption potential will disappear if the tip is fractured, but the diffusion potential will remain.) Finally, the difference in the concentration of the two electrolytes in contact with the two silver chloride pellets (one in the bath, one in the pipette) will produce a further DC potential adding to the tip potentials.

In fact, all amplifiers designed for use with micropipettes have a facility for offsetting to zero or 'nulling' these non-biological voltage sources. In practice, the microelectrode is lowered into the tissue until it is just outside a cell. The voltage is then nulled off until a zero reading is obtained. The cell is then penetrated by a small forward movement of the electrode. This can be done in a variety of ways. Some manipulators have a facility for a short 'tap' type movement. Some preamplifiers have means of rupturing the cell membrane by an electrical zap through the micropipette. In the end, it means that whichever way works for you, is the way you should do it.

2. A BRIEF NOTE ON PATCH CLAMPING

Because the technology is very similar, most of the commercial suppliers of patch clamp instrumentation are the same companies as those who supply equipment for intracellular work. Patch clamp electrodes are glass micropipettes, made in pullers closely related to those used for intracellular electrodes. The same manufacturers should be contacted for patch electrode pullers as for intracellular electrode pullers. However, the tips of patch electrodes are usually wider than those of intracellular electrodes (about 1 micron), and instead of being sharp at the end they are deliberately smoothed off using a 'microforge'. (These are also available commercially from the puller manufacturers.) Resistances of patch electrodes are typically considerably lower than those of intracellular electrodes; for example, 10 MΩ as compared with 50–100 MΩ. The idea in patch clamping is to fuse the polished glass at the end of the electrode on the cell membrane, not to penetrate it. Gentle suction is applied to the electrode until the cell membrane is firmly attached to the tip. If the fusion is successful the lipid of the membrane wets the glass to form a circular hydrophobic 'gigaseal' where the glass meets the membrane; the resistance through this seal is of the order of gigohms. Coating the electrode with a silicon resin ('Sylgard') is often used to improve seals. When the seal is formed the electrode amplifier is then used to measure the current through the electrode, not the voltage. Because of the seal, the current through the electrode is the same as the current through the membrane patch under the seal. Analysis of the current through the patch has led to an enormous increase in our understanding of the mechanism of ion channels in membranes.

The mechanical and physical requirements for patch clamping are similar to those for intracellular work, and the method of pulling the electrodes is also similar. However, because the patch clamp technique measures *current*, not voltage, a different kind of amplifier is needed. This is the patch clamp amplifier.

The two major players in the patch clamp amplifier business are undoubtedly Axon Instruments and List-Electronic (*Table 4* above). Each company makes a range of patch clamp amplifiers, and each firmly believes that their instruments are the best. (Sometimes with an almost messianic zeal!) Because there are several forms of patch clamping (for example, whole cell, patch detached, inside-in and inside-out patches) the advice given earlier is worth repeating here. Before buying anything, visit as many labs as will tolerate you, and see the different forms of apparatus actually in action. Some amplifiers are undoubtedly better for some tasks than others, and you should naturally choose the instruments that you know will work with the kind of cells you want to use.

3. EXTRACELLULAR RECORDING

3.1. Introduction

When an action potential occurs in the membrane of a neuron there is an (initially inward) current flow across the cell membrane. This transmembrane current generates a current flow in the volume of conducting tissue around the neuron. This extracellular current will in turn generate a voltage transient in the tissue, and this voltage transient can be detected as an extracellular 'spike'. Extracellular spikes from mammalian neurons in the central nervous system have a duration of about 0.2 to 2 milliseconds, depending on the type of neuron and the bandwidth of the spike recording system.

The great difference between intracellular and extracellular recording is that the extracellular spikes are usually between 10 and 100 times smaller than the intracellular spikes, and may go down to a few microvolts. So why bother with extracellular recording? The reason is that with extracellular recording you can record the activity of neurons without having to impale and thereby damage them. Also, because it is far less sensitive to movement, this technique can be used in the whole anesthetized brain, i.e. it is not restricted to isolated cells or brain slices.

On the other hand, with extracellular recording it is all too easy to record from two or more cells ('units') at the same time; this is technically 'multi-unit' recording. With extracellular recording the size of the spike is a matter of where the electrode is physically relative to the cell body and dendrites. So small spikes may be obtained from large cells, and large spikes from small cells if the electrode tip is placed optimally. Unfortunately, no-one knows what is the optimal placement for any given cell, so all extracellular recording is on the 'poke and hope' basis. In other words, you push the electrode into the tissue until a spike appears: then you stop and see what its characteristics are.

Extracellular recording can be done with glass micropipettes with relatively large tips (1–5 microns) filled with sodium chloride instead of potassium chloride. However, because these electrodes tend to be electrically noisy, most extracellular recording is done with metallic electrodes.

The two most popular forms of extracellular microelectrode are the tungsten electrode and the carbon fiber electrode.

3.2. Tungsten microelectrodes

At first sight, tungsten is an odd choice for an electrode. Copper, gold and silver are all very good conductors, and gold will not corrode in use. The reason tungsten was chosen was for its stiffness! This is not as silly as it seems. In order to get into the tissue without causing massive damage, microelectrodes need to have narrow shafts as well as small tips. They also need to be very stiff so that they can be pushed into nervous tissue without bending over. Most wire gets very floppy at a diameter of less than a hundred microns or so, and so is no use for a microelectrode, unless supported in some way (for example, being encased in a glass micropipette). Tungsten wire is remarkably stiff, much stiffer for example than stainless steel) and thus can be used to make rigid electrodes even when tapering down to diameters of a few microns. This is why it is used for electrodes.

Tungsten electrode manufacture is a black art! Until recently, researchers were expected to make their own electrodes, which could involve multiple etches, glass coating, and multistage electroplating of the tip, a far more frustrating and time consuming exercise than the manufacture of pipettes. Fortunately however all that has changed. You can now buy perfectly acceptable tungsten electrodes from a number of commercial suppliers (see *Table* 5). These electrodes have the great advantage of standardization and reproducibility, and will save you time and money in the long run. Making tungsten electrodes in the lab is a total waste of time, *unless* you have some very special requirement, for example for a super thin electrode, or one with a non-standard shape. (Ainsworth manufactures glass-coated tungsten electrodes to any feasible shape to special order; please phone 0181 520 3770.) But if you simply require tungsten electrodes for normal extracellular recording, go out and buy them.

This advice might raise eyebrows in some readers. Surely your own electrodes will be better than any commercial ones? But think about the other parts of the experimental rig that you use. Back in the 1950s neurophysiologists often had to make their own oscilloscopes and preamplifiers. It would be foolish to do that nowadays because excellent commercial products are available. The same applies to electrodes. Unlike glass micropipettes, which *have* to be made locally because of problems of tip fracture, electrostatic blocking of the tip etc., tungsten electrodes are excellent travellers. So, young scientist, by all means read about the classic tungsten electrode manufacturing methods of Hubel, Merrill, etc., but don't try to emulate them unless you really have to.

Table 5. Manufacturers of tungsten microelectrodes

A-M Systems Inc., 11627-A Airport Road, Everett, WA 98204, USA
Fax 206 353 1123

Ainsworth Specialist Electrodes, c/o MPB, Department of Physiology, Queen Mary College, Mile End Road, London E1 4NS, UK
Fax 0181 983 0467

Diamond General Corp., 3965 Research Park Drive, Ann Arbor, MI 48108, USA
Fax 313 747 9703

Micro Probe Inc., PO Box 87, Clarksburg, MD 20871, USA
Fax 301 349 2059

Rhodes Medical Instruments Inc. (for tungsten stimulating electrodes), 21044 Ventura Blvd, Woodland Hills, CA 91364, USA
Fax 207 729 1603

3.3. Carbon fiber microelectrodes

Carbon fibers are naturally very stiff and very strong, with particularly high tensile strength. They are also relatively good conductors. The carbon surface also makes a very low noise interface (see below) with a physiological saline solution. For these reasons they have become widely used for extracellular spike recording. Carbon fiber electrodes are commercially available from Axon Instruments and Diamond General (see *Tables 4 and 5*) and from the following suppliers:

Radiometer Analytical, 72 Rue D'Alsace, 69627 Villeurbanne, France
Fax 78 68 8812

MBP Electrodes, Department of Physiology, Queen Mary College, Mile End Road, London E1 4NS, UK
Fax 0181 983 0467

However, there is a problem with the commercial electrodes in that these electrodes are designed for micro electrochemical analysis (voltammetry) rather than extracellular recording. (Carbon fibers can be used for either voltage or current recording.) With the exception of MPB Electrodes (above) the carbon fiber tip designed for voltammetry may be either too long or too short for optimal spike recording.

For this reason it is useful for researchers to make their own carbon fiber electrodes. The technique is not much more difficult than that for making glass pipettes, and far simpler than that for making tungsten electrodes.

Industrial carbon fibers are normally used, and these come in a very restricted range of standard sizes. The most widely used fibers have an outside diameter of 8 microns.

3.4. Manufacture of carbon fiber electrodes

Carbon fibers suitable for use as microelectrodes are available from Goodfellow Metals in the UK (Goodfellow catalogue no. C005725), or from Amoco Performance Products (type P55) in the USA.

The fibers come as a multifiber 'tow'. To make the electrodes first cut a bundle of fibers about 15 cm long from the tow. This bundle can be stored in a fold of paper, arranged so that 1–2 cm of the tow protrudes from the end of the paper. The bundle is placed on a light box as used by photographers, so that the fibers are lit from below. A person with normal eyesight or myopia should be able to see single fibers clearly when they are backlit this way. Now fit plastic collars (a few mm of the insulation stripped from a small electrical signal wire makes a good collar) on to the tip of a pair of no. 5 jeweller's forceps (available from Fine Science Tools Inc, 202–277 Mountain Highway, North Vancouver, BC, Canada V7J 3P2; Fax 604 987 3299). The soft plastic collars prevent the forceps damaging the fibers during handling. Tease a single fiber from the tow using the forceps, and without letting go, pull the fiber out from the tow and then push it slowly into a 2 mm outside diameter glass electrode tube held almost horizontally in a test tube filled with acetone or absolute alcohol. The fiber should float down in the fluid until it fills the tube. The tube should then be carefully blotted to remove the acetone or alcohol, and then pulled in the normal way to make a glass micropipette with the carbon fiber embedded inside. Because the fiber has such high tensile strength, when the glass breaks in the puller to form the two pipettes, the intact fiber is pulled out of one of the pipettes, and held tightly by the constricted glass at the tip of the other one.

The pipette containing the fiber (which will have a long length of fiber protruding from the end) is then carefully removed and placed on a glass slide under a microscope. The protruding fiber is trimmed to size (for spike recording about 20 microns is usually optimal) with another pair of no. 5 forceps. These are mounted on a separate micromanipulator, and fitted with a screw control to push the tips together in a slow controlled way. When the tips are closed over the fiber it will snap apart. With practice this method can be used to control fiber length to better than plus or minus 5 microns.

The blunt end of the electrode can be filled with saline or KCl like a normal pipette, but for longer life it is better to make a 'dry' connection with the carbon fiber. This can be done by dipping a fine copper wire into conducting paint, and then thrusting the wire down into the electrode until the copper presses against the carbon fiber near the tip. The copper wire can then be cemented in place at the end of the pipette for mechanical strength.

3.5. Electrical considerations

Tungsten and carbon fiber electrodes are metallic conductors. When they are immersed in an electrolyte their surfaces exchange electrons and possibly ions to form a (voltage dependent) *capacitance* in the solution. What this means in practice is that these electrodes are very noisy if connected to a DC preamplifier suitable for intracellular electrodes. The DC level will drift in uncontrollable ways and will be responsive to any thermodynamic changes in the solution. Thus temperature, pH, solution ionic strength, all will affect the DC level. In other words these electrodes are no good for intracellular recording, where DC measurements are needed.

This is just as well, because the electrodes are also too big for intracellular recording. Typical tungsten electrodes have tips which are cones of 10–20 microns length tapering down to a submicron point, but starting from a diameter of 5–10 microns. These tips are obviously too big to place inside a cell, even if the electrode surface was suitable for DC recording. However, the tip capacitance can have quite a low impedance (AC resistance) at the frequency bandwidth needed to pick up the extracellular spikes from neurons. At 1 kHz, for example, a typical tungsten or carbon fiber electrode may have an impedance of 1 MΩ or less. This means that the irreducible intrinsic or 'Johnson' noise of the electrode is much less than that of a micropipette. For example, a high quality (glass insulated, platinum plated) tungsten electrode can have a peak-to-peak noise of less than 20 μV over the range 100 Hz–10 kHz. The best carbon fiber electrodes can have even less. This is much less than a tenth of the noise of a typical glass pipette. The amplitude of an extracellular spike will vary depending on how near the electrode tip is to the active area of membrane. Lower noise electrodes mean that a better signal-to-noise ratio of extracellular spikes is obtained, and so more different neurons can be sampled during a penetration of the brain. To take full advantage of these characteristics of extracellular electrodes, a different preamplifier should be used from that for intracellular electrodes. Preamplifiers for extracellular electrodes are AC coupled (i.e. the steady level of the signal is filtered out and discarded). They can have lower input impedances than those for intracellular work, and this means that they can have lower noise. Most of the manufacturers of intracellular amplifiers (*Table 4*) also make extracellular preamplifiers for use with tungsten or carbon electrodes.

With extracellular recording, however, the signal amplification is only half the story. The electrode tip may be near enough to several different cells to pick up extracellular action potentials from them. The different cells will each produce a spike but of a different amplitude or shape, depending on where it is relative to the electrode tip. How do you tell them apart? The answer is 'spike discrimination' units. These are devices which can sort out spikes into different logical channels depending on their amplitude and time course. More complicated techniques of spike discrimination use software and/or digital delay lines for more sophisticated sorting. It is pointless to buy an extracellular amplifier unless you are also going to invest in the spike discrimination machinery to go with it. Two manufacturers produce a complete range of spike discrimination and spike counting apparatus, and they also produce excellent extracellular spike preamplifiers. These are Bak Electronics in the USA and Digitimer Ltd in the UK (*Table 4*).

4. FURTHER READING

4.1. Intracellular recording

Brown, K.T. and Flaming, D.G. (1975) New microelectrode techniques for intracellular work in small cells. *Neuroscience*, **2**, 813.

Coombs, J.S., Eccles, J.C. and Fatt, P. (1955) Excitatory synaptic action in motoneurons. *J. Physiol.* **130**, 374.

Frank, K. and Fuortes, G.F. (1955) Potentials recorded from the spinal cord with microelectrodes. *J. Physiol.* **130**, 625.

Geddes, L.A. (1972) Electrodes and the measurement of biolelectric potentials. John Wiley, New York.

Ling, G. and Gerard, R.W. (1949) The normal membrane potential of frog sartorius fibers. *J. Cell. Comp. Physiol.*, **34**, 383.

Silinsky, E.M. (1992) Intracellular recording methods for neurons, in *Monitoring Neuronal Activity* (J.A. Stamford, ed.). IRL Press, Oxford, p. 29.

4.2. Extracellular recording: tungsten electrodes

Hubel, D.M. (1957) Tungsten microelectrode for recording from single units. *Science*, **125**, 549.

Levick, W.R. (1972) Another tungsten microelectrode. *Med. Biol. Eng.*, **10**, 510.

Levick, W.R. and Cleland, B.G. (1974) Selectivity of microelectrodes in recordings from cat retinal ganglion cells. *J. Neurophysiol.*, **37**, 1387.

Merrill, E.G. and Ainsworth, A. (1972) Glass-coated platinum-plated tungsten microelectrodes. *Med. Biol. Eng.*, **10**, 662.

Robinson, D.A. (1968) The electrical properties of metal microelectrodes. *Proc. IEEE*, **56**, 1065.

Wienman, J. and Mahler, J. (1964) An analysis of electrical properties of metal microelectrodes. *Med. Electron. Biol. Eng.*, **2**, 265.

4.3. Carbon fiber electrodes

Armstrong-James, M. and Millar, J. (1979) Carbon fiber microelectrodes. *J. Neurosci. Meth.*, **1**, 279.

Armstrong-James, M., Fox, K. and Millar, J. (1980) A method for etching the tips of carbon fibre microelectrodes. *J. Neurosci. Meth.*, **2**, 431.

Fox, K., Armstrong-James, M. and Millar, J. (1980) Electrical properties of carbon fibre microelectrodes. *J. Neurosci. Meth.*, **3**, 37.

Math, F. and Marianneau, G. (1994) A new method for manufacturing carbon-fiber microelectrodes *J. Neurosci. Meth.*, **52**, 149.

Millar, J. (1992) Extracellular single and multiple unit recording with microelectrodes, in *Monitoring Neuronal Activity* (J.A. Stamford, ed.) IRL Press, Oxford, p. 1.

CHAPTER 2
ELECTROPHYSIOLOGICAL RECORDING IN BRAIN SLICES

Z. I. Bashir and M. Vignes

Brain Slices

1. INTRODUCTION

The use of electrophysiological recording from brain slices has vastly increased our understanding of the mechanisms underlying synaptic transmission in the CNS. Some of the advantages/disadvantages of the use of *in vitro* preparations are outlined in *Table 1*. The hippocampus is often used as a model system for studying synaptic transmission in the CNS since it has the distinct advantage, over some other brain areas, of a relatively simple architecture and organization which can be viewed under low power magnification. This simplifies the placement of stimulating and recording electrodes. In addition, the simple circuitry of the hippocampus means that it is relatively easy to study monosynaptic excitatory postsynaptic potentials (EPSPs) (and inhibitory postsynaptic potentials (IPSPs)), without interference from polysynaptic transmission.

Table 1. Considerations when using *in vitro* preparations

Advantages	Disadvantages
• Mechanical stability • Direct visualization • Control of extracellular medium • Precise control over concentration of drugs	• Damage to neurons and glia • Loss of afferent and efferent connectivity; loss of vascular and hormonal regulation • The artificial conditions may alter cellular metabolism in many ways

Electrophysiological recording from brain slices is used to investigate a myriad of different aspects of function in a variety of different brain structures. In this chapter we will attempt to provide a user-friendly account of:

(i) commonly used methods to prepare hippocampal slices,
(ii) some of the electrophysiological recording techniques which are used in the study of synaptic function *in vitro*.

2. PREPARATION AND MAINTENANCE OF HIPPOCAMPAL SLICES

2.1. Cutting slices

When it comes to the finer details of slice preparation, there are as many different variations as there are laboratories using these methods (1); the protocols described here (*Table 2*) are based on personal experience. It is important, during preparation and subsequent storage of slices, that the aCSF solution is continually oxygenated. It is also usual, since this can take several minutes, to perform the whole procedure in ice-cold aCSF solution. In order to prevent excessive neuronal activity (which may be detrimental to slice health) during slice preparation, some workers use aCSF with no added calcium and/or high concentrations of magnesium (2). Generally speaking, transverse slices, that is to say across the longitudinal axis, are prepared in order to maintain the characteristic tri-synaptic circuitry of the hippocampus (*Figure 1*). In this way (most commonly) perforant path/dentate gyrus granule cell synapses, mossy fiber/CA3 synapses or Schaffer-collateral/CA1 synapses can be studied with some ease.

The thickness of the slice that is used is determined by the proposed experiment and can be anywhere from about 150 to 600 μm. Slices at the thinner end of the spectrum are used in order to visualize individual cells filled with fluorescent probes (e.g. calcium-sensitive dyes (3)). Slices of about 400 μm are generally used for most electrophysiological experiments (4), as these provide a reasonable degree of intact circuitry. Thicker slices are sometimes used by some workers (5). However, a potential problem is that the middle of such slices can become anoxic (6).

McIlwain chopper

The use of hippocampal slices for electrophysiological investigations came about from the original work of Henry McIlwain who cut slices of brain tissue to aid his studies on the metabolism of the CNS (see (7)). It was subsequently shown by Li and McIlwain (8) and Yamomoto and McIlwain (9) that they could record membrane potentials from neurons in slices of brain tissue prepared in this way. See *Table 2* for methods.

Vibroslice

Cutting slices using a vibrating blade has recently become very popular. This method appears to have several advantages over the McIlwain chopper:

(i) the ability to cut thinner slices;

Figure 1. Schematic representation of a transverse hippocampal slice illustrating the major synaptic pathways within the hippocampus. Electrical stimulation of the Schaffer collateral commissural fibers results in synaptic activity in the CA1 region which can be recorded by several different electrophysiological means either from the cell body or from the dendritic region.

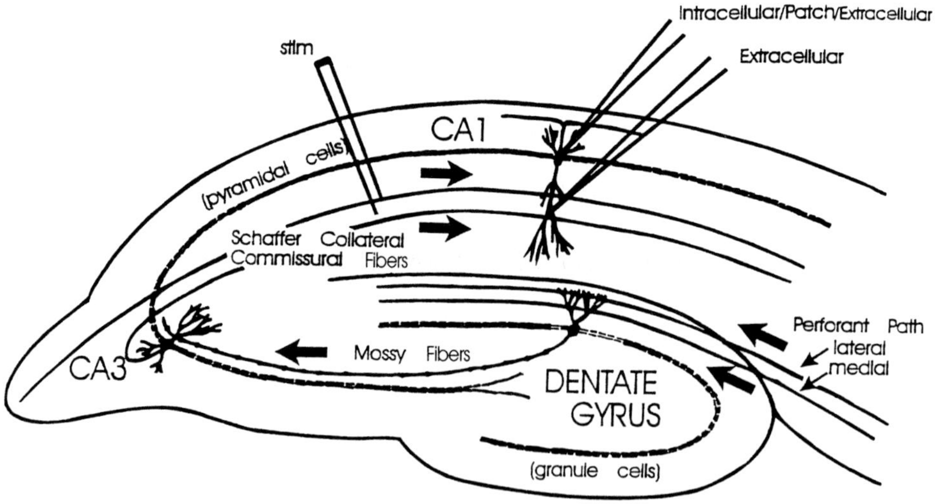

(ii) the relative ease with which slices of brain tissue from immature animals (which have less connective tissue) can be cut;
(iii) the health of slices, as judged by morphological and electrophysiological parameters, appears better than those cut with a McIlwain chopper (2).

The more common devices are the Vibroslice (Campden Instruments Ltd, UK) and the vibrotome (Oxford Laboratory, USA). The set up of a vibroslice (illustrated in *Figure 2A*) is such that the blade is moved forward horizontally through the tissue, whilst vibrating at right angles to this forward motion. Therefore, in order to cut transverse slices of hippocampus, the brain has to be orientated so that the longitudinal axis of the hippocampus is more or less vertical. See *Table 2* for methods.

Table 2. Cutting slices

- Animals are decapitated, according to local guidelines. Cut skin from skull. Cut across back of skull and then along midline. Prise away skull.
- Prise and cut away dura. Cut frontal cortex and cerebellum away.
- Scoop out brain. Place in ice-cold aCSF for a minute or two.

McIlwain	Vibroslice
Each hemisphere is placed with the medial face uppermost.	The outer edge of the cerebral cortex is cut and the brain placed on this new flat surface, with the medial side facing upwards.
The brain stem and remaining cerebellum is removed with a spatula and forceps to reveal the ventral surface of the hippocampus. A spatula is inserted underneath the hippocampus and the septal and temporal ends are severed to gently remove the whole hippocampus.	'Superglue' a block of sylgard or agar to the ventral side of the brain. The brain and the sylgard block are then 'superglued' to the stage of the slicer. The sylgard block provides support to the tissue and prevents it from being pushed over by the force of the cutting blade (*Figure 2*).
The hippocampus is placed on aCSF-moistened filter paper on the stage of the McIlwain chopper (Mickle Laboratory Engineering Co.), with the long axis at right angles to the cutting blade. The chopper drops the blade through the tissue, raises the blade and then moves the stage on by a pre-determined distance (the slice thickness) before dropping the blade once more. The hippocampus can be cut in about 15 s.	The stage is placed in the bath, which is filled with oxygenated, ice-cold aCSF. Cut uppermost 1 or 2 mm of the brain straight away. This will reveal the hippocampus in all its transverse glory. Cut slices of the desired thickness. Adjust cutting and vibration speed so that slices can be cut quickly but without pushing and forcing the tissue.
Transfer the filter paper and slices to a dish of oxygenated aCSF and gently separate the slices with a fine paint brush.	Once the hippocampus has been cut through its entirety, all the slices in the bath can be removed, using a wide-mouth (~0.5 cm) Pasteur pipette, to a dish of aCSF where the hippocampal slices can be dissected away from the surrounding tissue.

The slices are then stored in continually oxygenated aCSF and allowed to recover for at least one hour. See method described by Gibb and Edwards (25).

Figure 2. (a) Cutting hippocampal slices using a vibroslice. One hemisphere of the brain is 'superglued' to a block of Sylgard. This ensemble is glued in turn to the nylon stage of the vibroslice and bathed in continually oxygenated, ice-cold aCSF (not shown). The cortex faces the vibrating cutting blade. The stage is then slowly brought towards the blade. (b) The maintenance of a hippocampal slice in a submerged recording chamber. Nylon fibers are glued on a U-shaped piece of stainless steel, in order to form a mesh, and this keeps the slice stable in the chamber. Note that the distance between the fibers leaves enough room to position stimulating and recording electrodes in the CA1 area.

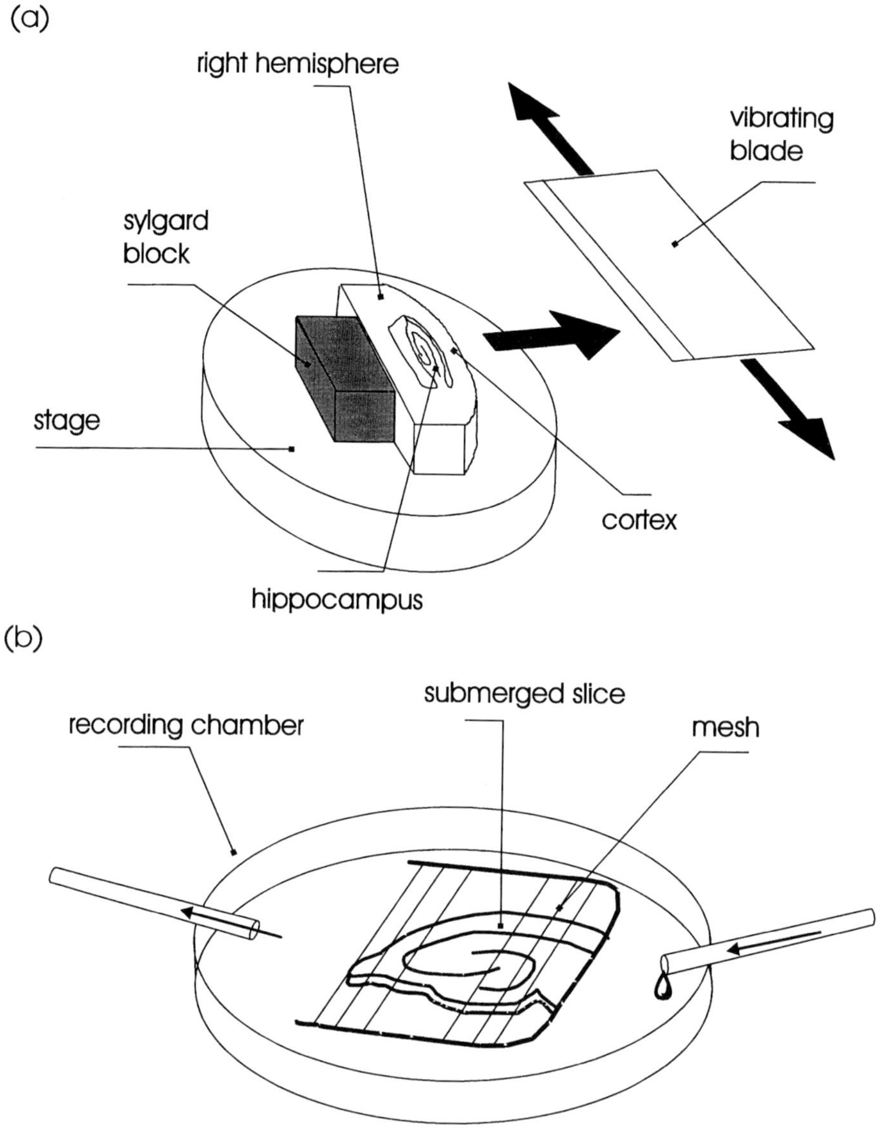

2.2. Recording chambers

There are basically two methods, 'submerged' and 'interface', of maintaining slices in order to record from them (see *Table 3*). These two methods have led to a multitude of different

Table 3. Submerged and interface chambers

Submerged	Interface
Slices are placed on the bottom surface of a small bath (with a volume between 1 and 1.5 ml) through which the bathing medium passes.	Slices are placed on a taut nylon net within a small bath and aCSF flows underneath the net. Capillary action ensures that the slice itself is bathed in a fine film of aCSF.
In order to stabilize the slice, various means can be used. One of the simplest is to use a U-shaped piece of stainless steel with nylon fibers glued across it. These and the weight of the steel satisfactorily hold the slice stationary (*Figure 2b*). The nylon fibers should be sufficiently separated to allow placement of recording and stimulating electrodes as desired.	To provide sufficient oxygen, 95% O_2/5% CO_2 is bubbled into distilled water around the slice chamber. The whole chamber has a lid to provide an O_2 rich environment. If O_2 supply is not sufficient, the slices become 'crinkly' with a dry looking appearance. 'Epileptiform' activity (multiple population spikes) and a deterioration in the viability of the slice also occurs. This can be resolved by increasing the rate of O_2, raising the water level through which the O_2 passes, or changing the height of the chamber lid.

variations of home-made and commercially-available recording chambers. For a fuller discussion of the different types of recording chambers, the reader should consult *Brain Slices*, edited by Raymond Dingledine (4). One problem that can occur, with both submerged and interface preparations, is the appearance of air bubbles in the recording chamber. This is because the aCSF is at a higher temperature in the recording chamber compared with where it was initially oxygenated; therefore keep the temperature of the aCSF the same, as far as possible, throughout its course.

Submerged versus interface

It is generally considered that there is no 'right or wrong' design or type of chamber to use; the experimenter must decide which system will be best for the experiments in question. However, it has recently become apparent that there may be fundamental differences between slices in interface and submerged chambers. Using extracellular recording it was demonstrated that the application of 1S,3R-aminocyclopentane dicarboxylic acid (1S,3R-ACPD), an agonist of metabotropic glutamate receptors (mGluRs), results in a slowly developing potentiation of synaptic transmission (10). This potentiation was shown not to occur in submerged slices. Indeed, if in the same experiment a submerged slice was then put in an interface configuration, the subsequent application of 1S,3R-ACPD resulted in synaptic potentiation (11). The reasons for this are not clear but it was demonstrated that there was an alkalinization of about 0.15 pH units, in the same slice, on going from a submerged to an interface configuration (11). It may be that there are many other differences between 'submerged' and 'interface' slices and more care should be taken in evaluating and considering what effect, if any, such differences may have on the properties of CNS tissue (2).

2.3. Bathing medium

The actual composition of the medium is obviously determined by the nature of the investigation. However, a 'normal' aCSF solution comprises (in mM) NaCl (124), KCl (3), $NaHCO_3$ (26), $NaHPO_4$ (1.25), $CaCl_2$ (2), $MgSO_4$ (1) and D-glucose (10), which is gassed with 95% O_2/ 5% CO_2.

2.4. Some common problems

If no evoked synaptic activity is detected in the slice, it is possible that:

(i) Distilled water instead of aCSF is being used.
(ii) Calcium has not been added to the aCSF. If this is the case a fiber volley, but no synaptic potential, may be observed with extracellular recording.
(iii) If all else appears in order, and there is still no activity, then it is just possible that the slices are dead. This can happen from time to time to even the most experienced of slicers.

Another problem that may occur is that multiple population spikes (epileptiform activity) are evoked in response to a single stimulus. This will occur if:

(i) magnesium is not added to the aCSF (12), or
(ii) if the slices are 'unhealthy', which can occur due to lack of oxygenation either within the recording chamber or to problems/delays during the dissection.

3. EXTRACELLULAR RECORDING TECHNIQUES

3.1. Glass electrodes

For a comprehensive account of the manufacture, use and some of the problems associated with glass electrodes the reader should consult Purves (13) and Millar (this volume). Extracellular recording *in vitro* is usually of field potentials evoked by electrical stimulation of the appropriate afferent pathway(s). Glass extracellular electrodes are filled with a concentrated (4M) solution of NaCl. The resistances of electrodes used are generally in the range of 2–10 MΩ. The electrode is connected via a Ag/AgCl wire, or pellet, within the electrode holder, to the headstage and hence to a DC amplifier. A bath or reference electrode made of chlorided silver wire is placed within the aCSF solution and connected, usually by a gold-plated connector, to the headstage. The electrical activity can be monitored on an oscilloscope, or more frequently nowadays, on a computer screen.

Electrode problems

If the electrodes are too blunt the leakage of a concentrated hypertonic salt solution into the tissue will result in the rapid loss of electrical activity. The leakage is characterized by the appearance of a bright white area centered around the tip of the electrode. The loss of activity should be reversible if the electrode is removed rapidly. This problem can be countered by using sharper, higher resistance electrodes (with an associated increase in electrical noise) or using less concentrated salt solutions (or even aCSF) in the electrode.

Stimulating electrodes

When stimulating an afferent pathway, such as the Schaffer collateral–commissural pathway (SCCP), a bipolar stimulating electrode is generally used. It is connected to an isolated stimulating unit which can deliver either constant current or constant voltage. Stimulating electrodes can be made by twisting together two strands of insulated nickel chromium wire (dia ~55 μm). When this is cut, the ends are exposed and this provides the 'business end' of the stimulating electrode. At the other end, the wires are separated and the insulation scraped away (using a scalpel blade) and these can then be soldered, or connected by some other means (e.g. conducting paint), to the output of the stimulating unit. The stimulating electrode can then be housed in plastic or glass tubing to provide some rigidity, and then mounted on a micromanipulator.

Synaptic responses

Extracellular recording from hippocampal slices is still widely used today because the interpretation of such recordings is relatively straightforward as monosynaptic excitatory potentials can be easily evoked in the absence of polysynaptic events. In addition, extracellular recording has the advantage that stable recordings can be made for many hours; this is very useful for studying long-term changes in synaptic efficiency.

For a given stimulating electrode position, the form of the evoked response is determined by the site of the recording electrode (*Figure 3*). It is generally accepted that the response recorded extracellularly is a reflection of intracellular EPSPs and that IPSPs play little part in the generation of the field potential. However, under certain circumstances, it has proved possible to record extracellular field IPSPs (14). With the recording electrode within the *stratum pyramidale* (cell body layer), and the stimulating electrode in the stratum radiatum, a positive-going field EPSP is recorded. With a sufficiently strong stimulus, a negative-going population spike may be triggered. Within the stratum radiatum, a negative-going field EPSP, which may trigger a positive-going population spike, will be recorded. Fine adjustment of the stimulating and recording electrodes through the depth of the slice can drastically alter the size and form of the evoked response.

Figure 3. Examples of synaptic activity evoked by single electrical stimuli delivered to the SCCP and recorded in the whole-cell, intracellular and extracellular modes. The top trace is an AMPA/NMDA EPSC recorded at –60 mV in the presence of picrotoxin to block $GABA_A$ mediated IPSCs and Cs^+ in the patch pipette to block $GABA_B$ mediated IPSCs. The middle trace is an intracellular EPSP/IPSP recorded at –55 mV. Note the biphasic $GABA_A$ and $GABA_B$ IPSP. Bottom traces are examples of extracellular synaptic responses recorded from the pyramidal (left) and radiatum (right) layers. Adapted from (36).

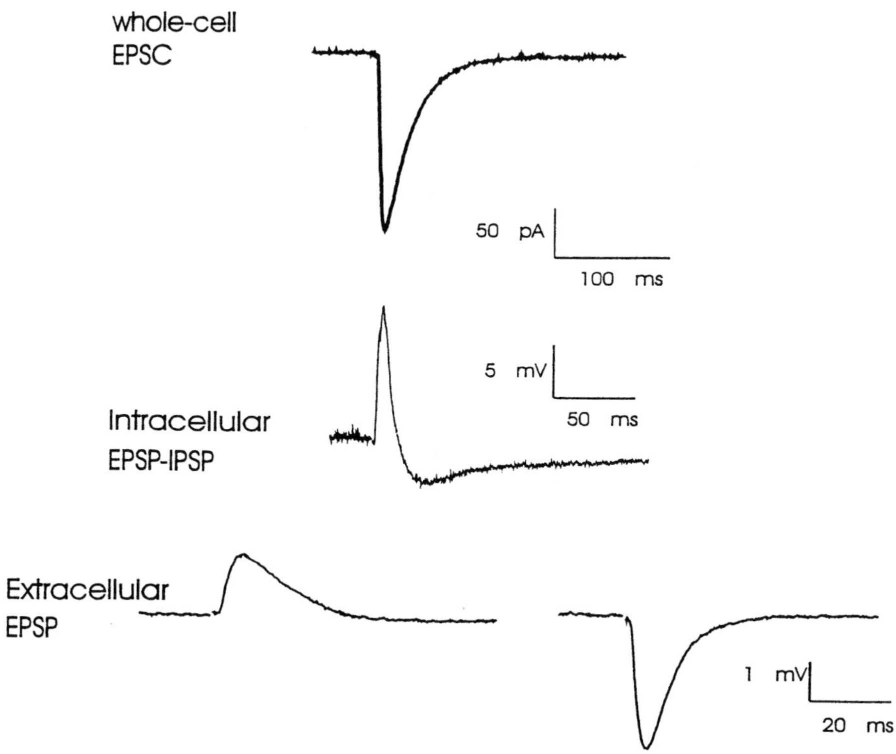

It is important not to stimulate the pathway at too high a frequency as this may result in activity-dependent changes in synaptic efficiency (e.g. long-term potentiation, long-term depression); a test frequency of between 0.1 and 0.033 Hz is therefore often used.

A frequently used method in hippocampal electrophysiology is to stimulate two converging but separate sets of afferent fibers; in this way one of the stimulated inputs can be used as the experimental pathway and the other can act as a control. This can be done in the SCCP by placing two stimulating electrodes within the stratum radiatum, but on each side of the recording electrode (15). Alternatively, two different, independent pathways synapsing onto the CA1 neurons can be stimulated (16, 17). In the SCCP, paired pulse facilitation (PPF) occurs when two stimuli (paired pulse interval of ~50 ms) are delivered to the same pathway. Therefore, when stimulating two separate sets of afferents, the independence of the fibers can be determined by ensuring that there is no paired pulse facilitation between the two pathways.

3.2. Grease-gap recording

The grease-gap preparation has been used to record extracellularly from different biological preparations for many years and has recently been applied to hippocampal slice preparation (18). The essence of the technique is to place the recording electrode in contact with the slice but to isolate the recording electrode, and the part of the slice to which it is connected, from the bathing medium (see *Figure 4* and *Table 4*). In this way the potential difference between a reference electrode in the bathing medium and the isolated recording electrode, in contact with an adjacent part of the slice, is measured. This technique is very useful for some agonist/antagonist pharmacology since dose-dependent changes in the DC potential are seen upon the application of agonists which activate ligand gated ion channels. In addition, synaptic responses can also be evoked and this method provides a relatively stable means of

Table 4. Slice preparation for grease-gap recording

1. Hippocampal slices are prepared in the normal way (see *Table 2*). The alveus is partially dissected (carefully, under magnification) from the stratum oriens (without impinging on stratum pyramidale) using a no. 15 scalpel blade.
2. The slice is placed, with the aid of a fine paint brush, on a plastic platform (e.g. an upside down blackened petri dish), the surface of which has been lightly smeared in grease. Gently remove aCSF from the slice with tissue paper and sweep out the alveus at right angles to the slice.
3. The recording electrode is placed on the alveus, at a short distance from the slice, and both the alveus and electrode are covered in grease (see below). The grease may just go over onto the edge of the slice but not onto the stratum pyramidale (*Figure 4*).
4. To aid the flow of aCSF over the slice, four small pieces of tissue paper (these are not shown in the figure) are wetted in aCSF and placed over the greased side, and the other three sides of the slice. ACSF is delivered (~2 ml min^{-1}) through a heated water jacket (28–30°C) and dripped onto the slice, via a piece of tissue paper to minimize any pulsatile flow. Since the platform is inclined at an angle (~20°), the aCSF flows over the slice and drops off the platform to be caught in a receptacle from where it can be sucked away.
5. A reference electrode is placed in contact with the aCSF solution bathing the slice and an earth electrode, to minimize electrical noise, is placed in contact with the aCSF, where it drips off the slice.

Figure 4. The grease-gap recording technique. A schematic representation of the hippocampal slice, illustrating the partially dissected alveus and the placement of electrodes and the grease barrier. See text for details.

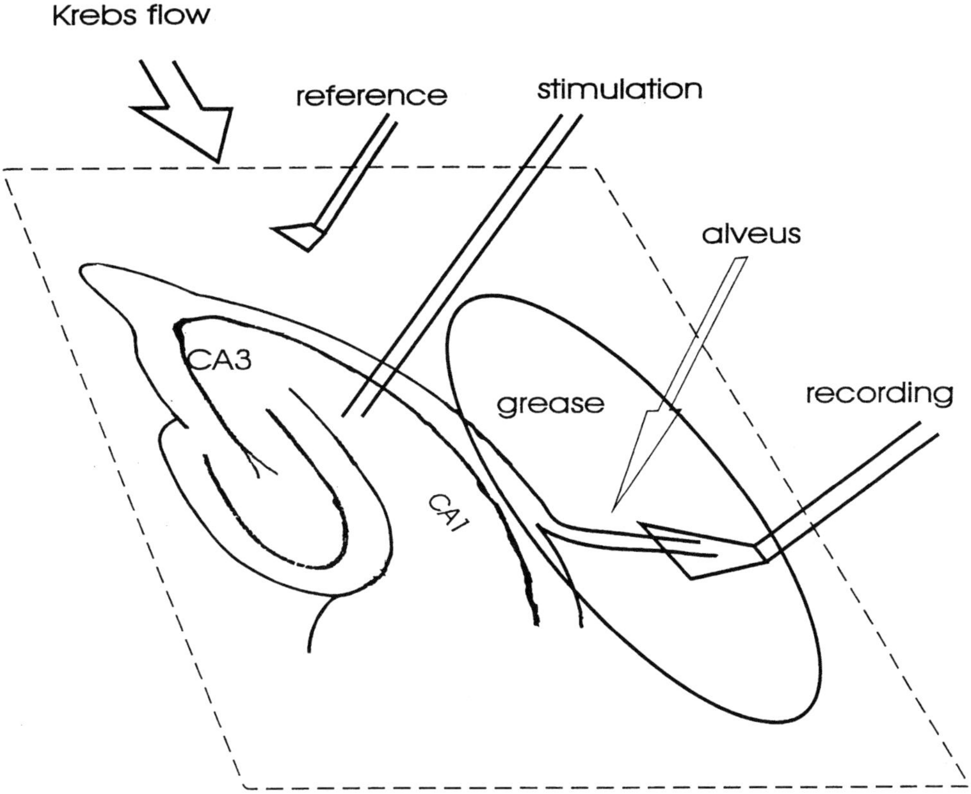

long-term extracellular recording. Indeed, stable recordings have been carried out for over 24 hours (19).

Electrodes and grease

The grease used is a mixture of petroleum jelly and liquid paraffin in a ratio of between 60:40 and 70:30, by weight. The grease is applied over the alveus and recording electrode through a fine hypodermic needle attached to a 2 ml syringe. The recording and reference electrodes are wick electrodes made from tissue paper (lens paper/nappy liner) which is passed through a Gilson pipette tip. The pipette tip is then filled with aCSF and attached by a rubber joint to a glass tube which contains a Ag/AgCl wire or pellet which is connected appropriately to the input of the amplifier. The electrodes can be used for many weeks without problems, providing that air bubbles are not allowed to build up inside the tips. Therefore, they should be stored with the tips submerged in aCSF. Once the electrodes have been set up on the slice, there is often an initial 'settling' time during which there may be a slow shift in DC potential between the recording and reference electrode.

Electrophysiological responses

Synaptic potentials can be evoked by a stimulating electrode in the stratum radiatum (see above) and agonist-induced depolarizing DC potentials (*Figure 5*) induced by delivery of

Figure 5. Electrophysiological responses that can be recorded in the grease-gap preparation. The left panel illustrates dose-dependent changes in the DC level induced by the application of 2 ml of (i) 5 μM and (ii) 10 μM AMPA. The right hand panel illustrates synaptic responses evoked by stimulation of the SCCP at (i) low and (ii) higher stimulus intensity.

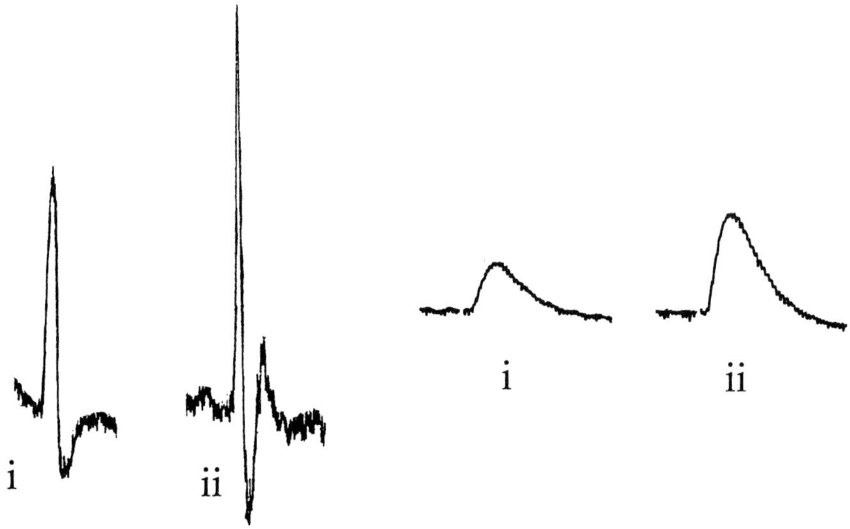

(usually 2 ml) of a given concentration of agonist (e.g. ionotropic glutamate receptor agonists α-amino-3-hydroxy-5-methylisoxazole-4-propionate (AMPA), *N*-methyl-D-aspartate (NMDA) etc.). This technique has proved very useful in investigating the interactions between mGluRs and NMDA receptors as well as determining the actions of some glutamate receptor agonists and antagonists. See refs 20–22 for some recent reports based on this technique.

4. INTRACELLULAR RECORDING

Until relatively recently, intracellular recording in brain slices was confined to using high resistance, sharp electrodes which pierce the cell membrane and thus enter the cell. Since the advent of blind whole-cell patch-clamp recording in slice preparations, much of the voltage-clamp analysis of macroscopic whole-cell currents is now undertaken using this technique, because of the low noise and high resolution provided. However, it should not be thought that sharp electrode recording has in any way been made obsolete (27, 28). Indeed, under certain circumstances, it is an advantage not to dialyse the intracellular contents of the neuron, and sharp electrode techniques are ideal for this situation. For a discussion of sharp intracellular electrodes see Millar, this volume. Patch-clamp recording also enables one to study the activity of single channels in the cell-attached, outside-out or inside-out modes. However, these uses of patch-clamping are outside the scope of this chapter and will not be discussed further. The interested reader should see refs 23–25. We will describe methods of whole-cell recording.

4.1. Patch-clamp recording

Patch-clamp recording, a technique developed by Neher and Sakmann (23), is based on the use of large diameter (1–3 μm) tipped, low resistance (1–10 MΩ) glass electrodes. Until recently, this technique was used exclusively to record from cultured cells, or specially prepared tissue, in which the recording electrode could be placed on individual, visualized cells. It was considered that the electrode tip had to be maintained in a very clean condition otherwise the electrode would not seal sufficiently tightly with the cell membrane to produce a gigaseal; a seal of resistance of at least 10^9 Ω. Indeed this seal is vital to patch-clamp recording. Firstly, the leak current and noise are minimized allowing small, single channel currents to be recorded. Secondly, the seal allows rupturing of the patch immediately underneath the electrode without the electrode and cell becoming dissociated, which permits intracellular (whole-cell) recording with a low resistance electrode.

Patch-clamping in slices

The method of blind whole-cell recording from intact slices was first described by Blanton *et al.* (26), and is a relatively straightforward procedure (see *Table 5* and *Figure 6* for methods). It has recently been demonstrated (37–39) that is possible, upon visualising cells within a slice, to make patch-clamp recordings from dendrites (and whole-cell recording from the cell body of the same neuron). This has the potential advantage of reducing dendritic filtering and space-clamp artefacts (see below).

Synaptic responses

In whole-cell voltage-clamp experiments, the flow of current resulting from either synaptic, agonist or voltage activation of ligand-gated receptors or voltage-gated channels can be measured. Activation of afferents in most regions of the mammalian CNS results in 'fast' synaptic responses which are mediated by glutamate and γ-amino-butyric acid (GABA) receptors. The separation of a complex EPSP/IPSP sequence into its constituent parts can be achieved by the use of selective antagonists of the different components of the response. In the case of an EPSP/IPSP sequence it can be shown that the EPSP is made up of a fast AMPA receptor-mediated and a slow NMDA receptor-mediated component and that the IPSP comprises a fast $GABA_A$ receptor-mediated and a slow $GABA_B$ receptor-mediated component (*Figures 7 and 8*).

Figure 9 illustrates the use of voltage-clamping in the study of NMDA receptor-mediated synaptic transmission: a current–voltage relationship for an NMDA receptor-mediated EPSC is plotted. It shows the characteristic negative slope conductance (at hyperpolarized membrane potentials), which is due to the voltage-dependent Mg^{2+} block of the NMDA receptor-channel (29).

Filling solutions

With whole-cell recording, the contents of the pipette will diffuse throughout the cell. In addition, the soluble contents of the cell will also be diluted in the large pipette volume. Thus, in order to maintain the capacity for cellular metabolism, it is common practice in some laboratories to routinely add ATP and GTP to the intracellular solution. Recording solutions used are commonly potassium or caesium based with acetate, methylsulfonate, gluconate or chloride as the major anions. Thus, a solution that we and others have used successfully is (in mM): $KMeSO_4$ or $CsMeSO_4$ (130), NaCl (1), MgCl (1), HEPES (5), EGTA (0.1). HEPES is a pH buffer and EGTA a calcium buffer.

There are some important points to consider:

(i) The use of a low EGTA concentration allows the investigation of calcium-dependent phenomena, such as the induction of long-term potentiation and depression. It is the experience of some workers that higher EGTA concentrations (5–10 mM) help to

Table 5. Blind patch-clamping in slices

1. Voltage steps (~1 mV) are applied to the electrode (ideally faster than 2 Hz). The current responses, indicative of the electrode resistance, are monitored on a chart recorder or computer screen (*Figure 6a*). Junction potentials should be offset. Apply and maintain positive pressure (a light blowing action) to the electrode via a 1 ml syringe connected via a two-way tap to the electrode holder. This prevents the electrode tip from becoming excessively clogged with debris. The electrode is positioned and then advanced through the tissue, using a fine control micromanipulator, until the seal resistance is seen to increase.
2. The increase in seal resistance (this may be gradual or sudden, may be only a small increase through to a 100% or greater increase) indicates that the electrode tip is in the proximity of a cell (*Figure 6b*). Advance the electrode slowly (not more than a few microns) and a further increase in the resistance should occur. The positive pressure is released by opening the tap. This often results in a further increase in resistance.
3. Gentle suction is applied to the electrode (by mouth, via the syringe) and negative pressure maintained by closing the tap. This should result in the complete (or almost complete) abolition of any remaining current. The negative pressure is crucial in producing a gigaseal, the formation of which is aided by changing the electrode holding potential to –60 mV (*Figure 6c*). At about –30 mV, release the negative pressure; this should help the sealing process. The resistance of the seal can be calculated knowing the current required to hold the patch of membrane under the electrode at the holding potential. Once this has reached 60 pA (at –60 mV), a gigaseal, which will often gradually increase further, has been achieved (*Figure 6d*). However, if, as is sometimes the case, the membrane has become somewhat perforated then the holding current will not provide an accurate measure of the seal resistance. This can be calculated by applying a 10 mV (or greater) step and measuring the size of the current response.
4. To obtain whole-cell mode increase the voltage step to +10 mV. The ZAP function (on Axon Instruments Axopatch), or a variant of this, delivers a 1.5 V step of variable duration (0.1 to 10 ms) to the electrode. This, in combination with brief negative pressure, breaks down the cell membrane. Success is achieved when a burst of Na^+ and Ca^{2+}-dependent spikes is seen arising from a large current step (which is a function of the input resistance and will be several hundreds of MΩ) in response to the 10 mV step. Large capacitance transients, representing the charging of electrode and cell membrane, are seen on the rising and falling phases of the step. Whole-cell access can often be improved by the gentle application of positive pressure and/or gradual withdrawal (of only a few microns) of the electrode from the cell.
5. Once the whole-cell mode has been obtained it is possible to record evoked (and spontaneous) synaptic responses in either voltage-clamp mode (*Figure 6e*) or current-clamp mode.

maintain whole-cell access and this should be considered if changes in Ca^{2+} concentration are not important for the investigation.

(ii) The use of caesium ions (Cs^+) in the filling solution blocks cellular potassium channels and results in a gradual drift, in a depolarizing direction, of the cell membrane potential. Therefore, the presence of Cs^+ increases membrane resistivity which reduces electrotonic filtering, especially of slower membrane events. In addition, the presence of

➤p. 28

Figure 6. Whole-cell patch clamp recording methods in slices. (a) Voltage steps are applied to the electrode (in track mode) and the resulting current steps monitored (upper panel) on a fast and (middle panel) on a slow time base. (b) The seal resistance increases as a cell is approached, as indicated by a decrement in size of the current responses. (c) A seal begins to form between the cell and the electrode and the electrode potential is hyperpolarized to –60 mV. (d) The current required to clamp the cell at –60 mV decreases as the seal becomes tighter. (e) Following breakdown of the membrane underneath the electrode evoked synaptic currents can be recorded in response to electrical stimulation of the SCCP. The cell characteristics (input resistance and electrode/cell charging) are monitored by the injection of a voltage step (+10 mV), which precedes the synaptic response.

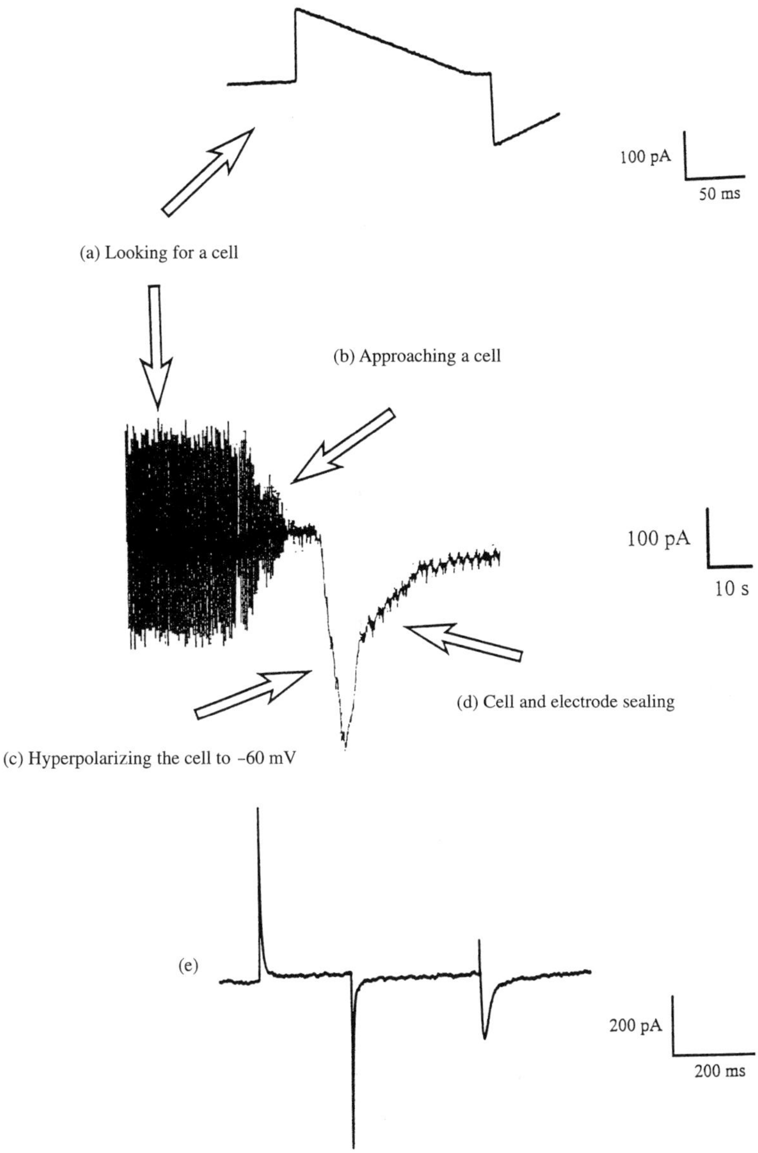

Figure 7. Some of the different components contributing to synaptic transmission at the SCCP synapses, recorded under whole-cell voltage clamp. All synaptic responses, unless otherwise indicated, were recorded at –60 mV. The synaptic response normally comprises a biphasic EPSC and biphasic IPSC. However, (a) with a caesium based intracellular solution the slow component of the IPSC is abolished, leaving the biphasic EPSC and the fast IPSC. (b) The fast IPSC is $GABA_A$ receptor-mediated and is abolished by the antagonist picrotoxin, leaving the biphasic EPSC. (c) The fast component of the EPSC is mediated by AMPA receptors and can be blocked by the additional presence of the antagonist, CNQX. This reveals the slow NMDA receptor-mediated component (recorded at a holding potential of –35 mV to maximize the NMDA response).

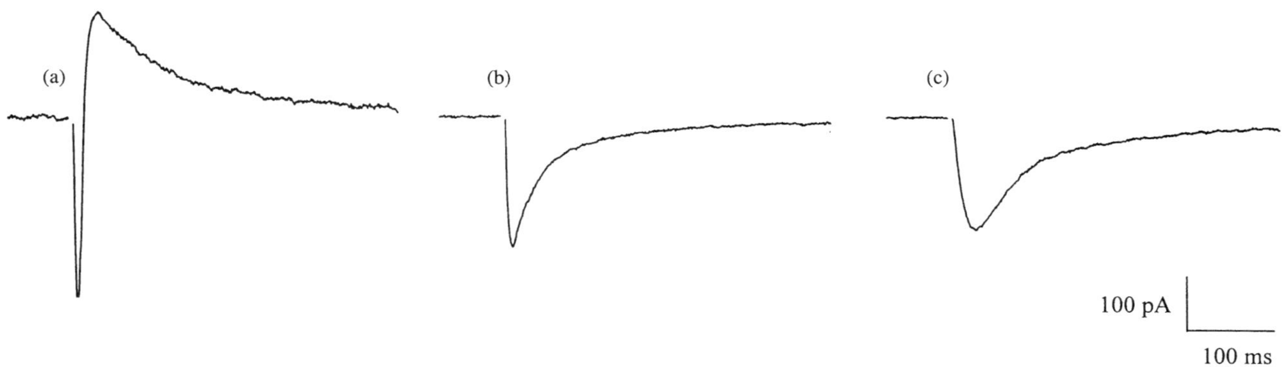

Figure 8. Attenuation of synaptic responses due to resealing of the cell membrane. The example illustrates an NMDA receptor-mediated EPSC recorded at –35 mV. The dotted line illustrates the cell characteristics and evoked synaptic current under conditions of good whole-cell access. The solid line shows how, in the same experiment, sealing of the cell alters both the cell characteristics and the evoked synaptic current. There is a significant slowing of the membrane charging characteristics and a considerable attenuation and slowing of the synaptic current. The time constant of the rising phase of the current is slowed from 12 ms (dotted line) to 20 ms (solid line). The time constant of the decay phase is not significantly affected.

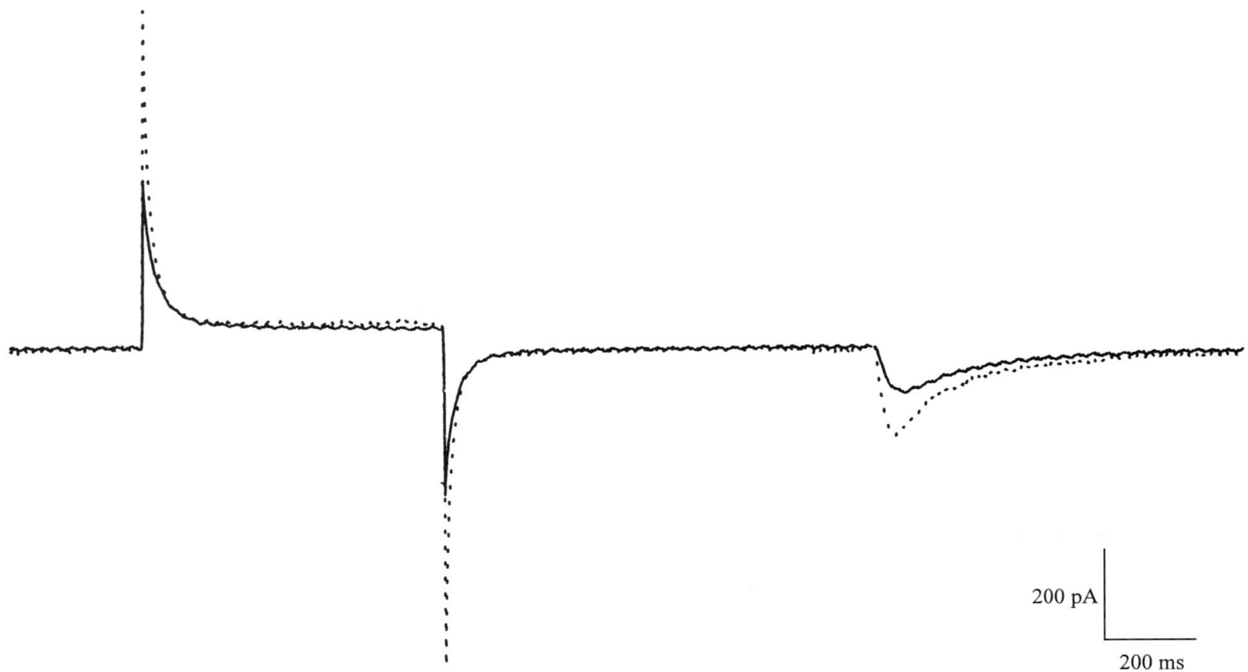

Figure 9. Voltage dependency of the NMDA receptor-mediated EPSC. Synaptic responses were evoked by stimulation of the SCCP. The traces illustrated were evoked at the holding potentials indicated on the graph. The graph shows the current–voltage relationship of NMDA receptor-mediated synaptic transmission.

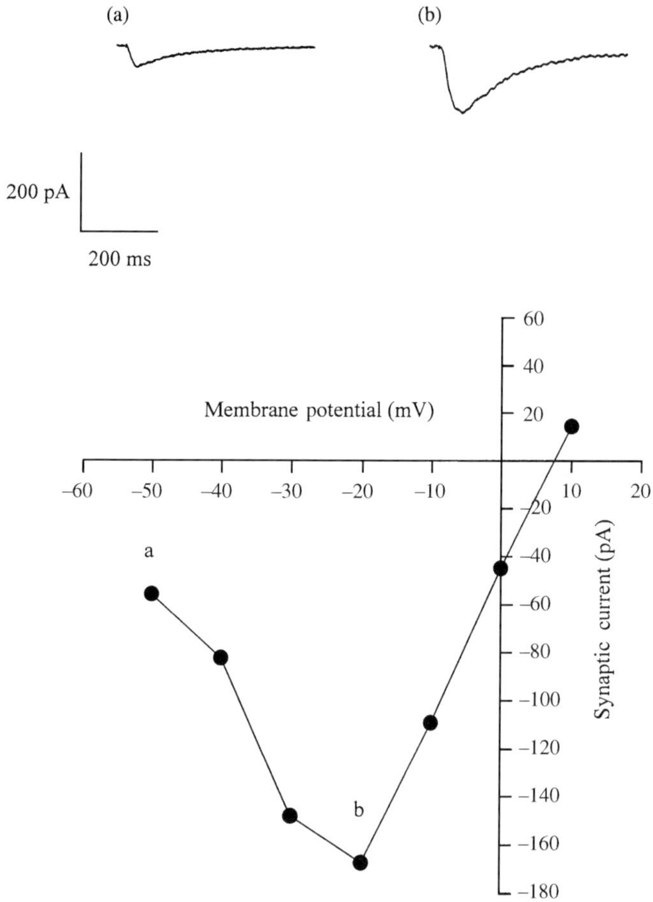

Cs^+ will result in a block of synaptic events (e.g. $GABA_B$ responses) dependent on K^+ channel activation.

(iii) It is also fairly common to add an intracellular blocker of Na^+ channels, such as *N*-(2,6-dimethyl-phenylcarbamoylmethyl)-triethylammonium bromide (QX-314), to the filling solution. This prevents action potential firing and thus decreases the likelihood of voltage breakthrough and helps to maintain the integrity of the voltage clamp.

(iv) Cs or KCl based intracellular solutions are sometimes used in order to study spontaneous IPSCs as this produces large 'depolarizing' IPSCs at 'normal' resting potentials.

(v) It should be borne in mind that cell dialysis is not necessarily a disadvantage and intracellular perfusion (30) of selected compounds can be a very useful technique. For an account of intracellular dialysis, see Kay (31).

4.2. Artefacts and problems

We will review here some common problems encountered in slice patch-clamping which can sometimes make this technique very frustrating. Whole-cell patch-clamping in slices demands a great deal of patience, but can be very rewarding!

Glass electrodes

Problems that can occur are invariably associated with the glass recording electrodes. For whole-cell recording from CA1 neurons we have found that electrodes of *about* 5 MΩ generally work best. These electrodes are pulled on a Flaming/Brown horizontal puller (Sutter Instrument Co., California, USA). This puller allows manipulation of some of the different factors that will influence the electrode, such as the current delivered to the heating coil, duration and strength of the pull, at each of the three stages of the pulling process. Altering these parameters to change electrode characteristics has sometimes been a hit and miss affair, but it seems that the recent introduction of a computer program to aid this process *may* save a lot of frustration. If unsatisfactory electrodes (i.e. electrodes which do not produce gigaseals, whole-cell mode cannot be achieved or stability of such is poor) are being pulled and the problem cannot be cured, then it may be time to change the heating coil. The resistance of the electrode is not necessarily an indicator in predicting a 'good' recording. Other factors, such as the angle and length of the taper and tip size, will influence the behaviour of the electrode. In addition, similar electrodes may not behave similarly when used to record from slices prepared from young animals compared with slices from adult animals. In slices from adult animals connective tissues (e.g. glial cells) tend to clog the electrode more easily than in slices from young animals where these cell types are less represented and the access to neurons is easier.

Stability of the recording

In patch clamping one of the major difficulties is in maintaining a constant whole-cell access over long time periods. Generally, re-sealing of the membrane underneath the electrode causes a decrease in the whole-cell access. This is seen as a diminuation of the amplitude of the capacitive transients evoked by the voltage step. The changing characteristics of the whole-cell access reduces the size and increases the rise time of the synaptic currents (*Figure 8*). In some instances, sealing can be overcome by re-applying a positive pressure to the electrode. However, this procedure must be handled with great care, since in many cases, it leads to a 'loss' of the cell if the pressure re-applied is too high. To avoid sealing, the resistance of the electrode should not be too high. It is imperative therefore that whole-cell access is constantly monitored throughout the experiment.

Series resistance errors

Series resistance (*Rs*) is the combined electrode and access resistance. When current flows across this resistance, there is a drop in potential proportional to $Rs*I$ which results in a discrepancy between the actual and clamped membrane potentials. Since neither *Rs* nor *I* are ever likely to be zero, it is clear that there is always going to be a series resistance error. In most experiments with relatively low resistance electrodes and small evoked or spontaneous responses, *Rs* is not usually considered to be a problem and tends to be ignored. The greatest problems arise when large responses are being recorded and/or large voltage steps are applied and/or *Rs* is large. In these instances care is needed in the interpretation of such results, especially when comparing between different experiments. *Rs* can be measured in current-clamp mode by repetitively applying a current step. The error voltage due to *Rs* can be seen as a fast step component preceding the slower charging of the membrane. The error voltage is then cancelled by the series resistance control which adds a voltage signal proportional to the membrane current, and scaled appropriately to the command voltage of the clamp amplifier.

Series resistance can be compensated for (only partially in practice) and the appropriate manual for the particular amplifier in use should be consulted. In practice it is preferable to try to keep *Rs* as small as possible, by using as low resistance electrodes as is practicable (32).

Space clamp errors

Since the voltage control is delivered to the cell body, it is pertinent to ask whether dendrites (and spines), which may be some millimetres from the clamp source are isopotential with the cell body, and hence the clamp potential. Thus, there is a potential source of error in that the synaptic potential, generated at a distance from the cell body, is not in fact being subjected to the voltage clamp to the same extent as the cell body. In practice, when recording *in situ* from cells with synapses at a distance from the clamp source, there is little that can be done to overcome any theoretical or practical errors that may occur as a consequence of space clamp problems. It should always be borne in mind however that space-clamp and electrotonic filtering can severely attenuate recorded synaptic responses (33).

5. STIMULATION, DATA ACQUISITION AND ANALYSIS

For a comprehensive account of this subject, see ref. 34. When studying evoked synaptic activity it is usual to use the computer system to trigger the stimulus and to record the activity. This is necessary to time-lock the stimulus, data acquisition and any on-line analysis. Therefore, the computer triggers the stimulator and starts data acquisition. The signal from the electrode is amplified and then fed to the computer via an A/D converter and stored on the hard disk.

There are many different means of storage of electrophysiological recordings. Hard copies, such as chart recordings, are often used where a continuous record of relatively slow activity, which may not necessarily require any extravagant analysis, is required. For example, DC potentials evoked by agonists in the grease gap preparation, or potentials or currents evoked by iontophoretic agonist application are generally recorded on a continuous chart record. On the other hand, fast events such as evoked and spontaneous potentials and currents, which could not be satisfactorily recorded by the above means, are recorded on magnetic medium such as video or audio tape, or directly on computer disk. Whichever of these is used there is the advantage that these responses can be subjected to analysis either on-line, or off-line at the end of the experiment.

Generally speaking, computer programs for electrophysiological recording are a combination of stimulation, data acquisition and analysis programs. The program, or suite of programs, that one uses is determined firstly by the amplifier(s) and interfaces that one has, and whether these have dedicated programs, and secondly the needs of the experimenter. Probably the most widely used amplifiers/interfaces are those manufactured by Axon Instruments (Axoclamp and Axopatch). This company also provides a fairly exhaustive set of programs (pClamp) to deal with stimulation, acquisition and analysis. There is now a comprehensive publication by Axon Instruments (35) which provides a useful background text to electrophysiology and Axon amplifiers/software. However, one of the main drawbacks of the Axon software, from a personal point of view, is the relative lack of on-line analysis and means of viewing any such analysis. When conducting electrophysiological experiments in which synaptic responses are being monitored, it is of great value to be able to ascertain accurately on-line what effect a certain manipulation (be it drug application, some form of stimulation etc.) has on the recorded response. Another downside has been the lack of a simple means of obtaining alternating outputs to drive two separate stimulating electrodes.

Because of the above, and other limitations of some of the commercially available data acquisition and analysis software, an in-house program for extra and intracellular recording has recently been developed by Dr William Anderson (Department of Anatomy, University of Bristol). This program interfaces with a variety of hardware and is currently being used very successfully in our own and several other laboratories. The major advantages of this program are the on-line analysis and graphing of the experiment and automated off-line analysis.

Brain Slices

6. ACKNOWLEDGMENTS

We wish to thank Dr Nicola Beretta and Dr Tim Benke for helpful comments and discussion. Z.B. is supported by the Wellcome Trust and M.V. by the European Community.

7. REFERENCES

1. Lipton, P., Aitken, P.G., Dudek, F.E., Eskessen, K., *et al.* (1995) *J. Neurosci. Meth.*, **59**, 151.

2. Aitken, P.G., Breese, G.R., Dudek, F.F., Edwards, F., *et al.* (1995) *J. Neurosci. Meth.*, **59**, 139.

3. Alford, S.T., Frenguelli, B.G., Schofield, J.G. and Collingridge, G.L. (1993) *J. Physiol*, **469**, 693.

4. Dingledine, R. (1984) *Brain Slices*, Plenum Press, New York and London.

5. Stasheff, S.F., Anderson, W.W., Clark, S. and Wilson, W.A. (1989) *Science*, **245**, 648.

6. Fujii, T., Baumgartl, H. and Lübbers, D.W. (1982) *Pflügers Arch.*, **393**, 83.

7. Collingridge, G.L. (1995) *J. Neurosci. Meth.*, **59**, 5.

8. Li, C-L. and McIlwain, H. (1957) *J. Physiol.*, **139**, 178.

9. Yamomoto, C. and McIlwain, H. (1966) *J. Neurochem.*, **13**, 153.

10. Bortolotto, Z.A. and Collingridge, G.L. (1993) *Neuropharmacol.*, **32**, 1.

11. Bortolotto, Z.A., Bashir, Z.I., Davies, C.H., Taira, T., Kaila, K. and Collingridge, G.L. (1995) *J. Neurosci. Meth.*, **59**, 19.

12. Coan, E.J. and Collingridge, G.L. (1985) *Neurosci. Lett.*, **53**, 21.

13. Purves, R.D. (1981) *Microelectrode Methods for Intracellular Recording and Ionophoresis*, Academic Press, London.

14. Lambert, N.A., Borroni, A.M., Grover, L.M. and Teyler, T.J. (1991) *J. Neurophysiol.*, **66**, 1538.

15. Bashir, Z.I. and Collingridge, G.L. (1992) *Eur. J. Neurosci.*, **4**, 485.

16. Andersen, P., Sundberg, S.H., Sveen, O. and Wigstrom, H. (1977) *Nature*, **266**, 736.

17. Lynch, G., Dunwiddie, T. and Gribkoff, V. (1977) *Nature*, **266**, 737.

18. Blake, J.F., Brown, M.W. and Collingridge, G.L. (1988) *Br. J. Pharmacol.*, **95**, 291.

19. Blake, J.F. (1989) *A Novel Technique for Investigating Excitatory Amino Acids in Area CA1 of the Rat Hippocampal Slice*, PhD Thesis, University of Bristol.

20. Frenguelli, B.G., Blake, J.F., Brown, M.W. and Collingridge, G.L. (1991) *Br. J. Pharmacol.*, **102**, 355.

21. Harvey, J. and Collingridge, G.L. (1993) *Br. J. Pharmacol.*, **109**, 1085.

22. Bashir, Z.I. and Collingridge, G.L. (1994) *Exp. Brain Res.*, **100**, 437.

23. Sakmann, B and Neher, E. (1983) *Single Channel Recording*. Plenum Publishing, New York.

24. Hamill, O.P., Marty, A., Neher, E., Sakmann, B. and Sigworth, F.J. (1981) *Pflug. Arch.*, **391**, 85.

25. Ogden, D. (ed.) (1994) *Microelectrode Techniques: The Plymouth Workshop Handbook*, Company of Biologists Ltd, Cambridge.

26. Blanton, M.G., LoTurco, J.J. and Kreigstein, A.R.J. (1989) *J. Neurosci. Meth.*, **30**, 203.

27. Larkman, A., Stratford, K. and Jack, J. (1991) *Nature*, **350**, 344.

28.Voronin, L.L., Kuhnt, U., Hess, G., Gusev, A.G. and Roschin, V. (1992) *Exp. Brain Res.*, **89**, 248.

29. Nowak, L. Bregestorski, P., Ascher, P., Herbert, A., and Prochiantz, A. (1984) *Nature*, **307**, 462.

30. Neher, E and Eckert, R. (1988) in *Calcium and Ion Channel Modulation* (D. Armstrong and M.B. Jackson, eds), Plenum Press, New York, p 371.

31. Kay, A.R. (1992) *J. Neurosci. Meth.*, **44**, 91.

32. Armstrong, C.M. and Gilly, W.F. (1992) in *Methods in Enzymology* **207**, Academic Press, San Diego, p. 100.

33. Spruston, N., Jaffe, D.B. and Johnston, D. (1994) *TINS*, **17**, 161.

34. Dempster, J. (1993) *Computer Analysis of Electrophysiological Signals*. Academic Press, San Diego.

35. Sherman-Gold, R. (ed.) (1993) *The Axon Guide for Electrophysiology and Biophysics*, Axon Insts, 1101 Chess Drive, Foster City, CA, USA.

36. Collingridge, G.L. (1992) *Exp. Physiol.*, **77**, 771.

37. Stuart, G.J. and Sakmann, B. (1994) *Nature* **367**, 69–72.

38. Magee, J.C. and Johnston, D. (1995) *Science* **268**, 301–304.

39. Spruston, N., Jonas, P. and Sakmann, B. (1995) *J. Physiol.* **482**, 325–352.

CHAPTER 3
NEUROANATOMICAL TRACING BASED ON CELLULAR TRANSPORT: HORSERADISH PEROXIDASE, *PHASEOLUS VULGARIS*-LEUCOAGGLUTININ AND BIOTINYLATED DEXTRAN AMINE

F.G. Wouterlood, P.H. Goede, B. Jorritsma-Byham and M.P. Witter

1. INTRODUCTION

Mapping of the complex connectivity in the central nervous system by means of experimental neuroanatomical tract-tracing methods began in 1850 when Waller (1) published his work on degenerating nerve fibers in the spinal cord. An amazing array of tracing methods and tracers has been introduced since then. Even today a continuous search is still going on for new neuroanatomical tracers.

A gross distinction can be made between *physiological* and *physicochemical* neuroanatomical tracing methods. Physiological are all those methods that exploit the uptake of a tracer substance by a *living* neuron and its subsequent transport into the cell's processes. The present chapter deals with several physiological methods. All other tracing methods fall into the category of physicochemical techniques. The latter category comprises techniques that capitalize on phenomena as diverse as Golgi silver-impregnation, diffusion of dyes in intracellularly injected neurons in fixed brain slices, and the enhanced argyrophilia of degenerating fibers and axon terminals that occurs following a surgical or chemical lesion or after ischemia.

A second classification of neuroanatomical tracing methods is based on the direction in which transport, diffusion, or degeneration occurs. This classification is most commonly used. *Anterograde* and *retrograde* tracing methods are distinguished. Anterograde transport means that the label is taken up by the dendrites and the cell body of a neuron. From here, the tracer is carried via cytoplasmic transport through the axon to the terminal axonal arborization including its terminal boutons (*Figure 1*). Retrograde transport occurs in the opposite direction. Typical anterograde tracing methods are the autoradiographic method which utilizes the uptake, incorporation in proteins and the subsequent anterograde transport of [^{3}H]-amino acids, and the techniques which utilize the transport of lectins, carbohydrates and biotinylated compounds. Retrograde methods comprise those that exploit the transport from the axon terminals back to the cell bodies of substances like fluorescent compounds, horseradish peroxidase, viruses, bacterial toxins and their fragments, gold-conjugated colloidal particles, latex microspheres, and so forth.

Both physiological and physicochemical methods can be of the anterograde or retrograde type. Difficult to classify is Golgi silver-impregnation since the silver salt deposits 'grow' in all the processes of neurons until these are completely filled (2).

Figure 1. Scheme indicating the two ways of transport distinguished in neurons. Anterograde = uptake by dendrites and/or the neuronal cell body and transport towards the terminal arborization and boutons terminaux. Retrograde = uptake by terminal boutons and transport along the axon towards the cell body.

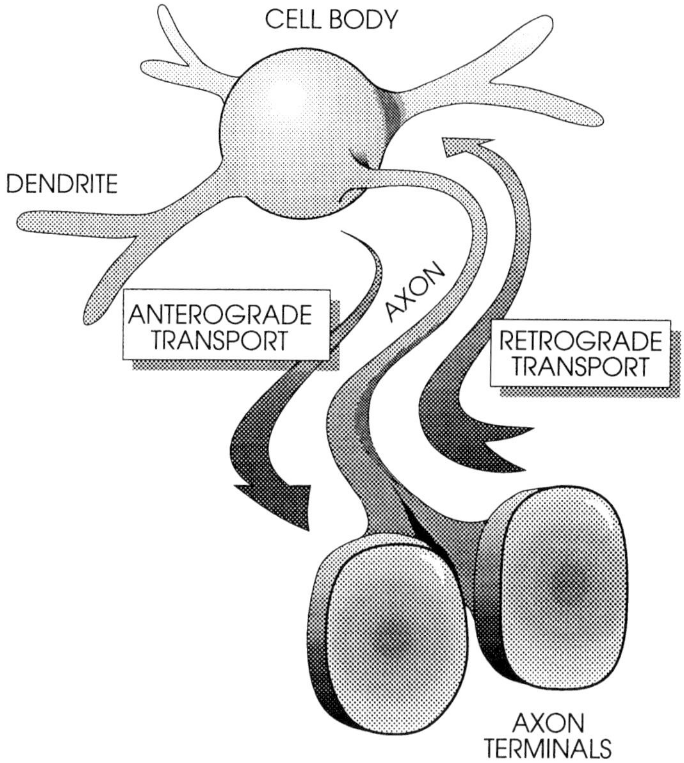

This chapter focuses on three widely used physiological neuroanatomical tracing methods. One is considered to be classical while the others are more recent. All three methods are based on the uptake and transport of tracer substances that are deposited in the extracellular space around the neurons of interest: horseradish peroxidase (HRP), *Phaseolus vulgaris*-leucoagglutinin (PHA-L), and biotinylated dextran amine (BDA). The uptake and transport occur during a 'survival time'. Afterwards, the location of the transported tracers inside the neurons is histochemically or immunocytochemically visualized. In this chapter we deal only with three of the most widely used methods since it is impossible to deal with all available methods. In the textbooks by Heimer and Robards (3), Heimer and Záborszky (4), Hökfelt *et al.* (5) and Bolam (6), extensive information can be found on nearly all tract tracing methods presently available.

An 'ideal' tracer should meet the following demands. First, it should visualize one neuron or a population of neurons to their full extent, including the longest fiber collaterals, the entire terminal fiber arborizations and all the specializations. Second, the area of application should unequivocally be demarcable. Third, to verify into what direction a fiber projects, the label should be transported exclusively in one direction, that is either anterogradely or retrogradely (*Figure 1*). Fourth, the marker should only be taken up by neurons and not by undamaged or damaged fibers traversing the area where the tracer has been deposited. Fifth, the transport should be restricted to one neuron of a chain of neurons, and not continue into subsequent neurons (so-called transsynaptic transport). Sixth, the detection should be uncomplicated and also be modest as regards investments in equipment, time and labor. Finally, the method should be compatible with other methods, for instance with the immunocytochemical detection of neuroactive substances, and with electron microscopy for the visualization of synaptic contacts.

None of the three methods described hereafter fulfills all the above criteria of an ideal tracing method. Yet, because many of their characteristics approach the ideal and because of their versatility, they are widely used in studies on connectivity in the peripheral and central nervous systems of a variety of animal species.

2. CLASSICAL RETROGRADE TRACING: UPTAKE AND TRANSPORT OF HORSERADISH PEROXIDASE (HRP)

The ability of horseradish peroxidase (HRP) to act as a catalyst for the conversion of diaminobenzidine into an insoluble reaction product was discovered only thirty years ago (7). In 1971, Kristensson and Olsson (8) discovered that HRP is taken up by peripheral nerves and retrogradely transported to the parent spinal motoneurons. Shortly afterwards, LaVail and LaVail (9) reported that HRP is also retrogradely transported in the central nervous system. In both studies the transported HRP was detected with diaminobenzidine (DAB) according to Graham and Karnovsky's (7) original method. The transported tracer is present in numerous small puncta in the cell bodies (lysosomes, ref. 10). The simplicity of the tracing method, and the fact that the DAB-precipitate is sufficiently dense to act also as an identifying label in the electron microscope, has led to its universal acceptance. Thus, a way was introduced to use the physiological transport within nerve cells for identifying the connectivity of these cells.

Later it was discovered that HRP is transported anterogradely as well. The anterograde component is made visible when the uptake is enhanced by using HRP conjugated to wheatgerm agglutinin (11) or cholera toxin subunit B (CTB). These conjugates bind to receptors in the neuronal cell membranes and are thus more easily internalized than free HRP which is taken up by pinocytosis only (10). The use of an immunohistochemical detection procedure instead of histochemistry further improves the sensitivity and results in the visualization of thin fibers and their axon terminals (12). Unfortunately, the tracer appears to be taken up and transported by fibers passing the site of injection that are damaged during the injection procedure.

Today, HRP tracing is used in a variety of protocols, mostly in a typical retrograde tracing setting, but also in combined anterograde–retrograde tracing. In addition to DAB, a variety of peroxidase-compatible chromogens has been introduced: benzidine dihydrochloride (BDHC, 13), *o*-dianisidine (14); *o*-tolidine (15), tetramethylbenzidine (TMB, 16), colloidal gold (17) (see also ref. 12), and vector-VIP substrate (66).

2.1. Anesthesia, surgery and injection

Rats are appropriately anesthetized and placed correctly in a stereotaxic apparatus. HRP can be applied in several ways. For the tracing of fibers of peripheral nerves into the CNS, use can be made of pieces of sterile, HRP-drenched gel-foam (18) or HRP-crystals (12). In the CNS itself, foam pieces are usually too big, and injections with HRP are common here, applied either mechanically or iontophoretically. Typically, a 35% HRP solution is prepared in saline (0.9% NaCl) prior to injection. The type of HRP used is Sigma type VI (Sigma, St Louis, MO) or Boehringer (Boehringer, Mannheim, FRG). Mechanical injections – that is, of volumes in the 0.1–0.5 μl range via a syringe – tend to be much larger than those made iontophoretically, through glass pipettes. Alternatives are the implantation of HRP-pellets, or HRP-containing polyacrylamide gels (19).

The skull over the brain area of interest is opened, and the injection instrument lowered to the desired stereotaxic location. The latter can be derived from several atlases of the rat brain. Commonly used today are the stereotaxic atlases by Paxinos and Watson (20) and Swanson (21). When mechanical injections are made it is advisable to reduce the outflow of the syringe by injecting the desired amount (rats: 0.1–0.3 μl) over a period of 10 minutes. Then, the syringe should be kept in place for another 10 minutes and after this period slowly retracted to allow the brain tissue to close over the injection site. Otherwise, labeling along the injection track may become prominent, or leakage may occur of HRP into the subarachnoidal space. It should be noted here that a disadvantage of the HRP-method is the difficulty in accurately delineating the *effective* site of injection. At the time that the animal is sacrificed, most HRP has already been taken up by terminals in the injection area and transported away, at least from the peripheral zone of the site of injection.

2.2. Survival time

The speed of retrograde transport of HRP is estimated at 48–120 mm per day (12). After uptake, the HRP is transported in small vesicles from the periphery to the cell body and ends its service life in lysosomal organelles where the enzyme is metabolically degraded (10, 12). This implies that during the survival period the enzymatic activity slowly diminishes. Although the best survival time for each new experiment should be determined empirically, as a rule of thumb the survival time depends on the animal species used, the system under study, and the length of the involved fibers and the density of the terminal boutons at the site of injection. In the early publications, relatively short survival times were used (24–48 h).

Today, a safe survival time as a starting point for fine-tuning of the optimum is considered 4 days in mammals, and twice as long in cold-blooded animals. Mesulam (12) advises against survival timer longer than 5–7 days, since a considerable amount of the HRP may then have undergone metabolic degradation.

2.3. Fixation

The most common way of fixing the central nervous system is via transcardial perfusion. The animals are anesthetized with a large dose of sodium pentobarbital (e.g. Nembutal, Ceva, France; 60 mg/kg bodyweight). When strong stimuli produce no reflexes any more (end-of-tail pinching, forepaw pinching), the animals are placed in a supine position and the thorax is opened. A cannula is inserted into the left ventricle and moved upwards into the ascending aorta. A clamp is placed on the descending aorta, and an incision made in the right atrium. Rinse fluid (a warm saline solution with some heparin) is allowed to flow (200 ml/min; hydrostatic pressure about 50 mmHg) through the animal until the fluid flowing out of the right atrium is clear from erythrocytes. Good vascular rinsing is an important step since erythrocytes contain strong peroxidase activity that may seriously compete with that of the transported HRP during the final histochemical staining.

A debate has been going on for a length of years what fixative would be the best for maximizing the yield of HRP-labeling. This dispute has never been satisfactorily settled, and the glutaraldehyde–paraformaldehyde–sucrose fixative prescribed by LaVail and LaVail (9) still gives good results in sections used both in light microscopy and in electron microscopy. We currently use the solution shown in *Table 1* which contains phosphate buffer at a osmolarity of 125 mM to compensate for the absence of any sucrose.

Table 1. Fixative for demonstration of transported HRP

1.5% glutaraldehyde
1% freshly depolymerized paraformaldehyde
125 mM phosphate buffer, pH 7.6
Together in a volume of 1000 ml, temperature 38°C, pH 7.6.

2.4. Dissection and sectioning

The brain is removed from the skull immediately after the perfusion and immersed in a 30% sucrose solution. Upon equilibration, sections are cut on a freezing microtome. For EM purposes, sections can be cut immediately on a Vibratome. The sections are collected and stored at 4°C in 125 mM phosphate buffer, pH 7.6. Prior to the histochemical procedure, the sections are rinsed in 50 mM Tris brought at pH 7.6 with 1N HCl.

2.5. Histochemical procedure

(i) Rinse 3 × 10 minutes in 50 mM Tris–HCl (pH 7.6).

(ii) Preincubate 15 minutes in 3-3′ diaminobenzidine tetrachloride (DAB, Sigma; 5 mg DAB in 10 ml of 50 mM Tris–HCl, pH 7.6). Filter the DAB solution prior to use. The tissue will be soaked with DAB that, because of the lack of electron donor, cannot react with the peroxidase.

(iii) Start the histochemical reaction by adding the electron donor, usually 3.3 μl of 30% H_2O_2 per 10 ml of incubation medium. The formation of (brown) DAB-precipitate can be monitored by taking sections out at intervals, rinsing them in Tris–HCl buffer, and inspecting them in a microscope. When the histochemical reaction goes too fast, a slow-releasing glucose oxidase-based electron donor system can be used instead of H_2O_2 (22).

(iv) When sufficient precipitate is formed, the reaction is terminated by repeated washing in 50 mM Tris–HCl, pH 7.6. The sections are subsequently mounted on glass slides from a solution of 0.2% gelatin in 50 mM Tris–HCl, pH 7.6, dried, counterstained if necessary, dehydrated, and coverslipped. Sections destined for electron microscopy are rinsed in 100 mM sodium cacodylate buffer, pH 7.6, postfixed for 1 h with 1% osmium tetroxide in 100 mM cacodylate buffer, pH 7.6, dehydrated in graded ethanol, and embedded in epoxy resin via a propylene oxide step.

Notice. Diaminobenzidine is a potential carcinogen and should be treated with care, that is, in a container placed in a fume hood. Always use gloves and decontaminate glassware, instruments and disposables that have been in contact with DAB with household chlorine bleach. To avoid contamination of microscopes, always rinse test sections first in 50 mM Tris–HCl, pH 7.6 or 8.0, before inspecting the progress of the staining in the microscope. This procedure should also be followed with the visualization of PHA-L and BDA (below).

3. ANTEROGRADE TRACING USING A LECTIN, PHASEOLUS VULGARIS-LEUCOAGGLUTININ (PHA-L)

PHA-L was first isolated from red kidney bean extracts, *Phaseolus vulgaris* (20). Two types of subunit have been identified: the leucoagglutinating (L) and the erythrocyte-agglutinating (E) varieties. One molecule of *Phaseolus* lectin contains four subunits that can be of any five possible combinations of L- and E-subunits. The 115 000 molecular weight PHA-L consists of exclusively L-subunits (24, 25), and only this lectin is taken up and transported by nerve cells (26). The uptake of PHA-L is receptor-mediated. Tomimoto *et al.* (27) reported the uptake and transport in CNS neurons of *Phaseolus vulgaris*-agglutinin P-form (PHA-P), but this variant most probably is a mixture of PHA-L and other isolectins composed of L- and E-subunits in various proportions.

The advantage of PHA-L over other tracers is that it fulfills nearly all criteria of an ideal tracer. There is a small, often negligible, retrograde transport component (see *Table 2*). The immunocytochemical processing of the brain tissue takes several days. This is relatively time- and labor-consuming compared with the histochemical processing of HRP-labeled brain tissue. Visualization of labeled neurons and neuronal processes is however complete and includes all cellular details such as dendritic spines and varicosities on the fibers. Light microscopy of PHA-L labeled neurons can be followed by electron microscopy of the same preparations (28, 29); electron microscopy of PHA-L labeled fibers and axon terminals has also successfully been combined with postembedding immunogold cytochemistry (30).

Table 2. Direction and degree of transport

Tracer	anterograde transport	retrograde transport
HRP[a]	–	++
HRP[b]	+++	+++
PHA-L	+++	–/+
BDA	+++	+

[a] unconjugated, [b] WGA- or CTB-conjugated
Degree of transport: – no transport, –/+ small amount of transport, + some transport, ++ good transport, +++ almost all transport in this direction

3.1. Application of the tracer

PHA-L is commercially available as a lyophilized powder (Vector, Burlingame, CA). It is dissolved at 2.5% in 50–100 mM sodium phosphate-buffered saline, pH 7.4 (26, 31, 32) or in 50 mM Tris–HCl buffered saline, pH 7.4 (e.g. refs. 28, 33, 34). Aliquots of 2.5% PHA-L solution can be deep frozen (–20° to –40°C) and stored for years. Repeated thawing and freezing has an adverse effect on the amount and 'crispness' of the labeling. The tracer is iontophoretically applied via micropipettes with a tip diameter of 10–30 µm. Tip diameter is critical. When larger diameters are used, no uptake and transport occurs. Several authors however claim good results with mechanically injected PHA-L (35–37).

The animals are appropriately anesthetized and placed in a stereotaxic apparatus. The skull over the brain area of interest is removed, and the injection instrument lowered to the desired stereotaxic location. A positive, pulsed DC current is applied to the micropipette. The most commonly used frequencies range between 0.14 Hz (e.g. 26, 34) and 1.0 Hz (38). The strength of the current is in the 2–10 mA range (e.g. refs 26, 39), and the application time 15 minutes. Notice that the small diameter of the pipette tip produces an impedance in the megohm range. The current source must be powerful enough to maintain a relatively high voltage (hundreds of volts) against this impedance. This is necessary to produce the desired current of 3–10 mA. Any power source that satisfies these demands is usable. Most authors use a Midgard CS3 constant current source (Stoelting, Chicago, IL).

The size of an injection strongly depends on the architecture of the brain region where the PHA-L is injected. The sites of injection are usually small, and can be made larger only within limits. Since extended injection times or increased injection currents produce only slighty larger injection sites, large areas to be injected can best be covered by multiple 'standard' injections close to each other. A consequence of a higher current is that the PHA-L may start to precipitate at some point. This then results in a clogged pipette. Very small injection sites are obtained by using pipettes with a small tip diameter, or by applying a low current. Reducing the duration of the injection usually produces less well stained neurons and less transport.

Reports on mechanical injections of PHA-L are scarce and are limited to rat (35), cat (37) and monkey (38). Following mechanical injections of (5–10%) PHA-L in rats, injection sites consisted of a central, darkly stained zone with less darkly stained neurons in the periphery (35).

3.2. Survival time

According to Gerfen and Sawchenko (26), the appearance of the injection site changes very little from 20 minutes up to seven days after the injection. A few hours of survival already produces anterograde labeling at some distance from the injection site. A transport rate can be anticipated of 4–6 mm per 24 hours (26). Thus the length of the pathway to be traced is an important factor. Also notice that the rate of transport may vary in different neuronal systems and animal species. In the rat, a survival time to begin experimenting with could be one week. PHA-L remains detectable up to 6 weeks after injection (40).

3.3. Fixation and immunocytochemical procedure

PHA-L is relatively tolerant with respect to fixatives. PHA-L 'withstands' fixatives which contain up to 2.5% glutaraldehyde. Thus, in tracing/immunofluorescence studies, use can be made of buffered 4% formaldehyde to reduce background autofluorescence (41–43). A small amount of glutaraldehyde (0.1–0.2%) is used when the objective is single PHA-L tracing or PHA-L tracing combined with oligopeptide- or enzyme immunocytochemistry (44, 45). Neurotransmitter immunocytochemistry combined with PHA-L tracing may require a higher amount of glutaraldehyde in the fixative; for electron microscopy of PHA-L labeled fibers and axon terminals, as well as for the detection of transported PHA-L in conjunction with GABA, we have used fixatives containing up to 2.5% glutaraldehyde.

The following is a common procedure in the rat. Use of another animal species, or any combination in rat brain with immunocytochemistry for a second antigen, may require modification.

(i) The animals are anesthetized and perfused with a vascular rinse fluid and a fixative as described in section 2.2. The vascular rinse fluid consists of 8.5 g NaCl, 0.25 g KCl, and 0.20 g $NaHCO_3$ dissolved in 1000 ml distilled water (body temperature) (46).

(ii) The pH is adjusted to 6.9 by bubbling through a mixture of 95% oxygen and 5% carbon dioxide. One young adult rat (200 g) requires about 100 ml of rinsing fluid (thoracic aorta closed with an artery clamp).

3.4. Fixative

We use per animal 1 liter of 125 mM phosphate buffer, pH 7.6, which contains 4% freshly depolymerized paraformaldehyde and 0.1% glutaraldehyde. The fixative may additionally contain 0.2% of a saturated solution of picric acid to improve the fixation of myelin.

The brain is usually dissected immediately after the perfusion, but can also be dissected later (e.g. two hours after perfusion). Brain sections can be cut immediately with a Vibratome or immersed in a cryoprotectant and sectioned on a freezing microtome. Although 30% sucrose as a cryprotectant suffices, a mixture of 2% DMSO–20% glycerin in 125 mM phosphate buffer (47) diffuses faster into the brain and produces in our hand better results. Sections of 40 μm thickness are subjected to a free-floating immunocytochemistry procedure.

3.5. Visualization procedure

PHA-L can be visualized with immunofluorescence-, immunoperoxidase- or avidin-biotin-peroxidase (ABC) methods. Immunofluorescence is routinely used for the purpose of quick inspection of injection sites, or used in combination with retrograde fluorescent tracing. Both the ABC- and the immunoperoxidase procedures have been extensively employed, with comparable results. We describe here the immunoperoxidase method.

3.6. Peroxidase–antiperoxidase (PAP) visualization

Unless explicitly stated otherwise, incubation and rinse buffers are as follows: 50 mM Tris, 0.85% NaCl, and 0.5% Triton X-100, pH 8.0 (TBS-TX; Triton trademark of Sigma, St Louis, USA), room temperature. All steps are carried out with gentle agitation. Rinse 3 × 10 minutes with TBS-TX before and after each step. Pre-incubation for one hour in 5–10% normal goat or swine serum is optional.

(i) Incubate in rabbit anti-PHA-L (Sigma; 1:2000), 60 hours, in a refrigerator.
(ii) Incubate in goat or swine anti-rabbit IgG (GAR or SwAR; Dakopatts A/S, Copenhagen, Denmark), 1:50–1:100, 90 minutes, room temperature.
(iii) Incubate in peroxidase–antiperoxidase complex raised in rabbit (rabbit-PAP; Dakopatts A/S, Copenhagen, Denmark), 1:800, 90 minutes, room temperature.
(iv) Rinse in TBS-TX and in 50 mM Tris–HCl buffer, pH 8.0, and react with 0.04% 3-3′ diaminobenzidine and 0.015% H_2O_2 in 50 mM Tris–HCl buffer, pH 7.6. Filter the solution prior to adding the hydrogen peroxide. The progress of staining can be monitored with a light microscope (rinse the sections in Tris–HCl before viewing them in the microscope).
(v) Mount the sections on slides, dry, dehydrate, and coverslip with Entellan (Merck, Germany).

3.7. Dual label immunocytochemistry

We present here a double-PAP immunocytochemical procedure for visualizing transported PHA-L in conjunction with the immunostaining for a second antigen (in this example vasopressin, cf. ref. 45).

Unless explicitly stated otherwise, the following buffer is used for all incubations: 50 mM Tris, 8.75 g NaCl/l, and 0.5% Triton X-100 (TBS-T) (pH 8.0, room temperature). During all steps the sections are gently agitated. Rinse 3 × 10 minutes in TBS-T before and after each step.

(i) Incubate in a cocktail of goat anti-PHA-L (1:8000; Vector, Burlingame, CA) and rabbit anti-vasopressin (1:4000; gift by Dr F. Vandesande, University of Leuven, Belgium), 60 hours, 4°C.
(ii) Incubate in a cocktail of donkey anti-goat IgG (DoAG; 1:50; Dakopatts A/S, Copenhagen, Denmark) and swine anti-rabbit IgG (SwAR, 1:50; Dakopatts), 2 hours, room temperature.
(iii) Incubate in goat peroxidase–antiperoxidase complex (goat-PAP; Dakopatts), 1:800, 2 hours, room temperature.
(iv) Rinse in 50 mM Tris–HCl, pH 8.0, and react with nickel-enhanced diaminobenzidine. Composition: 7.5 mg DAB, 200 mg nickel-ammoniumsulfate and 10 μl 30% H_2O_2 in 50 ml 50 mM Tris–HCl buffer, pH (exactly) 8.0. PHA-L labeled fibers will stain blue-black.

(v) Rinse thoroughly in 50 mM Tris–HCl, pH 8.0.
(vi) Incubate in rabbit peroxidase–antiperoxidase (1:800).
(vii) Rinse in 50 mM Tris–HCl, pH 7.6 and react with 3-3′diaminobenzidine (5 mg DAB in 10 ml 50 mM Tris–HCl, pH 7.6, 3.3 µl 30% H_2O_2) for 25 minutes. Filter the solution prior to adding the hydrogen peroxide. Vasopressin-immunoreactive structures will stain amber.
(viii) Mount the sections on slides from a 0.2% solution of gelatin in 50 mM Tris–HCl, pH 7.6, dry, dehydrate, and coverslip with Entellan.

4. ANTEROGRADE TRACING USING BIOTINYLATED DEXTRAN AMINE (BDA)

Dextran amine conjugated with the fluorescent dye, rhodamine, was introduced as a neuroanatomical tracer in 1986 (48). It proved to satisfy the majority of the criteria of an ideal anterograde tracer (see above), and since then several other conjugates have been proven good tracers as well (dextran amine–tetramethylrhodamine or fluoro-ruby, ref. 49; dextran amine–Lucifer yellow, ref. 50).

A disadvantage common to all these tracers is that their fluorescent label is not permanent, although the transported label can be made permanent via photoconversion (51) or immunocytochemistry (52). These treatments are however labor- and time-intensive. Another disadvantage is the considerable amount of retrograde transport, especially in areas of the cerebral cortex, and the possibility of false anterograde tracing via the collaterals of originally retrogradely labeled neurons.

Biotinylated tracers such as biocytin (53, 54) or biotinamide (55, 56) can be visualized in the fluorescence microscope via a short and simple one-step reaction with fluorophore-conjugated avidin, or can be made permanent via treatment with an Avidin–biotin–peroxidase complex. In retrospect it is no surprise that the power of dextran amine tracing and the ease of the visualization of biotinylated compounds were combined by Reiner and co-workers who introduced biotinylated dextran amine as neuroanatomical tracer (BDA; ref. 57; see also refs. 58 and 59). An additional advantage of BDA appeared to be its easy visualization in the electron microscope (60) and the ease with which it can be combined with a wealth of light and electron microscopic neuroanatomical methods (61, 62). Thus, we are currently applying BDA whenever anterograde tracing must be combined with electron microscopy, and in experiments in which two anterograde tracers must be combined in order to study convergence of fiber projections onto a common termination area.

4.1. Application of the tracer

Biotinylated dextran amine conjugated to lysine (BDA: molecular weight 10 000, lysine fixable) is commercially available (Molecular Probes, Eugene, OR). It is used as a 5% or 10% solution in 10 mM phosphate buffer, pH 7.25. This solution can be stored in aliquots in a freezer. Iontophoretical application of BDA is achieved through glass micropipettes and using injection parameters similar to those used for the application of PHA-L (see section 3.1). We prefer iontophoretic injections over mechanical injection since the injection site in an iontophoretic approach is much smaller than in a mechanical approach (see also ref. 59). Survival time is 6 days. BDA remains detectable in rats with survival periods up to 4 weeks.

4.2. Fixation and histochemical procedure

(i) The rats are anesthetized and transcardially perfused with vascular rinse fluid (composition: see section 3.3.), followed immediately by 1000 ml of a mixture of 4% formaldehyde, 0.1% glutaraldehyde and 0.2% picric acid in 125 mM phosphate buffer, pH 7.4 (room temperature). BDA 'withstands' the same range of fixatives as PHA-L (see sections 3.3 and 3.4).

(ii) Dissect the brain immediately after the perfusion.

(iii) Section the brain at a thickness of 40 μm on a Vibratome, using chilled 125 mM phosphate buffer as the bath fluid.

(iv) (optional) For electron microscopy: cryoprotect the sections by immersion in a mixture of 10% dimethylsulfoxide (DMSO) and 1% glycerin in 125 mM phosphate buffer, pH 7.4 (10 minutes, refrigerator), and subsequently in buffered 2% DMSO 20% glycerin (10 minutes). Then freeze the sections by dropping them in liquid nitrogen-cooled isopentane or by placing them on solid carbon dioxide. Allow the sections to thaw at 4°C (for details see ref. 62). Repeat freezing-thawing if necessary.

(v) Rinse the sections three times in 50 mM Tris, 0.875% NaCl, 0.5% Triton X-100, pH 8.0 (TBS-T; no Triton for electron microscopy), and react for 90 minutes at room temperature (overnight at 4°C for electron microscopy) with ABC complex (Vectastain, Vector, Burlingame, CA, USA) prepared according to the manufacturer's recommendations. Apply gentle agitation.

(vi) Rinse with TBS-T, pH 8.0 and 50 mM Tris–HCL buffer, pH 7.6, and react for 20 minutes in a solution of 0.04% 3.3′ diaminobenzidine tetrahydrochloride (DAB, BDH Chemicals, England) and 0.015% H_2O_2 in the Tris buffer (pH 7.6). Filter the solution prior to adding the hydrogen peroxide.

Terminate the reaction by rinsing three times in the Tris buffer, and collect sections for electron microscopy in 100 mM cacodylate buffer (pH 7.4) and postfix these for one hour in 1% OsO_4 at pH 7.4, block-stain for one hour with 2% cold, aqueous uranyl acetate and dehydrate through an ascending alcohol series. Embed the EM-sections via propylene oxide in Epon–Araldite on slides between polyethylene foils. Mount the sections for light microscopy on slides, and counterstain with cresyl violet if necessary.

4.3. Tracing with BDA combined with immunocytochemistry for a second marker

Light microscopy

We present here an ABC-PAP immunocytochemical procedure for visualizing transported BDA in conjunction with the immunostaining for a second marker (in this case parvalbumin, a calcium-binding protein). The fixative used in this particular combination is the same as above, except that 0.1% glutaraldehyde is used.

The rinse and incubation buffers are composed of 50 mM Tris, 0.85% NaCl, and 0.5% Triton X-100 (TBS-T) (pH 8.0). Sections are gently agitated during the incubations and rinses. Rinse 3 × 10 minutes before and after each step.

(i) React with Avidin–biotin–peroxidase complex (standard mixture) in TBS-T, pH 8.0 (Vectastain kit, Vector, Burlingame, USA), 90 minutes, room temperature.

(ii) Rinse in 50 mM Tris–HCl buffer, pH 8.0, and react the sections for 5–10 minutes in a reaction medium which consists of 50 ml 50 mM Tris–HCl, pH 8.0, 7.5 mg DAB, 200 mg nickel-ammoniumsulfate and 10 μl 30% H_2O_2. BDA-labeled structures stain blue-black.

(iii) Incubate in mouse anti-parvalbumin, 1:2000 (Sigma), 60 hours, 4°C.

(iv) Incubate in goat anti-mouse IgG (GaM; Dakopatts), 1:100, 90 minutes, room temperature.

(v) Incubate in mouse peroxidase–antiperoxidase complex (mouse-PAP), 1:100, 90 minutes, room temperature.

(vi) Rinse in 50 mM Tris–HCl buffer, pH 8.0, and react with 0.04% 3-3′ diaminobenzidine and 0.015% H_2O_2 in 50 mM Tris–HCl buffer, pH 7.6. Filter the solution prior to adding the hydrogen peroxide. The progress of staining can be monitored with a light microscope (rinse the sections first in Tris–HCl buffer). Parvalbumin-immunoreactive cells and their processes are stained brown.
(vii) Mount the sections on slides, dry, dehydrate, and coverslip with Entellan.

Electron microscopy
BDA-tracing can easily be combined with the detection of a second marker at the electron microscope level. For this purpose it is necessary to label neuronal processes which contain the tracer or the second marker with different electron dense precipitates. For this purpose we first immunodetect the second marker with DAB as the first precipitate, then conduct a silver–gold enhancement step to deposit metallic particles in all neuronal processes which contain DAB, and finally proceed with an ABC-DAB reaction to detect the transported BDA. Thus, in the electron microscope it is possible to differentiate BDA-labeled axon terminals (fuzzy DAB precipitate only) from the cell bodies, dendrites and other neuronal processes that contain the second marker (discrete metallic particles embedded in a DAB-precipitate). An example is provided here in which we have combined BDA-tracing with the detection in the electron microscope of parvalbumin. The fixative is described in section 4.2.

All rinse and incubation steps are carried out with TBS (same as above, *no* Triton X-100), pH 8.0, at 4°C and with continuous gentle agitation. Prior to the procedure, the sections are cryoprotected and freeze-thawed twice (see section 4.2). In between the incubations the sections are rinsed 3 × 10 minutes with TBS.

(i) Incubate 60 hours in mouse anti-parvalbumin (Sigma,1:2000).
(ii) Incubate 6–8 hours at 4°C in goat anti-mouse IgG, 1:100.
(iii) Incubate overnight at 4°C in peroxidase–antiperoxidase complex raised in mouse, 1:100.
(iv) Rinse with 50 mM Tris–HCl, pH 7.6, and react with 0.05% diaminobenzidine (DAB) in 50 mM Tris–HCl buffer (pH 7.6) and 0.015% H_2O_2 (room temperature). Filter the solution prior to adding the hydrogen peroxide. The reaction is terminated after 10–20 minutes by transferring the sections to fresh Tris–HCl buffer.
(v) Silver intensify the DAB reaction product (see below).
(vi) Incubate overnight at 4°C in ABC-solution (prepared according to the manufacturer's instructions).
(vii) React with DAB (see step (iv)). The reaction is stopped after 30–60 minutes by transferring the sections to fresh Tris–HCl buffer.
(viii) Process for electron microscopy as indicated in section 4.2.

Silver–gold intensification (according to van den Pol and Görcs, refs. 63, 64)
(All steps carried out at room temperature)

(a) Treat the sections 60 minutes with 10% thioglycolic acid (pH 2.0).
(b) Wash 2 × 10 minutes with 2% sodium acetate buffer (pH 8.6).
(c) Silver intensify sections for 3 minutes.
Preparation of reagent: slowly add 40 ml of a 37% stock solution of formalin to 10 ml of a mixture of 0.2% ammonium nitrate, 0.2% silver nitrate and 1% tungstosilicic acid while stirring the solutions constantly. The solution must remain clear. Then add 10 ml of 5% sodium carbonate (while stirring).
(d) Rinse 10 minutes in 1% acetic acid, pH 2.8.

(e) Rinse 2 × 10 minutes in 2% acetate buffer, pH 8.6.
(f) Convert the formed silver into metallic gold by treating sections for 10 minutes with 0.05% $HAuCl_4$ (gold chloride) made up freshly with distilled water (no agitation).
(g) Rinse one minute in 2% acetate buffer, pH 8.6.
(h) Treat with 3% sodium thiosulfate for 5 minutes.
(i) Rinse in 2% acetate buffer (2 × 10 minutes).

Silver-gold intensification (according to Derouiche and Frotscher, ref. 65)
(All steps carried out at room temperature)

(a) Rinse at least five times in 200 mM phosphate buffer, pH 7.2–7.4.
(b) Make up 0.2% paraformaldehyde in distilled water (from a stock solution of 10%), and add in this order: 1.0% tungstosilicic acid, 0.2% $AgNO_3$, and 0.2% NH_4NO_3.
(c) Add slowly and under continuous agitation an equal volume of 5% Na_2CO_3 in distilled water.
(d) Transfer the sections to the developer (same as in step (iv) of the protocol of van den Pol and Görcs) and react in the dark for 5–15 minutes or until the desired degree of intensification is reached. The extent of the intensification can be controlled by examining the sections periodically to verify color change from brown to black. Do not overdevelop since this results in aspecific silver deposits.
(e) Rinse the sections briefly in 200 mM phosphate buffer, pH 7.2–7.4, and place them in fixer (2.5% $Na_2S_2O_3$ in distilled water) for 1–3 minutes. This treatment significantly reduces the aspecific background.
(f) Convert the formed silver into metallic gold by incubating the sections for 45 minutes (4°C) in 0.05% $HAuCl_4$ (gold chloride) made up freshly with distilled water (no agitation).
(g) Rinse the sections briefly in 200 mM phosphate buffer, pH 7.2–7.4, and fix again in 2.5% $Na_2S_2O_3$ for precisely 5 minutes.
(h) Rinse in 200 mM phosphate buffer, pH 7.2–7.4.
(i) Rinse again in TBS.

5. REFERENCES

1. Waller, A.V. (1850) *Phil. Trans. Roy. Soc., Ser. B.*, **140**, 423.

2. Spacek, J. (1992) *Micr. Res. Techn.*, **23**, 264.

3. Heimer, J. and Robards, M.J. (eds) (1981) *Neuroanatomical Tract-Tracing Methods*, Plenum, New York.

4. Heimer, J. and Záborszki, L. (eds) (1989) *Neuroanatomical Tract-Tracing Methods: Recent Progress*, Plenum, New York.

5. Björklund, A., Hökfelt, T., Wouterlood, F.G. and van den Pol, A.N. (eds) (1990) *Handbook of Chemical Neuroanatomy*, vol. 8, Elsevier, Amsterdam.

6. Bolam, J.P. (ed.) (1992) *Experimental Neuroanatomy*, Oxford University Press, Oxford.

7. Graham, R.C. and Karnovsky, M.J. (1966) *J. Histochem. Cytochem.*, **14**, 291.

8. Kristensson, K. and Olsson, Y. (1971) *Brain Res.*, **32**, 399.

9. LaVail, J.H. and LaVail, M.M. (1972) *Science*, **176**, 1416.

10. Trojanowski, J.Q., Gonatas, J.O. and Gonatas, N.K. (1981) *J. Neurocytol.*, **10**, 441.

11. Schwab, M.E., Javoy-Agid, F. and Agid, Y. (1978) *Brain Res.*, **152**, 145.

12. Mesulam, M-M. (1982) in *Tracing Neuronal Connections with Horseradish Peroxidase* (M-M. Mesulam, ed.) IBRO Handbook Series: *Methods in the Neurosciences* (A.D. Smith, ed.) Wiley, Chichester, p. 1.

13. Levey, A.I., Bolam, J.P., Rye, D.B., Hallanger, A.E., Demuth, R.M., Mesulam, M-M. and Wainer, B.H. (1986) *J. Histochem. Cytochem.*, **9**, 1449.

14. Warr, B.B., De Olmos, J.S. and Heimer, L. (1981) in *Neuroanatomical Tract-Tracing* (L. Heimer and M.J. Robards, eds), Plenum, New York, p. 207.

15. Somogyi, P., Hodgson, A.J. and Smith, A.D. (1979) *Neuroscience*, **4**, 1805.

16. Holland, V.R., Saunders, B.C., Rose, F.L. and Walpole, A.L. (1974) *Tetrahendon*, **30**, 3299.

17. Ménétrey, D. (1985) *Histochemistry*, **83**, 391.

18. Escher, G., Schönenberger, N. and Van der Loos, H. (1983) *J. Neurosci. Meth.*, **9**, 87.

19. Griffin, G., Watkins, L.R. and Mayer, D.J. (1979) *Brain Res.* **168**, 595.

20. Paxinos, G. and Watson, C. (1986) *The Rat Brain in Stereotaxic Coordinates.* 2nd edn, Academic Press, Sydney, p. 218.

21. Swanson, L.W. (1992) *Brain Maps: Structure of the Rat Brain*, Elsevier, Amsterdam.

22. Itoh, K., Konishi, A., Nomura, S., Mizuno, N., Nakamura, Y. and Sugimimoto, T. (1979) *Brain Res.*, **175**, 341.

23. Räsänen, V., Weber, T.H. and Gräsbeck, R. (1973) *Eur. J. Biochem.*, **38**, 193.

24. Leavitt, R.D., Felsted, R.L. and Bachur, N.R. (1977) *J. Biol. Chem.*, **252**, 2961.

25. Felsted, R.L., Egorin, M.J., Leavitt, R.D. and Bachur, N.R. (1977) *J. Biol. Chem.*, **252,** 2967.

26. Gerfen, C.R. and Sawchenko, P.E. (1984) *Brain Res.*, **290,** 219.

27. Tomimoto, H., Kamo, H., Araki, M. and Kimura, H. (1987) *J. Neurosci. Meth.*, **22**, 1.

28. Groenewegen, H.J. and Wouterlood, F.G. (1990) in *Handbook of Chemical Neuroanatomy*, vol. 8 (A. Björklund, T. Hökfelt, F.G. Wouterlood and A.N. van den Pol, eds), Elsevier, Amsterdam, p. 47.

29. Wouterlood, F.G. and Groenewegen, H.J. (1985) *Brain Res.*, **326**, 188.

30. Freund, T.F. (1993) in *Immunohistochemistry II* (A.C. Cuello, ed.), Wiley, Chichester, p. 329.

31. Köhler, C. (1986) *J. Comp. Neurol.*, **246**, 149.

32. Sugiura, Y., Lee C.L. and Perl, E.R. (1986) *Science*, **234**, 358.

33. ter Horst, G.J., Groenewegen, H.J., Karst, H. and Luiten, P.G.M. (1984) *Brain Res.*, **307**, 379.

34. Groenewegen, H.J., Vermeulen-Van der Zee, E., te Kortschot, A. and Witter, M.P. (1987) *Neuroscience*, **23,** 103.

35. Russchen, F.T. and Price, J.L. (1984) *Neurosci. Lett.*, **47**, 15.

36. Russchen, F.T., Amaral, D.G. and Price, J.L. (1985) *J. Comp. Neurol.*, **242**, 1.

37. Matsuyama, K., Ohta, Y. and Mori, S. (1988) *Brain Res.*, **460**, 124.

38. Woolf, N.J., Hernit, M.C. and Butcher, L.L. (1986) *Neurosci. Lett.*, **66,** 281.

39. Groenewegen, H.J. and Berendse, H.W. (1990) *J. Comp. Neurol.*, **299**, 187.

40. Wouterlood, F.G., Goede, P.H. and Groenewegen, H.J. (1990) *J. Chem. Neuroanat.*, **3**, 11.

41. Gerfen, C.R. and Sawchenko, P.E. (1985) *Brain Res.*, **343**, 144.

42. Groenewegen, H.J. (1988) *Neuroscience*, **24**, 379.

43. Wouterlood, F.G., Goede, P.H., Arts, M.P.M. and Groenewegen, H.J. (1992) *J. Histochem. Cytochem.*, **40**, 457.

44. Wouterlood, F.G., Bol, J.G.J.M. and Steinbusch, H.W.M. (1987) *J. Histochem. Cytochem.*, **35**, 817.

45. Staiger, J.H. and Wouterlood, F.G. (1990) *Cell Tiss. Res.*, **261**, 17.

46. Friedrich, V.L. and Mugnaini, E. (1981) in *Neuroanatomical Tract-Tracing Methods* (L. Heimer and M.J. Robards, eds), Plenum Press, New York, p. 345.

47. Rosene, D.L., Roy, N.J. and Davis, B.J. (1986) *J. Histochem. Cytochem.*, **34**, 1301.

48. Glover, J.C., Petursdottir, G. and Jansen, J.K.S. (1986) *J. Neurosci. Meth.*, **18**, 243.

49. Schmued, L., Kyriakidis, K. and Heimer, L. (1990) *Brain Res.*, **526**, 127.

50. Chang, H.T. (1991) *Brain Res. Bull.*, **26**: 813.

51. Lübke, J. (1993) *Neurosci. Prot.*, **93**-050-06-01.

52. Dolleman-van der Weel, M.J., Wouterlood, F.G. and Witter, M.P. (1994) *J. Neurosci. Meth.*, **51**, 9.

53. King, M.A., Louis, P.M., Hunter, B.E. and Walker, D.W. (1989) *Brain Res.*, **497**, 361.

54. Izzo, P.N. (1991) *J. Neurosci. Meth.*, **36**, 155.

55. Kita, H. and Armstrong, W.E. (1991) *J. Neurosci. Meth.*, **37**, 141.

56. Huang *et al.* (1992) *J. Neurosci. Meth.*, **41**, 31.

57. Reiner, A., Veenman, C.L. and Honig, M.G. (1991) *Abstr. Soc. Neurosci.*, **17**, 1510.

58. Brandt, H.M. and Apkarian, A.V. (1992) *J. Neurosci. Meth.*, **45**, 35.

59. Veenman, C.L., Reiner, A. and Honig, M.G. (1992) *J. Neurosci. Meth.*, **41**, 239.

60. Wouterlood, F.G. and Jorritsma-Byham, B. (1993). *J. Neurosci. Meth.*, **48**, 75.

61. Jorritsma-Byham, B., Witter, M.P. and Wouterlood, F.G. (1994) *J. Neurosci. Meth.* **52**, 153.

62. Meredith, G.E. and Wouterlood, F.G. (1993) *Micr. Res. Techn.*, **24**, 31.

63. van den Pol, A.N. and Görcs, T. (1986) *J. Comp. Neurol.*, **252**, 507.

64. van den Pol, A.N. and Decavel, C. (1990) in *Handbook of Chemical Neuroanatomy*, vol. 8 (A. Björklund, T. Hökfelt, F.G. Wouterlood and A.N. van den Pol, eds), Elsevier, Amsterdam, p. 199.

65. Derouiche, A. and Frotscher, M. (1990) *Eur. J. Neurosci.*, **suppl. 3**, 49.

66. Zhou, M. and Grofova, I. (1995) *J. Neurosci. Meth.*, **62**, 149.

CHAPTER 4
PREPARATION OF TISSUE FOR MICROSCOPY

H.A. Davies and M.G. Stewart

Microscopy

1. INTRODUCTION

For conventional light or transmission electron microscopy protocols are essentially similar to those used for examination of non-nervous tissue (1, 2), but the main problem, especially with CNS tissue, arises because its structural integrity is so dependent on a stable oxygen supply. Since the key requirement for optimal preservation in microscopy (and especially electron microscopy) is rapid fixation, this almost invariably necessitates perfusion of the brain of the living (but anesthetized) animal's circulatory system, normally via the heart. In many countries it will be necessary to ensure that appropriate authorization has been obtained to administer anesthesia to an animal, and to carry out the subsequent perfusion. In any case, animal surgery requires skill and experience and should not be attempted without training or adequate supervision. Suitable courses are available via a number of institutions to ensure that the required level of competence exists prior to attempting perfusion of an animal.

2. FIXATIVES FOR LIGHT AND ELECTRON MICROSCOPY

The description provided here is for electron microscopy; for light microscopy, similar primary fixatives are used except that glutaraldehyde 2% is omitted and the paraformaldehyde concentration of 2% is raised to 4%. The secondary fixative, osmium tetroxide (OsO_4), is used only for electron microscopic procedures. See *Table 1* for a list of solutions.

Table 1. Solutions required for preparation of tissue for light and electron microscopy

Solution	Preparation	Precautions/notes
0.2M phosphate buffer	$NaH_2PO_4.2H_2O$ (MW=137.99) Na_2HPO_4 (MW=141.96) For each 100 ml buffer: 0.497 g $NaH_2PO_4.2H_2O$ 2.328 g Na_2HPO_4	Check the formula Otherwise use a table (ref. 1)
Paraformaldehyde (4% for light microscopy, 2% for electron microscopy)	Add powder to distilled water plus 6 drops of 1M NaOH per 20 ml Heat until dissolved (60°C). Cool and make up to half-volume with distilled water Add 0.2M PO_4 buffer (or double strength buffer)	Filter before use in perfusion fixation Keeps less than 24 h at 4°C
Paraformaldehyde + glutaraldehyde (2% for electron microscopy only)	Add glutaraldehyde after cooling. Make up as above	Filter before use in perfusion fixation Keeps less than 24 h at 4°C

Table 1. Continued

Solution	Preparation	Precautions/notes
Osmium tetroxide	1 g crystals in 50 ml distilled water	Stock 2% solution, keep at 4°C Allow 12–24 h to dissolve
Toluidene blue	1 g toluidene blue + 5 g borax + 100 ml distilled water; stir, with heat, to dissolve; filter	Stain 'matures' with storage Microfilter to remove particles before every slide is stained
Epoxy resins	See (ref. 1)	
Reynolds lead citrate	See (ref. 3)	
Uranyl acetate (aqueous or methanolic)	See (ref. 3)	
Cresyl violet (ref. 4)	0.1 g in 100 ml of 95% ethanol, acidified by adding 0.7 ml of 10% acetic acid pH 3.5 to 3.8	Filter before use
Saline	0.9 g in 100 ml distilled water	Physiological concentration
Glycerol/egg albumin	50:50 glycerol:egg albumin Leave to dissolve at RT	Apply to slide with a paint brush or cotton bud
Chrome alum	0.1 g chromic potassium sulphate 1.0 g gelatine 200 ml distilled water	Heat gently to dissolve Filter Dip racks of slides to sub them
Poly-L-lysine	0.01% in distilled water	It is possible to use a 1 in 10 of commercially available solution of 0.1%

3. PERFUSION FOR LIGHT AND ELECTRON MICROSCOPY

3.1. Anesthesia

Pentobarbitone (Sagatal), urethane, halothane, ether, or other anesthetic is used as required by the demands of the experiment. The effect of an anesthetic on ultrastructure needs to be considered, since there is some evidence that certain anesthetics may affect the final appearance of the fine structure of the tissue. In practice we have found little difference in our studies between tissues from animals anesthetized with urethane, pentobarbitone or ether, but if one is doing a quantitative study it may be helpful to check, and most certainly one should use the same anesthetic throughout. For small vertebrates (birds, rodents) pentobarbitone is the anesthetic of choice and can be administered intraperitoneally at a dose of 1.0 ml/kg. Once the animal is deeply unconscious, it should be placed on a dissecting board and the forelimbs gently stretched out.

3.2. Surgery and fixation

This procedure is outlined in *Table 2*. With larger adult rodents the success rate for perfusions should be almost 100% for an experienced investigator, but for smaller vertebrates, such as young chicks, success rates in perfusions can be as low as 60% owing to the more delicate nature of the blood vessels.

Table 2. Procedure for perfusion fixation

Procedure	Rationale/notes
Make a longitudinal cut in the ventral body surface in the midline Cut away the top part of the rib cage	To expose the thoracic cavity
Remove the pericardium and insert a cannula (attached to a peristaltic pump) into the left ventricle	The tip of the cannula should be filed away and protected with a 'blob' of rapid setting resin Gravity feed can be used instead of pump
Push cannula from the ventricle through the ventricular valve into the aortic arch	Tie suture thread around the aortic arch to secure or clamp with artery forceps
Move heart and lungs to one side Clamp dorsal aorta with artery forceps	To ensure only the head is perfused
Cut the right atrium	To allow drainage of blood and perfusate
Run peristaltic pump at 5.6 ml/min with 15–30 ml saline	To flush out the blood
Run primary fixative at either 37°C or 20–22°C	Use approx 150 ml fixative/300 g animal The animal should go rigid, followed by hardening of the tissues

4. TISSUE DISSECTION

The whole brain or a dissected area may be used. For light microscopy, the tissue should be washed under running tap water overnight after primary fixation to remove traces of fixative and then dehydrated and finally embedded in paraffin wax (*Table 4*). Sections are cut and can be stained with cresyl violet (*Table 5*).

For electron microscopy, an area of brain must be dissected or Vibratome sections used, which are then postfixed (*Table 3*), dehydrated and embedded (*Table 4*).

Table 3. Tissue dissection and secondary fixation for electron microscopy

Procedure	Rationale/notes
After primary fixation, leave animal for 2 hours	To ensure adequate fixation
Remove brain from skull	Tissue must be kept moist Avoid physical trauma
Dissect out the region of interest	May be appropriate to cut 40–300 μm sections with a razor blade or vibrating microtome A brain mould (either commercial or home-made) may be used Keep a record of the orientation[a]

Table 3. Continued

Procedure	Rationale/notes
Lay the slice on a sheet of dental wax Whilst viewing with a binocular microscope, cut out the area of interest	Use a cold light source to avoid heating the tissue
Cut 1 mm^3 blocks of tissue and place in small vials with approx 2 ml fixative	To allow good penetration of secondary fixative at the later stage below Keep a record of the orientation[a]
Wash tissue in buffer for 3 × 5 min	To remove all traces of aldehyde fixative
Replace buffer with 1% OsO_4 in buffer (approx 1 ml)	Leave for 1 h at RT
Wash in buffer for 3 × 5 min Store at 4°C if required.	To remove all traces of OsO_4

[a] In the rat hippocampus, to examine CA1 pyramidal cells and the granule cells of the dentate gyrus, tissue can be cut in a topless diamond shape where the CA1 cells are at the top

Table 4. Dehydration and embedding

	Light microscopy	Electron microscopy
Dehydration	70% ethanol for 1 h 80% ethanol for 1 h 90% ethanol for 1 h 100% ethanol for 3 × 1 h	50% ethanol/acetone for 10 min 70% ethanol/acetone for 10 min 90% ethanol/acetone for 10 min 100% ethanol/acetone 2 × 20 min 100% ethanol/acetone (dried over molecular sieve) for 20 min
Embedding	50:50 mixture of chloroform and xylene for 1.5 h or overnight (sections sink when fully infiltrated) OR 50:50 mixture of chloroform and 100% ethanol for 1.5 h or overnight (sections sink when fully infiltrated)	50:50 mixture of epoxy resin (e.g. Epon and Araldite) and 100% acetone overnight
	100% xylene (or Histoclear) for 2 × 1 h or overnight	100% epoxy resin 2 × 4 h
	100% wax – 1 h in vacuum oven	
	100% wax – 2 h in vacuum oven	
	Embed in fresh wax in plastic moulds containing a layer of partially solidified wax Allow to cool	Embed in fresh epoxy resin in rubber moulds or polythene capsules with paper labels with codes written in pencil
Polymerization	Harden in tray of cold water	Polymerize at 60°C for 24–48 h
	Mount onto wooden blocks	
	Cut 8–20 μm sections	Cut 70–100 nm ultrathin sections
	Collect on glycerol/egg albumin coated slides	Collect on appropriate grids, with or without support films.

Table 5. Nissl staining of sections for light microscopy: the fast cresyl method (4)

Procedure	Notes
100% xylene for 2 × 3 min[a]	To dewax the paraffin sections Can also be used for cryostat sections but it is more common to start with the distilled water step below
100% ethanol for 1 min	
90% ethanol for 1 min	
80% ethanol for 1 min	
70% ethanol for 1 min	
50% ethanol for 1 min	
Distilled water	**The start for cryostat sections**
Cresyl violet for 20–30 min at RT	
Wash in water	View in microscope If staining is too heavy, destain in 95% ethanol until the background is clear Adjust staining times as appropriate for subsequent sections
70% ethanol for 3 min	
95% ethanol for 3 min	
100% ethanol for 3 min	
100% ethanol for 1 min	
100% xylene (or Histoclear) for 2 × 3 min	To clear the section in readiness for the mountant If longer times are used, the sections may dislodge
Mount under a coverslip with Canada balsam or DPX	

[a]These times are for 20 μm sections and may need to be adjusted if different thicknesses are used

5. PREPARATION OF UNFIXED MATERIAL FOR LIGHT MICROSCOPY

This is used for routine light microscopy (e.g. to determine the location of electrode tracts). It is unnecessary to fix the brain, and rapid processing of the tissue can be achieved by freezing and subsequent sectioning in a cryostat (*Table 6*). This is also the route if brain tissue is required for *in vitro* receptor autoradiography where distribution of ligand binding in the whole brain is to be mapped (5). Cryostat sections are also used for metabolic mapping studies with 2-deoxyglucose (6).

Table 6. Preparation of cryostat sections for light microscopy

Procedure	Rationale/notes
Kill animal swiftly by cervical dislocation or decapitation	Speed is important to prevent post-mortem changes
Remove the brain from the skull and separate into fore- and hind-brain	Smaller pieces freeze better A mold may be used to give the same orientation for comparison between animals
Place brain into a beaker of isopentane cooled in solid CO_2 or liquid nitrogen	Place beaker in polystyrene box for safety Isopentane cryoprotects and prevents fracturing of the brain
Wrap in cooled aluminum foil with a code and store at –80°C until required	Allows accumulation of many brains from one experiment before sectioning
To section, place on cryostat chuck and mount with OCT compound (Tissue Tek)	
Cut 10–20 µm sections	For 12 µm sections, –18°C is suggested Thicker sections require slightly lower temperatures
Collect sections on chrome alum or poly-L-lysine subbed slides	To ensure good adhesion of section to the slide
Dry sections at room temperature	
100% ethanol for 4 × 60 s and through a descending series of ethanol to water (see *Table 4*)	To defat the sections
Stain with cresyl violet	See *Table 5*

6. VIEWING OF WHOLE CELLS AND PROCESSES

Ultrathin sectioning allows parts of the cells and processes (synapses, axons and dendrites) to be viewed in the electron microscope, but without recourse to extensive and time-consuming serial sectioning and reconstruction methods it is difficult to discern the organization and structural relationships of the parts of a nervous tissue, and indeed of particular neurons. However, a variety of methods are now available to enable examination of individual neurons. These can fall into two types, namely selective and non-selective methods.

Selective methods involve filling identified neurons. They are usually viewed in the light microscope in living or even fixed tissue, or in physiologically identified living cells in anesthetized animals. A very complete review of methodology for PHAL, HRP, virus tracing etc. has been published by Bolam (7).

The best known non-selective method is the Golgi procedure (8) which results in impregnation of neurons and their processes with silver (*Table 7*). The non-selective nature of the process is often cited as a disadvantage but it does enable a number of neurons to be visualized, as opposed to only single neurons following cell filling procedures. In addition the Golgi method has been elaborated by a number of authors (9), and if the first attempt at Golgi impregnation fails to produce a sufficient numbers of filled cells, it is possible to repeat the

Table 7. Golgi impregnation for light microscopy and gold-toning for electron microscopy

Staining	Procedure	Notes
Golgi impregnation After Valverde (8)	Place aldehyde and osmium-fixed block in aqueous 3.5% potassium dichromate overnight at RT	Tissue up to 3–4 mm thick is used for this procedure
	Place tissue in aqueous 0.7% silver nitrate overnight in the dark	
	Embed tissue in a drop of fresh 5% gelatine on a 4.5 mm filter paper	To secure the tissue for future chopping
	Place under the blade of a tissue chopper (e.g. Sorvall) and cut 80–150 μm thick sections.	
	Transfer sections with a paint brush wetted in glycerol to a slide and view in a microscope	To check for impregnation If not present, repeat the reaction
	Dehydrate and mount for light microscopy as in *Table 4*	
Gold-toning of Golgi-stained sections		For examination in the electron microscope
	Place an impregnated section in glycerol under a strong cooled light source for 20–30 min	A 100 W bulb of a light microscope is an appropriate light source
	Incubate sections in vials in 0.07% gold chloride in 0.25% glycerol for 30 min at 4°C	
	Wash sections in distilled water	
	Wash sections in 0.2% oxalic acid for a few minutes	
	Wash sections in distilled water	
	1% sodium thiosulfate 3 × 15 min at RT	To fix the reaction
	Rinse in distilled water and prepare for elecron microscopy (*Table 4*)	

procedure several times until a cell or cells that one wishes to examine are eventually impregnated.

Once Golgi impregnation is complete, one may wish to view the tissue in the electron microscope. Unfortunately the silver creates holes in the ultrathin sections when cut and is affected by the electron beam so it must be replaced by electron dense gold. This procedure ('gold-toning') is described in *Table 7*. The gold-toned tissue is embedded in Epon resin as described above (*Tables 3 and 4*) and can then be sectioned and viewed in the electron microscope. It is usually unnecessary to enhance contrast further with lead and uranium salts, since neural structures are normally readily visible in gold-toned tissue. The tissue can either be thin-sectioned (at 60–80 nm) or cut at a thickness of up to 7 μm, provided an accelerating voltage of 200 kV is used on unstained sections. However, we have been able to examine sections of 1–2 μm thickness at 100 kV. Such sections produce a 3-dimensional image which can be fully appreciated on an electron microscope with a gonoimeter stage in which it is possible to view dendritic spines in fine detail (10).

7. SECTIONING OF EMBEDDED TISSUE FOR LIGHT MICROSCOPE ANALYSIS PRIOR TO ELECTRON MICROSCOPY

Semi-thin sections of resin embedded tissue are cut using an ultramicrotome, initially with glass knives at approximately 1 μm thickness, followed by staining with 1% toluidine blue in 5% borax on a hotplate to enable localization of the tissue components, in a light microscope. Further semi-thin sectioning may be necessary before the tissue of interest is located and drawings or photomicrographs may be taken to record the area and ensure that a correlated study can be made when viewing an ultrathin (~60 nm) section in the electron microscope. Blocks are trimmed to give a smaller block-face and 50–100 nm ultrathin sections collected.

For nervous tissue, particularly if a quantitative morphometric study is to be undertaken, it is often necessary to collect a series of sections, for which support films of formvar or pioloform (which may be reinforced with carbon) on 2 × 1 mm slot grids are necessary, and if serial sections are to be cut it is preferable to use a diamond knife since glass knives can lose their edge after a limited number of sections are cut.

To examine the tissue in an electron microscope it is usually necessary to improve tissue contrast by staining with lead and uranium salts (3), such as Reynold's lead citrate and uranyl acetate. An autostainer (e.g. the Leica Ultrastainer) may be used and stains are available commercially in ready made packs. For sections of 60–80 nm thickness stained in this way, an accelerating voltage of 60 or 80 kV is ideal to allow viewing of the tissue components in most modern transmission electron microscopes.

8. Immunocytochemical methods at the electron microscope level

Immunocytochemistry (ICC) is the localization of intracellular protein or carbohydrate targets in tissue using specific antibodies or lectins. It is used at the EM level to allow fine structural localization and can be quantitated so that comparisons can be made in research projects between experimental and control subjects. Examples of the types of targets that have been studied in neurobiology are substance P (11), ionotropic and metabotropic receptors (12–14), neurotransmitters (15, 16) and cell adhesion molecules (17). The technique is a balance between the retention of immunoreactivity of the target in the tissue and preservation of the ultrastructure.

Factors affecting the success or otherwise of the method are the preparation of material, either by pre-embedding or post-embedding methods, or the immunocytochemistry methodology itself. *Table 8* lists the solutions used.

8.1. Pre-embedding immunocytochemistry

The ICC reaction is carried out before the tissue is embedded. The problem with the pre-embedding methodology is access of the reagents into the tissue because the reaction is

Table 8. Solutions for EM immunocytochemistry (ICC)

Solution	Preparation	Precautions/notes
Glutaraldehyde + paraformaldehyde + picric acid	See *Table 1* for glutaraldehyde + paraformaldehyde preparation Add picric acid	
Periodate/lysine/ paraformaldehyde (PLP)	To 10 ml 0.2 M DL lysine monohydrochloride, add 0.1 M Na_2HPO_4 until pH 7.4 reached Dilute to a final lysine concentration of 0.1 M with 0.1 M phosphate buffer pH 7.4 (volume =20 ml) Make an 8% paraformaldehyde aqueous solution Add 3 volumes of lysine buffer to 1 volume of paraformaldehyde Add solid Na m-periodate to give 0.1 M	Fix tissue without readjusting pH pH is 6.4–7.4 Preserves carbohydrate moieties
PBSA	8 g NaCl 0.2 g KCl 1.15 g Na_2HPO_4 0.2 g KH_2PO_4 Make up to 1 litre	pH is 7.4 Sterilize at 115°C for 38 min (max) in 500 ml aliquots Improved antigenicity with an extra 4 g/l of NaCl
BSA/PBSA	Add weighed amount of bovine serum albumin to PBSA Leave overnight to dissolve Add NaN_3 (1 in 200)	Microfilter and store at 4°C
Diaminobenzidene (DAB)	Make up required concentration Aliquot and freeze OR Use commercially available tablets which can be made for each use	Discard into 20-40% sodium hypochlorite (bleach) to inactivate

mainly on the surface (2–3 μm) of the tissue section. The advantage is that the result can be monitored under the light microscope and is particularly useful to locate a brain area prior to examination at the electron microscope level.

Table 9 lists the choices to be considered in the design of a pre-embedding protocol. *Table 10* is an example protocol.

8.2. Post-embedding immunocytochemistry

Tissue prepared in this way has the ICC reaction performed on the ultrathin section on a grid. The problem with this methodology is the retention of the immunoreactivity of the target whilst preserving good ultrastructure. The main advantage is the discrete localization, usually with a particulate marker, for precise localization that can also be quantitated. It is essential to use well-characterized antisera and to determine the fixation sensitivity of the antigens under study. See *Tables 11–13*.

➤p. 61

Table 9. Pre-embedding immunocytochemistry: choice of method

Method	Choice	Rationale
Fixative	Glutaraldehyde + paraformaldehyde	Differing concentrations
	Glutaraldehyde + paraformaldehyde + picric acid (18)	Improved immunoreactivity
	1% monomeric glutaraldehyde (18)	Improved immunoreactivity
	Periodate–lysine–paraformaldehyde (PLP) (19)	Retains 45% antigenicity The periodate oxidizes carbohydrate moieties to aldehyde groups which are then cross-linked by the lysine Paraformaldehyde stabilizes proteins and lipids (20)
	Perfusion fixation	Good penetration with no concentration gradient across the tissue block (21)
Permeabilization or not		To allow penetration of reagents into the block; may not be required
	Aqueous 1% sodium metaperiodate + aqueous 1% sodium borohydride	Often used (22)
	Freeze-thawing tissue block	Preferred by some workers (23)
Primary antibody	Type and specificity	Depends on what is available, either from friendly laboratories or commercially
	Use whole immunoglobulin	Penetration into tissue may be a problem
	Use divalent or monovalent Fab fragments	To aid penetration into the tissue
Second antibody, substrate and chromogen	Anti-species antibody	Dependent upon the species of the primary antibody
	Second antibody conjugated to horseradish peroxidase (HRP)	Detected by peroxidase–antiperoxidase, reacted with DAB, giving a coloured product that is also electron-dense
	Second antibody conjugated to biotin	Detected by the Avidin–biotin complex (ABC) from Vector laboratories; gives a similar product
	Second antibody conjugated to colloidal gold	Directly visible in the EM

Table 9. Continued

Method	Choice	Rationale
Final processing	Post-fixation in OsO_4, dehydration and embedding for EM as *Table 4* These can also be viewed in the light microscope directly	To keep Vibratome sections flat, osmicate and dehydrate between 2 sheets of Postlip Histology Paper After resin infiltration, place sections onto silicon- or PTFE-coated slides and cover with a silicon-coated coverslip Wrap in foil and keep weighted during polymerization

Table 10. Example protocol of pre-embedding processing, after Martini and Schachner (22)

Procedure	Rationale/alternative
Dissect tissue from perfusion-fixed animal	
Fix for 2–24 h at 4°C	
Embed in agar if tissue is small	1.6% for fresh tissue, 6% for glutaraldehyde-fixed tissue
Cut 50–100 μm Vibratome sections and place into PBS in a microtitre, 24- or 6-well plate	
Permeabilize in 0.1 M sodium m-metaperiodate in PBS for 15 min at RT and wash in PBS for 15 min	Can freeze-thaw instead
Treat with sodium borohydride (10 mg/ml in PBS) for 10 min at RT and wash in PBS for 15 min	To reduce NH–CO bonds to increase immunoreactivity
Soak in 10% DMSO (dimethyl sulfoxide) in PBS for 10 min at RT and wash in PBS for 15 min	To permeabilize cell membranes
Incubate in 0.1–1% bovine serum albumin in PBS for 2 h at RT	To block non-specific binding sites % BSA is determined by experiment to block non-specific binding but not to reduce the specific reaction
Incubate in primary antibody diluted in BSA/PBS 12–24 h at 4°C and wash several times in BSA/PBS over 60 min	To wash out the unbound primary antibody
Incubate with the diluted second antibody (HRP-conjugated) at 4°C overnight and wash several times in BSA/PBS over 60 min	To wash out the unbound second antibody

Table 10. Continued

Procedure	Rationale/alternative
Sections treated with diaminobenzidene (DAB) 2 mg in 10 ml 0.04 M Tris, pH 7.6, for 30 min in the dark at RT, and then DAB + 0.005% H_2O_2 for 15 min in the dark at RT	To produce a colored and electron-dense reaction product
Postfix in 2% OsO_4 in 0.01 M phosphate buffer, dehydrate and embed for EM (*Table 4*)	

Table 11. Post-embedding immunocytochemistry: choice of method

Procedure	Choice of procedure	Rationale
Fixation	Either glutaraldehyde alone, or Paraformaldehyde alone, or Glutaraldehyde + paraformaldehyde, or Glutaraldehyde + paraformaldehyde+picric acid, or no fixation	To characterize the primary antibody LM ICC can be performed on tissue fixed under different conditions; a dot blot can be perfomed where brain homogenate is dotted onto nitrocellulose membrane, then cut into strips; each strip is fixed in a different fixative, washed, reacted with antibodies and the reaction developed Perfusion fixation for rapid penetration (24)
Postfixation or not	Uranyl acetate	Reduces lipid loss during dehydration whilst retaining immunoreactivity (18)
	Osmium tetroxide	Reacts with unsaturated lipids and stabilizes membranes Reacts with proteins Much reduced immunoreactivity UVL polymerization of osmicated blocks leads to poor polymerization
Choice of solvent and temperature of dehydration	Ethanol	Less lipid extraction during progressive lowering of temperature during dehydration (PLT) Can be used for partial dehydration into LR White resin or Lowicryl K4M (25)
	Progressive lowering of temperature during dehydration (PLT)	Less extraction of material Slows down molecular diffusion within tissue during processing

Table 11. Continued

Procedure	Choice of procedure	Rationale
Unmasking antigenic sites	Non-fixation sensitive antigens, postfixed in OsO_4, embedded in epoxy resin	Etch resin from sections with sodium metaperiodate and/or sodium ethoxide (26, 27) Reveals antigens on section surface Reduce NH–CO bonds with sodium borohydride (26) Increases immunoreactivity Hydrogen peroxide to remove OsO_4 (28) or etch the resin (29)
	Labile antigens embedded in acrylic resins, e.g. Lowicryl[a]	Unmask antigens enzymatically from tissue with proteolytic enzymes, e.g. 0.1% trypsin in Tris, pH 7.8, for 30 min at 37°C (30)
Resins		
Stable antigens	Epoxy resins	Usually need to be etched
Less stable antigens	LR White or Lowicryl K4M	After partial dehydration at RT
	Lowicryl K4M and K11M	Polar, hydrophilic low viscosity acrylic resins Use after progressive lowering of temperature (PLT)
	Lowicryl HM20 and HM23	Non-polar, hydrophobic low viscosity acrylic resins
	Lowicryl K4M and HM20	Retain low viscosity to –35°C and –70°C
	Lowicryl K11M and HM23	Retain low viscosity to –20°C to –30°C
Polymerization of resin	Epoxy resins at 60°C	Deleterious to most antigens
	LR White at 50°C or chemically	Lower temperature may improve immunoreactivity Chemical polymerization must be kept cold to avoid overheating (18)
	LR White	Must be anoxic, use gelatin capsules or exclude air
	Lowicryls	Ultraviolet light polymerization at low temperatures (see above)

[a] When sectioning Lowicryls, particularly K4M, reduce the height of water in boat to avoid wetting the block face

Table 12. Example protocol of post-embedding labeling for stable antigens (etching), after Petrusz and Rustioni (26)

Procedure	Rationale/comments
Cut ultrathin sections approx 100 nm thick	Use sections thicker than normal
Collect on carbon/formvar coated mesh grids	Film should be thicker than normal Cu grids for phosphate buffer, Ni or Au for Tris buffer
Etch by floating grids on 20 µl drops of saturated sodium metaperiodate$_{aq}$ for 30 min at RT	Microcentrifuge the metaperiodate solution to remove any undissolved salt Place drops on Parafilm in the lid of a box which can be covered during incubation to prevent drying out
Blot the edges of the grids and wash under a stream of distilled water from a wash-bottle	
Blot and place onto 1% Na borohydride$_{aq}$ for 5 min at RT	Borohydride solution should be freshly made and produce many hydrogen bubbles to ensure activity (CARE)
Blot the edges of the grids and wash under a stream of distilled water from a wash-bottle	
Proceed with on-grid ICC as in *Tables 16 or 17*	

Table 13. Example processing schedule for post-embedding labeling of labile antigens (progressive lowering of temperature during dehydration), from Rusakov *et al.* (31)

Procedure[a]	Rationale/comments
Fix as appropriate and wash in buffer	
Soak in 0.5 M NH_4Cl for approx 2 h	To bind excess aldehyde groups
30% ethanol at 0°C for 30 min	Start of progressive lowering of temperature
50% ethanol at –25°C for 30 min	
70% ethanol at –25°C for 30 min	
90% ethanol at –25°C for 30 min	Partial dehydration
1:1 Lowicryl K4M:90% ethanol at –25°C for 30 min	Infiltration
2:1 Lowicryl K4M:90% ethanol at –25°C for 1 h or overnight	
Neat Lowicryl K4M for 4 × 1 h minimum at –25°C	
Neat Lowicryl K4M overnight at –25°C	

Table 13. Continued

Procedure[a]	Rationale/comments
Embed in fresh Lowicryl K4M in polythene capsules	Degas the Lowicryl using a water pump to ensure consistent polymerization Fill the capsules to exclude air
Polymerize at –25°C under 360 nm UVL for 24–48 h	

[a] This protocol is designed for Lowicryl K4M but if the temperatures are changed could be used with any of the Lowicryls

9. CRYOTECHNIQUES FOR HIGHLY LABILE ANTIGENS

These can be used on unfixed tissue and are applicable for antigens that are sensitive to fixation or that cannot be stabilized by fixation. In neurobiology they are also used on fixed material as the immunoreaction is improved and is an appropriate choice once the other techniques have been explored.

The tissue must be rapidly cryofixed to prevent ice crystal damage and many techniques are available for optimal freezing. Impact freezing at –197°C is currently the cryofixation method of choice for brain material, usually with a spacer to prevent squashing. If unfixed tissue is to be frozen, it must be dissected rapidly to avoid post mortem changes. Once frozen, the tissue can be ultrathin cryosectioned and the ICC reaction performed on thawed sections. Alternatively, cryosubstitution can be employed on either fixed or unfixed brain areas after dissection where the material is placed directly after freezing, into substitution fluid (usually pure methanol with or without uranyl acetate) at –80°C for 48 h minimum before embedding in one of the Lowicryls (*Table 14*).

Table 14. Example processing schedule for highly labile antigens: cryosubstitution (from our own laboratory)

Procedure	Rationale/notes
After perfusion, the brain area is fixed for a further 5–10 min at RT	Avoid prolonged exposure to fixative
Wash in buffer for 3 × 5 min	
10% Analar glycerol in buffer for 1 h at RT	Increasing concentrations of cryoprotectant
20% Analar glycerol in buffer for 1 h at RT	
30% Analar glycerol in buffer for 1 h at RT	
Impact-freeze on cooled copper metal block	Highly polished and cleaned and cooled to –197°C with liquid nitrogen in a Leica MM80E impact freezer Use a spacer of approx 1 mm thick to prevent squashing the brain tissue
Transfer quickly to containers of Analar methanol cooled to –80°C in a Leica Auto FS freeze substitution equipment	Dispense stock methanol into 100 ml aliquots to prevent water accumulation each time the stock bottle is opened Use a cooled transfer device to avoid condensation on the blocks

Table 14. Continued

Procedure	Rationale/notes
Hold temperature at –80°C for 48 h	Substitution of the frozen water with methanol
Increase temperature to –50°C for 22 h	
1:1 Lowicryl HM20:methanol for 1 h at –50°C	Low temperature infiltration with acrylic resin
2:1 Lowicryl HM20:methanol for 1 h at –50°C	
Neat Lowicryl HM20 for 2 h at –50°C	
Neat Lowicryl HM20 overnight at –50°C	
Neat Lowicryl HM20 for 1 h at –50°C, prior to embedding	This step ensures good polymerization
Embed in polythene capsules or the flat moulds supplied with the equipment in fresh Lowicryl HM20	
Polymerize in the freeze substitution equipment under 360 nm ultraviolet light for 48 h at –50°C	

[a]Ensure all solutions are pre-cooled before transfer

The freeze substitution equipment ensures the tissue is maintained at the required temperature and all manipulations are carried out at that temperature

10. ON-GRID IMMUNOCYTOCHEMISTRY METHODOLOGY

Similar principles are applicable to both pre-embedding and post-embedding labeling. The problem to overcome is to block the non-specific labeling whilst maintaining the specific labeling. The specificity of the label can be checked firstly by ensuring that the same structures are labeled in two serial sections, and secondly by checking areas of resin that do not contain tissue (e.g. the lumen of a capillary) and the film on the grid.

Table 15 lists the choices that have to be considered when designing an experiment. *Table 16* is an example protocol of labeling for a single antigen and *Table 17* labeling for two antigens on the same ultrathin section.

➤p. 65

Table 15. On-grid immuncytochemistry: choices to be made

Method	Choice	Rationale
Composition of grid	Copper mesh grids with support films	Formvar with or without carbon evaporation Pioloform is preferred because it is more mechanically stable than formvar
	Gold or nickel mesh grids with support films	Copper grid may be toxic If ammonium solutions are used If Tris buffer used If grids are to be submerged during the ICC reaction
Buffer system	Tris buffer pH 8.2 may be optimal	If tissue postfixed in uranyl acetate For some antigens
Blocking agents		Prevent non-specific binding and maintain specific binding
	Bovine serum albumin (BSA) Acetylated BSA Ovalbumin Fish gelatin Tween 20	Examples of blocking agents that may be used individually or in combination
	0.1%–10%	Optimal concentration is a balance of labeling between specific and non-specific This is established in a trial run where the concentration giving the highest signal-to-noise is the one to use
Primary antibody	Longer incubation times give increased labeling	Usually at 4°C but there may be an increase in non-specific label
Second antibody or protein A	Conjugated to horseradish peroxidase (HRP), biotin or gold	Is stated (by British BioCell) that the anti-species conjugate amplifies the signal
	Protein A may be similarly conjugated	The reaction is stoichiometric Colloidal gold is available in several discrete sizes so that double and triple labeling can be performed on the same section HRP and biotin conjugates produce diffuse electron dense products that are not quantitatable
Controls		All reactions should be performed on duplicate grids
	Single labeling for one antigen (*Table 16*)	Omit primary antibody Use pre-immune serum in parallel

Table 15. Continued

Method	Choice	Rationale
	Double labeling for two antigens on the same ultrathin section (*Table 17*)	Omit both test antisera Use pre-immune sera in parallel Omit primary antiserum but use both probes Omit secondary antiserum but use both probes

Table 16. Example protocol of post-embedding labeling for one antigen (single labeling) from Rusakov *et al.* (31)

Procedure[a]	Rationale
0.1%–5% BSA in PBSA for 30 min at RT	To block non-specific binding Microfilter before use
Antibody dilutions in BSA/PBSA for 1–2 h at RT or overnight at 4°C	Microcentrifuge before use to remove particulate debris Longer incubation may increase labeling density at the expense of increased non-specific binding
BSA/PBSA for 3 × 5 min at RT	To wash off unbound antibody
Gold conjugate (5, 10 or 15 nm) diluted in BSA/PBSA for 60 min at RT	Protein A/gold at 1 in 100 is in routine use
BSA/PBSA for 3 × 5 min at RT	To wash off unbound conjugate
Stream-wash with distilled water from a wash-bottle	To wash off salts
Stain sections with uranyl acetate and lead citrate	To contrast the tissue for electron microscopy

[a] All operations are performed by floating the grids section-downward on 20 μl drops of reagents on Parafilm in the lid of a sandwich box
Putting the bottom over the lid at each stage prevents evaporation and excludes dust from the grids

Table 17. Example protocol of post-embedding labeling for two antigens (double labeling), from Rusakov *et al.* (32)

Procedure[a]	Rationale
0.1%–5% BSA in PBSA for 30 min at RT	To block non-specific binding Microfilter before use
Primary antibody diluted in BSA/PBSA for 1–2 h at RT or overnight at 4°C	Microcentrifuge before use to remove particulate debris Longer incubation may increase labeling density at the expense of increased non-specific binding
BSA/PBSA for 3 × 5 min at RT	To wash off unbound antibody

Table 17. Continued

Procedure[a]	Rationale
5 nm gold conjugate diluted in BSA/PBSA for 60 min at RT	Protein A/gold at 1 in 100 is in routine use
BSA/PBSA for 3 × 5 min at RT	To wash off unbound conjugate
Protein A 100 μg/ml in BSA/PBSA for 90 min at RT	To block excess IgG sites not bound by the first gold conjugate
BSA/PBSA for 3 × 5 min at RT	To wash off unbound Protein A
Second antibody diluted in BSA/PBSA for 1–2 h at RT	Microcentrifuge before use to remove particulate debris Longer incubation may increase labeling density at the expense of increased non-specific binding
BSA/PBSA for 3 × 5 min at RT	To wash off unbound antibody
15 nm gold conjugate diluted in BSA/PBSA for 60 min at RT	Protein A/gold at 1 in 100 is in routine use
BSA/PBSA for 3 × 5 min at RT	To wash off unbound conjugate
Stream-wash with distilled water from a wash-bottle	To wash off salts
Stain sections with uranyl acetate and lead citrate	To contrast the tissue for electron microscopy

[a] All operations are performed by floating the grids section-downward on 20 μl drops of reagents on Parafilm in the lid of a sandwich box. Putting the bottom over the lid at each stage prevents evaporation and excludes dust from the grids.

If the two antisera are from different species, they can be reacted as a cocktail. Similarly, if anti-species second antibody probes are used, these can also be applied as a cocktail Controls are crucial, refer to *Table 15*

11. NOTES ON SUPPLIERS

General EM supplies: Agar Scientific Ltd, 66a Cambridge Road, Stansted, Essex CM24 8DA, UK.
Immunogold reagents and accessories: AurIon, Costerweg 5, 6702 AA Wageningen, Netherlands.
Immunogold reagents: British BioCell International, Cardiff Business Technology Centre, Senghenydd Road, Cardiff CF2 4AY, UK.
Histological equipment: Histological Equipment Ltd, 70 Davies Road, West Bridgford, Nottingham NG2 5JA, UK.
EM ancillary equipment: Leica UK Ltd, Davy Avenue, Knowlhill, Milton Keynes MK5 8LB, UK.
General laboratory chemicals: Merck Ltd, Hunter Boulevard, Magna Park, Lutterworth, Leicestershire LE17 4XN, UK.
Surgical instrument suppliers: Rocket of London Ltd, Imperial Way, Watford WD2 4XX, UK.
General laboratory chemicals: Sigma Chemical, Fancy Road, Poole, Dorset BH17 7BR, UK.

Immunocytochemistry reagents: Vector Laboratories, 16 Wulfric Square, Bretton, Peterborough PE3 8RF, UK.
Sagatal: Manufactured by Rhone Merieux (Ireland), is a 'prescription medicine only' in the UK and available only through veterinary surgeons.

12. ACKNOWLEDGMENT

We would like to thank Ms Jackie Brown for help in compiling the light microscopy section of this chapter.

13. REFERENCES

1. Glauert, A. (ed.) (1991) *Practical Methods in Electron Microscopy*, vol. 3, part I, Elsevier Science, Amsterdam.

2. Harris, N. (1992) in *Cell Biology Labfax*, p. 53.

3. Glauert, A. (ed.) (1991) *Practical Methods in Electron Microscopy*, vol 5, part I, Elsevier Science, Amsterdam.

4. Bancroft, J.D. and Stevens, A. (1990) *Theory and Practice of Histological Techniques*, Churchill Livingstone, Edinburgh.

5. Stewart, M.G. and Bourne, R.C. (1992) in *Quantitative Methods in Neuroanatomy* (M.G. Stewart, ed.), Wiley, Chichester.

6. Sokoloff, L., Reivich, M., Kennedy, C., Des Resiers, M.J., Patlak, C.S., Pettigrew, K.D., Sakrada, O. and Shinohara, M. (1977) *J. Neurchem*, **28**, 897.

7. Bolam, J.P. (ed.) (1992) *Experimental Neuroanatomy: A Practical Approach*, IRL Press, Oxford.

8. Valverde, F. (1970) in *Contemporary Research Methods in Neuroanatomy* (W.J.H. Nauta, and S.O.E. Ebbesson, eds), Springer Verlag, New York, p. 11.

9. Gabbott, P.L.A. and Somogyi, J. (1986) *J. Neurosci. Meth.*, **11**, 221.

10. Sojka, M., Davies, H.A., Rusakov, D.A. and Stewart, M.G. (1995) *J. Neurosci. Meth.*, **62**, 73.

11. Merighi, A., Polak, J. and Theodosis, D.T. (1991) *Neuroscience*, **40**, 67.

12. Nusser, Z., Mulvihill, E., Streit, P. and Somogyi, P. (1994) *Neuroscience*, **61**, 421.

13. Baude, A., Nusser, Z., Roberts, J.D.B., Mulvihill, E., McIlhinney, R.A.J. and Somogyi, P. (1993) *Neuron*, **11**, 771.

14. Molnar, E., McIlhinney, R.A.J., Baude, A., Nusser, Z. and Somogyi, P. (1994) *J. Neurochem.*, **63**, 683.

15. Csillag, A. (1991) *J. Neurocytol.*, **20**, 518.

16. Hamori, J., Takacs, J. and Petrusz P. (1990) *J. Histochem. Cytochem.*, **38**, 1767.

17. Martini, R. (1994) *J. Neurocytol.*, **23**, 1.

18. Newman, G.R. and Hobot, J.A. (1993) *Resin Microscopy and On-Section Immunocytochemistry*, Springer Verlag, New York.

19. Gotow, T. and Tanaka, J. (1994) *J. Neurosci. Meth.*, **37**, 691.

20. McClean, I.W. and Nakane, P.K. (1974) *J. Histochem. Cytochem.*, **22**, 1077.

21. Totterdell, S., Ingham, C.A. and Bolam, J.P. (1992) in *Experimental Neuroanatomy: A Practical Approach* (J.P. Bolam, ed.), IRL Press, Oxford.

22. Martini, R. and Schachner, M. (1986) *J. Cell Biol.*, **103**, 2439.

23. Somogyi, P. and Takagi, H. (1982) *Neuroscience*, **7**, 1779.

24. Ingham, C.A. (1992) in *Experimental Neuroanatomy: A Practical Approach* (J.P. Bolam, ed.), IRL Press, Oxford.

25. Hobot, J.A. (1990) *J. Struct. Biol.*, **104**, 169.

26. Petrusz, P. and Rustioni, A. (1989) in *Techniques in Immunocytochemistry*, vol. 4, Academic Press, San Diego.

27. Lane, B.P. and Europa, D.L. (1965) *J. Histochem. Cytochem.*, **13**, 579.

28. Baskin, D.G., Erlandsen, S.L. and Parsons, J.A. (1979) *J. Histochem. Cytochem.*, **27**, 1290.

29. Bendayan, M. and Zollinger, M. (1983) *J. Histochem. Cytochem.*, **31**, 101.

30. Doerr-Schott J. (1989) in *Colloidal Gold*, vol. I (M.A. Hayat, ed.), Academic Press, New York.

31. Rusakov, D.A., Davies, H.A., Krivko, I.M., Stewart, M.G. and Schachner, M. (1994) *Neuroreport*, **5**, 2469.

32. Rusakov, D.A., Davies, H.A., Stewart, M.G. and Schachner, M. (1995) *Neurosci. Let.*, **183**, 50.

CHAPTER 5
EXPERIMENTAL NEUROPSYCHOLOGY: THE LOGIC OF, AND LESSONS FROM, LESIONS

S.M. O'Mara and V. Walsh

1. INTRODUCTION

The lesion method is one of the oldest methods for studying brain function in humans and animals. To date, it has also been the single most informative technique available for understanding the neural bases of behavior. Lesions fall into two classes. The first is *accidental lesions*, which result from brain insults such as cerebrovascular accidents (CVA; stroke); penetrating head injuries, from gunshot or other incidents; focal contusions, usually resulting from closed-head injury; or neuronal sclerosis and neurodegeneration, where a widespread area of tissue dies or functions in a deeply abnormal manner because of widespread neuronal death or degeneration from, for example, Alzheimer's or Parkinson's disease. Disease states associated with neurodegenerative diseases have tended to prove less useful in the localization of brain function because there tends to be widespread degeneration associated with a variety of structures, making interpretation of the disrupted functions extremely difficult.

The second class of lesions are *deliberate lesions* which are usually, but not exclusively, performed on experimental animals. There are a variety of such lesions: tissue excision, where tissue is deliberately removed (usually by aspiration); tissue destruction, where tissue is destroyed electrolytically or chemically; tissue deactivation, where tissue is reversibly deactivated (most commonly via injections of local anesthetic or through cooling) tissue disconnection, where brain areas are left intact but fiber tracts which communicate between these areas are permanently cut or temporarily interrupted (by anesthetic or cooling); and finally, gene knockout, where the gene or genes that control the expression of receptors or other functional systems are disrupted, replaced or deleted and the effects of this manipulation are then assessed behaviorally. Neurosurgery involving the excision or aspiration of brain areas is often conducted on humans in order to relieve conditions such as drug-resistant epilepsy, relief from intractable pain or brain tumors (1–3).

The consequence of all these manipulations will be to permanently or transiently remove from the neural information processing chain a brain area that is hypothesized to be involved in a particular function. If this function can continue to be performed in the absence of this brain area, then it may be concluded that this area is not essential for executing this function. An area may of course provide some modulatory processing and may receive inputs from the irreducibly essential areas, while not being essential itself for the execution of a particular function or set of functions.

2. THE LOGIC OF LESION ANALYSES

At the heart of the lesion analysis is the notion of dissociation (4): given areas A and B in the brain and functions Y and Z, an experimenter may lesion area A and find that function Y is no longer possible whereas function Z is intact, whereas a lesion of area B produces a deficit in function Z, but function Y remains intact. Thus, functions Y and Z can be doubly dissociated. The logic may be laid out in the form of a truth table:

	Area A	**Area B**
Function Y	*Impairment*	No impairment
Function Z	No impairment	*Impairment*

The interpretation of the double dissociation is as follows (4): function Y depends on area A but not area B, whereas function Z depends on area B but not area A. Therefore, each function can proceed independently of the other, and therefore each function is mediated by an area with an input and an output to the rest of the system which is independent of the inputs and ouputs of the other area and its associated function.

What is immediately obvious from this sort of analysis is the importance of examining what the brain *can* do post-lesion as well as what it *can't* do. Not so obvious from the above table is the requirement to perform, if possible, control experiments involving sham lesions, but this is absolutely required in order to control for nonlesion-induced artifacts, such as the stress of the operation, extraneous tissue damage resulting from the presence of electrodes or cannulae, etc. Sometimes control lesions are conducted in sites remote from the lesion site of interest in order to control for the absolute amount of tissue removed. It is also important to try several different types of lesion technique, as *fibers of passage* provide the possibility of a major experimental confound. Fibers of passage are fibers that may pass through an area of interest but which do not contribute to the function of that area. Lesion effects may result from damage to these fibers and not from any damage intrinsic to the lesioned area. Electrolytic lesions of the amygdala strongly implicated the amygdala in conditioned taste aversions (see Holscher and O'Mara, this volume), but control experiments conducted using ibotenic acid, which does not affect fibers of passage but does destroy cell bodies, demonstrated that this was not so (5). There are actually three possibilities for dissociation in lesion analysis (4) – see *Table 1*.

Table 1. The theoretical and experimental analysis of dissociation

Type of dissociation	Experimental desirability	Theoretical and experimental rigor	Relationship to theory
No dissociation	Least desirable	'Fishing expedition' – asks what will happen if area A is damaged	Can be suggestive of further hypotheses; not definitive
Single dissociation	Moderately desirable	Reasonable control – lesion may be hypothesis or theory driven	Suggestive of further hypotheses, but not definitive
Double dissociation	Most desirable	Best of all, as lesion must be driven by directly opposed experimental questions Lesion analysis is driven by 'in principle' questions	Best available relationship to theory, providing definitive answer to well-posed questions

It is possible to extend the analysis further and look for triple or other higher-order dissociations. Such dissociations present formidable interpretative problems and are rarely used.

3. WHICH TYPE OF LESION TO USE?

The answer to this question will depend on the question being asked, as well as on technical considerations such as the ease or difficulty with which a particular structure can be lesioned. The hippocampus of the rat is notoriously difficult to lesion precisely as it is shaped like a standing cashew nut and it is located subcortically, so any attack on the hippocampus invariably involves damage to overlying cortex. The problem is even more difficult when tackling retrohippocampal structures such as the subiculum that present an even more convoluted morphology than the hippocampus itself. Furthermore, inferences regarding the effects of such lesions require controls where the overlying cortex is damaged, but the target structure remains intact, in order that any observed deficits be attributed unequivocally to particular areas and not to others. Dorsally located cortical areas present the least problem for lesioning, but they require the presence of visual landmarks which may not be obvious on the smooth cortical surface of the rat. This is not such a problem in the monkey, as monkeys have a well-gyrified cortical surface and individual gyri can be used as landmarks to delimit the maximal and minimal extent of an intended lesion.

Tables 2 and 3 present the relative merits and demerits of deliberate and accidental lesions, respectively. In all cases, appropriate controls must be used: there should be one unoperated normal control group and there should be a sham-lesioned group (see above).

Deliberate lesions are conducted under standard conditions: a standard neuroanatomical atlas of the particular species is used to delimit the relevant areas; plots are then made of the intended lesions and these are correlated with stereotaxic coordinates for the brain area(s) in question. The animal is then anesthetized and a burr hole is opened in the skull using an electric hand-drill; the lesion instrument is then lowered to the stereotaxic coordinates and the lesion procedure (for example, injection, aspiration) is carried out. The burrhole is then plugged with bonewax and the animal is allowed to recover before the behavioral experiments are conducted.

Transcranial magnetic stimulation

As noted above, deliberate lesions are only rarely conducted on humans and when they are it is for the relief of intractable epilepsy or other neurological conditions serious enough to warrant neurosurgery. One recently introduced technique, transcranial magnetic stimulation (TMS), allows the induction of transient lesions in humans (53). This non-invasive technique can be used to disrupt normal processing, usually in a cortical area, with millisecond precision. Single or repetitive pulses of current are applied to the skull overlying a cortical region of interest at a specified time during the performance of a task.

The advantages of the technique are that it is non-invasive, safe, reversible and designed for use with human subjects. Using TMS demands that the user has a hypothesis about the temporal sequence of events in the task being studied. For this reason it has been of most use so far in studies of the visual system and the motor system, i.e. those domains in which there is a long history of experimental psychology and single-unit neurophysiology. A typical experiment involves presenting subjects with a visual display and a discrimination task to perform. TMS is applied at n milliseconds after the onset of the array and reaction times are recorded. TMS effects can be measured as errors or increases in reaction time. TMS is

➤p. 73

Table 2. Deliberate lesions

Lesion type	Method	Advantages	Disadvantages	Sample references
Aspirative	Tissue suction of landmark- or stereotactically-defined area	Offers precise visual control over cortical area to be lesioned Straightforward to conduct	Can be less precise for subcortical structures Suffers from fibers of passage problem	5–8
Anesthetic	Injection of local anesthetic via cannula	Effects are reversible and thus each animal can be its own control Can be combined profitably with other techniques, e.g. single-unit recording	Area of diffusion may not be precisely defined Usually requires presence of indwelling cannula with its associated problems	9–11
Chemical	Injection of tissue-destroying agent	Does not suffer from fiber of passage problem Can be used to destroy a small but precisely defined tissue area, e.g. a small nucleus	Requires multiple injections to be made Time course of lesion induction (~days) may preclude certain behavioral preparations	12–16
Cooling	Thermal cooling coils used to cool defined area of tissue	Effects are reversible and thus each animal can be its own control The separate processes underlying a particular behavior can be fractionated into their components by using the cooling coils at carefully titrated times	Usually requires presence of indwelling thermal cooling coil with its associated problems	17–21
Disconnection	Knife cut of precisely defined fiber bundle between two communicating areas of interest	Structures are intact, but communication is no longer possible between them	Degeneration of associated target neurons in the associated structures may occur	3, 22–24

Electrolytic	Stereotactic electrode used to generate either a radio-frequency lesion or electrical lesion	Quick and simple lesion to conduct Technique of choice for preparations with time constraints, such as chicks	Suffers from fibers of passage problem	25, 26
Gene knockout	Most recently developed type of lesion: specific genes are deleted and their effect on biochemistry, physiology and behavior are assessed	Provides a high degree of experimental control over developmental changes	Expensive and recently developed technology, largely applied to fruitflies and mice Animals lacking a particular gene or set of genes are 'different' in their development – careful controls must be conducted	27–30, 31

Table 3. Accidental lesions

Lesion type	Insult origin	Advantages	Disadvantages	References
Cerebrovascular accident (CVA)	May result from either ischemic stroke or embolic stroke	Can be very precisely located, depending on the artery involved	May be very diffusely located, with many associated changes	32–38
Penetrating head injury	Gunshot or explosions are usual cause	Can be very precisely located	Often associated with diffuse damage Possible confounds resulting from collateral damage to body	39–41
Focal contusion	Usually result from falls in the elderly (65 yrs +) and road traffic accidents in young (< 30 yrs); bimodally distributed and preponderance of males	Can be associated with interesting syndromes, e.g. frontal executive dysfunction Lesion often resolves over a period of time, providing data on plasticity and recovery of function	May be difficult to visualize	42–46
Neurodegenerative/ sclerosis	Multiple causation: e.g. developmental (heatshock from influenza); excessive alcohol consumption	Can be very precisely located, e.g. hippocampal sclerosis		47–52

thought of as primarily a temporal tool but many studies report localization of effects within a resolution of 1 cm. For some areas of sensory systems analysis and cognitive neuroscience the technique holds much promise.

One drawback of TMS is that to date it has not proved possible to investigate higher level cognitive functions with single pulse TMS. Repetitive TMS has been used to investigate memory and language, but this precludes analyses of the temporal dynamics of the function under study. Anyone wishing to give repetitive TMS to human subjects may also find themselves constrained by ethical committees. Good examples of how to use the technique can be found in refs 54 and 55. Technical details are given in ref 56. The MagStim (TM) Co. (Whiteland Industrial Estate, Whitland, Dyfed SA34 OHR, UK) are also supportive and well-informed and try to keep an in-house list of publications reporting uses of the technique.

4. THE RESOLUTION OF THE LESION TECHNIQUE

The resolution of the lesion technique is limited by technical and behavioral considerations. Damasio (40) suggests the following considerations:

(i) the sophistication of the neuropsychological question being investigated;
(ii) the sophistication of the underlying theory regarding the question of interest;
(iii) the sophistication of the conceptualization of the nature of neuronal information processing (local circuits, systems, networks etc.);
(iv) the resolution of the procedures used to reconstruct the missing tissue.

A very important issue is the reconstruction of the lesioned area and its relationship to the areas that remain. The ideal for lesion analysis is the reasonably rapid postmortem examination of the brain after some question of interest has been investigated. Such rapid examination is important for several reasons. These include the prevention of the functional reorganization after the brain insult which may result in some degree of reorganization of structure and/or function to compensate for the tissue loss. Glassman (57) notes that there may be some redistribution of neurotransmitters as a result of the lesion; there may be some irritation of tissue adjacent to the lesion as a result of electrolytic treatment. Another important factor may be the age of the animal: whether or not it is still growing or if it has reached maturity, as the organization of neuronal distribution and associated neurotransmitters varies with age.

The most usual procedure will be to lesion the animal, conduct the behavioral assay of interest and then euthanize the animal with a standard anesthetic. The animal is then transcardially perfused with saline followed by a formalin and saline mixture. The brain is then removed from the cranial vault and placed in a variety of different solutions, depending on how it is to be sectioned (see Wouterlood *et al.*, this volume, for further details). It is usually helpful to photograph the brain from a number of standard positions and distances; other dimensions such as length and width should also be taken as should an estimate of the lesion size using calipers if the lesion is a cortical one. The brain undergoes shrinkage after the postmortem perfusion of formalin; a standard correction for this shrinkage used in the literature is 10% to 15%.

The brain should then be blocked and sectioned using a microtome and stained and counterstained for cell bodies and fibers (e.g. cresyl violet, nissl stain). The lesioned sections should be examined under the light microscope and drawings should be made using a camera lucida procedure of all the sectioned areas. In addition, the sections should be examined to determine if there is neuronal degeneration in areas other than the lesioned area. The nature and extent of the lesioned area should be reconstructed and delimited; inferences may then be drawn about whether the actual and intended lesion area correspond. Furthermore, an estimate must be made of the percentage volume of the actually lesioned area of the target area by comparison with sections from control animals or standard atlases. To some extent, the reconstruction of the lesioned area will be a matter of guesswork which must be undertaken in good faith. It is desirable that the lesion reconstructions be examined independently, if at all possible. Lesions that extensively invade areas other than the intended area should be correlated with the behavior of the individual animals concerned and it is probably desirable that this data be discarded but noted in any publications.

5. EXAMPLES OF LESION ANALYSES

This section presents examples of lesion analyses that have been particularly instructive for our understanding of brain function. Lesions of the hippocampus and its associated structures cause grave deficits in the allocentric representation of place and in allocentric spatial navigation (for reviews see refs 58 and 59, and Hölscher and O'Mara, this volume). Rats with hippocampal lesions show a variety of characteristic behaviors in water and dry mazes whereby they are unable to find hidden platforms in relation to distal cues in the environment (60). Rats who have sustained lesions to the amygdala do not show this pattern of spatial deficit at all, but do show deficits in cue-consequence learning in fear conditioning (61, 62). Lesions of the hippocampus do not affect acquisition or retention of fear conditioning; the hippocampus therefore does not participate in fear conditioning. We may therefore conclude that the hippocampus plays a crucial role in the allocentric representation of space and the amygdala plays a crucial role in fear conditioning; the inputs for these two different types of information processing are independent as are the outputs from these structures – a clear example of double dissociation of structure and function.

Another clear example of successful lesion analysis is represented in the program of experiments initiated by Iwai and Mishkin (64) and continued by Ungerleider and Mishkin (65). The consequence of the early stage of the program was to lay down the foundation of visual processing in the temporal cortex of the macaque monkey, while the later experiments generated the 'two streams' hypothesis of visual processing which has probably been the single most fruitful theory in understanding cortical vision in the last 25 years. Iwai and Mishkin (64) made strip lesions of areas of cortex in the extrastriate cortex and inferotemporal cortex of macaque monkeys. From the effects of these lesions they derived a gradient of severity of the effects of lesions on visual performance. The apparent inelegance of the approach was due to the combination of originality and thoroughness, and it formed the bedrock for dozens of subsequent experiments. Ungerleider and Mishkin (66, 67) later extended the work to lesions of temporal cortex and parietal cortex. Two major findings came from this work, one a single dissociation the other a double dissociation. Monkeys with large parietal cortex lesions were found to be impaired on the landmark task (a test of spatial ability) only if at least two subregions of parietal cortex were damaged (single dissociation). They were unimpaired on object discrimination problems however. Monkeys with lesions of the inferotemporal cortex were impaired on tests of object discrimination, but unimpaired on the landmark task (providing a clear double dissociation of the effects of temporal and parietal lesions on object and spatial tasks respectively). The papers underlying this work have been summarized, reviewed and absorbed into the literature. However, as case studies in lesion analysis they are worth reading again in their own right.

6. LESION ANALYSIS AND FUNCTIONAL IMAGING

Static imaging, as noted above, is extremely important for investigating the extent of lesions in human patients, and combined with careful neuropsychological analyses it has provided us with much important data regarding the localization of function. Now that fullblown functional imaging technologies such as PET and fMRI are available ((68) and journals cited in *Table 4* below) it may be asked if there is any point in continuing to conduct lesion analyses now. An acid reading of the last fifteen years of the functional imaging literature would lead one to suspect that the imaging literature is still largely concerned with the high-tech confirmation of the preceding 150 years of careful neuropsychological analysis. Furthermore, it can legitimately be asked if functional imaging technologies are predictive or simply provide *post hoc* confirmation of what we already know (69).

The subtractive logic of PET and fMRI studies does not leave us with a definitive conclusion about whether or not a particular area is absolutely required in a neura processing chain. The only way to definitively decide if function Y is dependent on area A is to remove that area from the processing chain; single-unit recording is subject to exactly the same critique and single-unit electrophysiologists usually accept that a lesion analysis must proceed hand-in-hand with a neurophysiological analysis.

Nadeau and Crosson explicitly compare the advantages and disadvantages of lesion analysis with functional imaging (70, *Table 2*). The particular advantages they cite in favor of the lesion strategy include the permanent removal of components of the processing chain, its persistence over time (which allows many systematic studies of the effects of the lesion on the processing chain) and, finally, the relative inexpensiveness of the procedure. Functional imaging complements lesion analysis by identifying necessary and contributing components of the processing chain; functional imaging also allows the identification of specific components of the processing chain.

We may conclude overall that the one strategy complements and underlines the other. The rationale for many imaging studies rests on years of neuropsychological analysis. Nevertheless, one could be forgiven, on reading some imaging papers, for thinking that lesion analysis has taken the place of phrenology as yesteryear's mistake. This is a dangerous error. When we have color pictures of every conceivable cortical 'network' in the literature, the question of what individual areas contribute to those networks will remain. The solution will be founded on an understanding of what is required to perform a task – the domain of classical experimental psychology and neuropsychological analysis. The question of the 'added value' of functional imaging must be confronted before initiating experiments with this technology: what will the use of this technology tell us that we didn't already know?

We finish up this section by pointing to the possible use of evoked potential measures for lesion analysis (71, 72). By comparing the performance of carefully selected brain damaged patients with the performance of normal controls on identical tasks such that the patients have selective deficits it is possible to compare the components of the response generated by normals with those generated by patients. This type of analysis, if conducted across a variety of patients, allows the identification of areas that are necessary for the generation of particular components of the overall evoked potential waveform. It also allows the identification of brain areas that are essential to the performance of particular tasks and by comparison to the performance of normals, differences in the waveform generated by lesioned subjects and of the waveform components that are essential to the successful performance in the tasks in question. A double dissociation analysis is also possible here: given waveform components A and B in normal subjects and lesioned areas Y and Z in lesioned subjects, what will be the outcome of the waveform and task analysis? The following may occur (by reference to the foregoing truth table):

(i) Task 1: waveform component A present in normals, absent in patients; Task 1 outcome for patients: normal;
(ii) Task 2: waveform component B present in normals, absent in patients; Task 2 outcome for patients: fail.

We may conclude therefore that although component A is missing, the activity of the area it represents is not essential for the performance of task 2; in contrast, component B is missing and patients show a deficit on task 2. Therefore, the inputs required for task 1 are independent of those required for task 2; the use of evoked potentials therefore allows the localization of damage to specific neural pathways that are independent of the pathways required for successful performance of task 1.

7. CONCLUSIONS

A meaningful lesion analysis is easiest to conduct if there is a reasonable degree of knowledge of the underlying neuroanatomy of the to-be-lesioned region and if there is a meaningful bedrock of psychological theory on which to build the lesion experiment. Such theory, informed by the neuroanatomy, should be capable of making some qualitative prediction, at the least, about the effect that the induced lesion will have. Another important but related issue is the reconstruction of the lesioned area and its relation to the brain areas that remain. For humans, the ideal – rapid postmortem examination after some question of interest has been investigated – is almost never attained. Humans may live for several decades after the lesion and subtle but hard-to-define reorganization may occur over this time. Other brain subsystems may come to take over functions that were previously the province of one area; Robertson and colleagues (e.g. 73, 74) have demonstrated that left-hemisphere language systems can be successfully entrained to overcome right-hemisphere sustained attention deficits after stroke. In humans, the next best thing will be to conduct high-resolution MRI combined with some form of volumetric analysis to get some idea of lesion volume and location (40); the current resolution of MRI is about 7 mm of tissue and this resolution can be expected to improve further as developments in imaging technologies progress. New techniques are becoming available for the co-registration of MRI, PET and electrophysiological data from evoked potentials and EEG (75) which promise to revolutionize further our understanding of the necessary components of the neural information processing chain. *Table 4* lists the more important journals that publish data related to lesions; these provide many useful tips on appropriate conventions for lesion reconstruction and interpretation.

The ideal for lesion analysis is approached in experimental animals: a discreet lesion can be induced, often in association with other techniques (e.g. anatomical tract-tracing, single-unit recording) and a well-defined behavioral question can be investigated. There exists excellent control over the animal's prelesion history and any required controls (within- and between-subject) can be easily conducted. The point of using animals then rests on the answer to a different set of questions: is animal work supposed to throw light on human brain function? If so, how secure are the homologies of structure and function between humans and the animal species (see (76, 77) for contrasting perspectives)? Or is the interest in how a brain solves a particular problem, without worries about whether or not this brain is homologous to the human brain? With humans, the question may have to be changed. Instead of looking for populations of subjects, with, say, discrete bilateral damage to a particular area, it is probably more appropriate to have a well-defined psychological or functional question relating to a particular domain, with a well-fractionated set of experimental tasks, and with these in hand to look for humans who have disruptions of these functions. In concert with data from imaging in these patients, functional imaging data from normal subjects and other sources of evidence (for example, evoked potentials) it will be possible to begin to ascribe functions to areas or networks of areas and to define essential circuits for particular types of functional operations. How these areas perform these functions is, of course, another question entirely.

Table 4. Important data sources

Journal names	Subject type
Brain *Cognitive Brain Research* *Cognitive Neuropsychology* *Cortex* *Human Brain Mapping* *Journal of Cognitive Neuroscience* *Nature* *Neurocase* *Neuropsychologia* *Neuropsychological Rehabilitation*	Mostly or exclusively work with humans, covering a variety of subjects from functional activation studies in normal humans, through assessment of deficit post-lesion, through to rehabilitation after damage has occurred
Behavioural Neuroscience *Behavioural Brain Research* *Brain Research* *European Journal of Neuroscience* *Experimental Brain Research* *Journal of Comparative Neurology* *Journal of Neuroscience* *Learning and Memory* *NeuroReport* *Psychobiology*	Mostly or exclusively work with animals, providing essential data on different types of lesions and on neuroanatomical connections

Neuropsychology

8. ACKNOWLEDGMENTS

We thank Dr John Gigg, Department of Psychology, Trinity College, Dublin, for reviewing the manuscript thoroughly for us.

9. REFERENCES

1. McMackin, D. and Staunton, H. (1991) *J. Irish Coll. Phys. Surg.* **20**, 193.

2. Jones-Gotman, M. *et al.* (1996) *Neuropsychologia*, in press.

3. McMackin, D. *et al.* (1995) *Acta Neurochir.*, **135**, 12.

4. Olton, D.S. (1991) in *Learning and Memory: A Biological View* (J.L. Martinez and R.P. Kesner, eds) Academic Press: San Diego.

5. Dunn, L.T. and Everitt, B.J. (1988) *Behav. Neurosci.*, **102**, 3.

6. Stevens, R. and Cowey, A. (1973) *Brain Res.*, **52**, 203.

7. Rawlins, J.N.P. *et al.* (1993) *Behav. Neurosci.*, **107**, 236.

8. Lawler, K.A. and Cowey, A. (1987) *Exper. Brain Res.*, **65**. 695.

9. Girard, P., Salin, P.A. and Bullier, J. (1991) *J. Neurophysiol.*, **66**, 1493.

10. Thinus-Blanc, C. *et al.* (1991) *Hippocampus*, **1**, 365.

11. Mizumori, S.J.Y. *et al.* (1989) *J. Neurosci.*, **9**, 3915.

12. Holscher, C. and Schmidt, W.J. (1994) *Behav. Brain Res.*, **63**, 187.

13. Tanaka, T. *et al.* (1992) *Prog. Neurobiol.*, **38**, 317.

14. Handelmann, G.E. and Olton, D.S. (1981) *Brain Res.*, **217**, 41.

15. Bender, D.B. and Baizer, J.S. (1990) *Brain Res.*, **79**, 467.

16. Levesque, F. *et al.* (1990) *Exp. Brain Res.*, **80**, 191.

17. Yamada, A. *et al.* (1992) *J. Neurosci.*, **12**, 729.

18. Horel, J.A. (1991) *Meth. Neurosci.*, **7**, 97.

19. Martin-Elkins, C.L. and Horel, J.A. (1992) *J. Comp Neurol.*, **3**, 178.

20. Stein, J.F. (1978) in *Active Touch* (G. Gordon, ed.) Pergamon Press, Oxford, p. 79.

21. Horel, J.A. (1996) *Behav. Brain Res.*, **76**, 199.

22. Gazzaniga, M.S. (1995) *Neuron*, **14**, 217.

23. Gaffan, D. (1993) *Neuropsychologia*, in press.

24. Gaffan, E.A., Gaffan, D. and Narrison, S. (1988) *J. Neurosci.*, **8**, 3144.

25. Purves, D., Bonardi, C. and Hall, G. (1995) *Behav. Neurosci.*, **109**, 366.

26. Sutherland, R.J., Whishaw, I.Q. and Kolb, B. (1983), **7**, 133.

27. DeZazzo, J. and Tully, T. (1995) *Trends Neurosci.*, **18**, 212.

28. Baringa, M. (1992) *Science*, July.

29. Grant, S.G.N. and Silva, A.J. (1994) *Trends Neurosci.*, **17**, 71.

30. Tully, T. *et al.* (1994) *Cell*, **79**, 35.

31. Mayford, M., Abel, T., and Kandel, E.R. (1995) *Curr. Opin. Neurobiol.*, **5**, 141.

32. Hom, J. and Reitan, R.M. (1990) *J. Clin. Exper. Neuropsychol.*, **12**, 644.

33. Kinsella, G. and Ford, B. (1980) *Med. J. Australia*, **2**, 663.

34. Blanc-Garin, J. (1994) *Neuropsychol. Rehab.*, **4**, 359.

35. Herrmann, M. *et al.* (1995) *Stroke*, **26**. 850.

36. Sturm, W. and Willmes, K. (1991) *Neuropsychol. Rehab.*, **1**, 259.

37. Zihl, J., Von Cramon, D. and Mai, N. (1983) *Brain*, **106**, 313.

38. Halligan, P.W., Marshall, J.C. and Wade, D.T. (1993) *J. Neurol., Neurosurg. Psych.*, **56**, 159.

39. Newcombe, F., Ratcliff, G. and Damasio, H. (1987) *Neuropsychologia*, **25**, 149.

40. Damasio, A.R. and Damasion, H.A. (1989) *Lesion Analysis in Neuropsychology*, Oxford University Press, New York.

41. Damasio, A.R. (1994) *Descartes' Error*, Picador, London.

42. Filley, C.M. *et al.* (1987) *Arch. of Neurol.*, **44**, 194.

43. Richardson, J. (1990) *Clinical and Neuropsychological Aspects of Closed Head Injury*, Taylor & Francis, London.

44. Azouvi, P. *et al.* (1996) *J. Clin. Exper. Neuropsychol.*, in press.

45. Mattson, A.J. and Levin, H.S. (1990) *J. Nerv. and Ment. Dis.*, **178**, 282.

46. Jaffe, K.M. *et al.* (1992) *Arch. Phys. Med. Rehab.*, **73**, 540.

47. Glickstein, M. and Stein, J. (1991) *Trends Neurosci.*, **14**, 480.

48. Lange, K.W. *et al.* (1992) *Psychopharmacology*, **107**, 394.

49. Sahakian, B.J. *et al.* (1988) *Brain*, **111**, 695.

50. Brunfaut, E. and d'Ydewalle, G. (1996) *Neuropsychologia*, in press.

51. Owen, A.M. (1995) *Neuropsychology*, **9**, 126.

52. Sagar, H.H. *et al.* (1988) *Brain*, **111**, 185.

53. Barker, A.T., Jalinous, R. and Freeston, I.L. (1985) *Lancet*, **1**, 1106.

54. Priori, A. *et al.* (1993) *Brain*, **116**, 353.

55. Pascual-Leone, A. *et al.* (1994) *Neurology*, **44**, 494.

56. Jalinous, R. (1991) *J. Clin. Neurophysiol.*, **8**, 10.

57. Glassman, R. (1978) *Recovery from Brain Damage Research and Theory* (S. Finger, ed.) Plenum Press, New York.

58. O'Mara, S.M. (1995) *Prog. Neurobiol.*, **45**, 253.

59. Rolls, E.T. and O'Mara, S.M. (1993) in *Brain Mechanisms of Perception: From Neuron to Behavior* (T. Ono, M. Raichle, D. Perrett and M. Fukuda, eds), Oxford University Press, New York.

60. Morris, R.G.M. *et al.* (1982) *Nature*, **297**, 681.

61. LeDoux, J.E. (1993) *Behav. Brain Res.*, **58**, 69.

62. Campeau, S. and Davis, M. (1995) *J. Neurosci.*, **15**, 2301.

63. Davis, M. (1991) *Int. Rev. Neurobiol.*, **36**, 225.

64. Iwai, E. and Mishkin, M. (1969) *Exp. Neurol.*, **25**, 585.

65. Ungerleider, L. and Mishkin, M. (1982) in *Analysis of Visual Behavior* (D.J. Ingle, M.A. Goodale and R.J.W. Mansfield, eds) MIT Press: Cambridge, MA.

66. Mishkin, M., Lewis, M. and Ungerleider, L. (1982) *Behav. Brain Res.*, **6**, 41.

67. Mishkin, M. and Ungerleider, L. (1982) *Behav. Brain Res.*, **6**, 57.

68. Posner, M. and Raichle, M. (1994) *Images of Mind*, Scientific American Library/W.H. Freeman, New York.

69. O'Mara, S.M. (1995) *Behav. Brain Sci.*, **18**, 365.

70. Nadeau, S.E. and Crosson, B. (1995) *Neuropsychiat., Neuropsychol. Behav. Neurol.*, **8**, 143.

71. Knight, R.T. *et al.* (1981) *Electroenceph. Clin. Neurophysiol.*, **52**, 571.

72. Knight, R.T. (1990) in *Event-Related Potentials: Basic Issues and Applications* (J.W. Rohrbaugh, R. Parasurman and R. Johnson, eds) Oxford University Press, New York.

73. Robertson, I.H. *et al.* (1995) *J. Clin. Exper. Neuropsychol.*, **17**, 416.

74. Robertson, I.H. *et al.* (1996) *J. Int. Neuropsychol. Soc.*, in press.

75. Wood, C.C. (1994) *Hum. Brain Imag.*, **1**(4), ii.

76. Preuss, T.M. (1995) in *The Cognitive Neurosciences* (M.S. Gazzaniga, ed.), MIT Press, Cambridge, MA.

77. Kolb, B. (1990) *Prog. in Brain Res.*, **85**, 501.

CHAPTER 6

MODEL LEARNING AND MEMORY SYSTEMS IN NEUROBIOLOGICAL RESEARCH: CONDITIONING AND ASSOCIATIVE LEARNING PROCEDURES AND SPATIAL LEARNING PARADIGMS

C. Hölscher and S. M. O'Mara

1. INTRODUCTION

The past fifteen years have seen an explosion of research interest in the biological bases of learning and memory. Particular interest has focused on maze procedures used to analyse spatial learning in the rat and on analysing the neurobiology of model systems derived from conditioning and associative learning procedures. There are now a huge number of procedures available to the investigator, depending on the question of interest; this chapter presents an overview of the more popular methods and techniques used to examine the neurobiology of learning and memory. The strategy in this field has generally been to choose a behaviour of interest and to examine its neural bases using a variety of techniques such as lesions of various types (see Chapter 6 by O'Mara & Walsh), neuroanatomical tract-tracing (Chapter 4 by Wouterlood), single-unit recording (Chapter 1 by Muller *et al.*), and more recently, computational modeling (Chapter 3 by Boddeley).

There have been major advances in the theoretical analysis of conditioning and associative learning at a psychological level over the past 15 years. In particular, there is now a greater understanding of what information the animal represents after undergoing a particular learning experience and how this representation itself is structured (see (1–4)). There have been equally major advances in theory and experiment in the neurophysiological, pharmacological and behavioral analysis of spatial navigation and spatial representation, a topic which has largely remained outside the purview of the more traditional associative learning and conditioning approaches. Much of the work on spatial navigation has been driven by the discovery of the extreme sensitivity of allocentric spatial learning to lesions of the hippocampal formation (for reviews, see refs 5–7).

2. CONDITIONING AND ASSOCIATIVE LEARNING PROCEDURES

Learning in this context refers simply to an enduring change in behavior as a result of experience, that it is not attributable to extraneous factors such as aging or drugs, etc. but is attributable to the interaction between the organism and environmental events that it experiences. However, a full consideration of the underlying psychological processes requires that we consider:

(i) the circumstances that produce learning;
(ii) the content of learning;
(iii) the way learning affects the behavior of the organism over time (8, 9).

In associative conditioning procedures animals learn which event predicts an outcome in which it has an interest (e.g. food availability or footshock avoidance). A large literature has developed using variants of these procedures (see, e.g., *Journal of Experimental Psychology: Animal Behavior Processes* or *Behavioral Neuroscience*) and the techniques have been applied with success to challenging behavior in clinical settings (see, e.g., *Behaviour Research and Therapy*).

In operant conditioning experiments, a rat is placed in a small chamber with a lever(s) present and is taught a simple rule – press the lever for food pellets. The animal can either discover the contingent relationship accidentally (by, for example, accidentally pressing the lever) or through deliberate training on the part of the experimenter ('shaping'). The number of presses per food pellet can be varied, as can the requirement to distribute lever-pressing per unit time between two or more levers ('contingencies of reinforcement'). Other events (referred to as discriminative stimuli) such as lights or tones can be used to signal periods of food availability or non-availability. These procedures can be used to develop long sequences of behavior that are very stable through time; stable baselines can be established, an intervention can be performed (e.g. a drug administered) and the effects of this intervention can be evaluated against baseline performance. A large and technical literature has developed using such procedures (see, e.g., *Journal of the Experimental Analysis of Behavior*) and such techniques have been applied with success in clinical settings (see, e.g., *Journal of Applied Behavior Analysis*). There has been little investigation of the neurobiological substrates of operant procedures and they will not be examined in any detail here.

3. ASSOCIATIVE CONDITIONING PROCEDURES

A great variety of procedures fall under the rubric of associative conditioning (1, 10). The phrase 'conditioned stimulus' (CS) refers to any stimulus which does not, in the absence of a manipulation, elicit a response by the animal (these can be events such as lights and tones, etc.). Differing CSs will be referred to as CS_A, CS_B, etc. The phrase 'unconditioned stimulus' refers to any event which does elicit a response from an animal, such as food presentation or footshock.

Figure 1 shows the possible temporal relationships between the various stimuli that may be presented to an animal.

Table 1 presents a summary of a variety of different associative conditioning paradigms. Procedures marked * have not been the subject of intensive neurobiological investigation to date.

Table 2 presents the paradigms for which the neural bases of learning have been worked out in any reasoanble detail.

The procedures defined in *Tables 1 and 2* have the great advantage that they offer precisely defined stimulus presentation parameters, well-defined assays of learning and an integrated theoretical analysis which provides the context for interpreting the results and predicting the outcome of further experiments. A great deal of work remains to be done, however, and the neurobiological analysis of learning requires approaches drawing on behavioral, anatomical, electrophysiological and computational neuroscience.

Table 1. Popular conditioning and associative learning paradigms

Procedure	CS manipulation	UCS manipulation	CS–UCS relation	Outcome
Habituation		UCS presentation only for *n* trials		Diminution in response over trials
Sensitization		Strong UCS presentation only for *n* trials		Increased responsivity to stimuli
Classical conditioning	CS–UCS pairing	Appetitive or aversive UCS	Any of the typical relationships illustrated in *Figure 1*	CS elicits UCS response over trials
Fear conditioning	CS_A–aversive UCS pairing	CS_A presentation during ongoing behavior (e.g. lever-pressing for food)	UCS onset typically precedes and overlaps CS onset in time	Suppression of ongoing behavior as measured by suppression ratio
Latent inhibition	Prior presentation of CS for *n* trials before UCS–CS pairing	Appetitive or aversive UCS	Typically forward pairing	Retarded acquisition of CS–UCS relationship
Blocking*	CS_A–UCS pairing; subsequent CS_B–UCS pairing	Appetitive or aversive UCS	Typically forward pairing	Acquisition of CS_A–UCS pairing; blocking of CS_B–UCS pairing
Second-order conditioning*	CS_A–UCS pairing; subsequent CS_B– CS_A pairing	Appetitive or aversive UCS	Typically forward pairing	CS_B also elicits CR
Sensory preconditioning*	CS_A–CS_B pairings for *n* trials; CS_B–UCS pairings	Appetitive or aversive UCS	Typically forward pairing	CS_A presentation will also elicit CR

Table 1. Continued

Procedure	CS manipulation	UCS manipulation	CS–UCS relation	Outcome
Overshadowing*	Compound CS_A–CS_B presentation: CS_A is an intense stimulus; CS_B is a weak stimulus	Appetitive or aversive UCS	Typically forward pairing	All conditioning accrues to the more intense CS_A; relatively little to CS_B
Taste aversion learning	CS is a novel taste; UCS is induced illness	Aversive UCS	Usually forward pairing, but CS–UCS interval can be up to 24 h	Robust learning; subject to usual variables in associative learning such as blocking, overshadowing, etc.

Table 2. Paradigms for which the neural bases of learning have been worked out in reasonable detail

Procedure	Method	Neurobiology	Key references
Habituation	Method described in *Table 1*	Habituation and sensitization have been little researched over the past two decades; neurobiology differs depending on the preparation and response system used	Ref. 11 provides a reasonable starting point for the historical literature
Sensitization	Method described in *Table 1*		
Classical defence conditioning	Method described in *Table 1*; best known preparation is classical defence conditioning of the nictitating membrane (3rd eyelid) in the rabbit. Rapid and robust response acquisition; control procedures have been worked out in detail	Physiology, pharmacology and neuroanatomy worked out in great detail by Thompson and colleagues; learned response depends on intact cerebellum; reflex response unaffected by cerebellar lesions	Refs 11–13
Fear conditioning (conditioned suppression)	Method described in *Table 1*; usually measured as disruption (suppression) of other ongoing behavior – measured by a 'suppression ratio'; rapid and robust response acquisition; control procedures have been worked out in detail	Physiology, pharmacology and neuroanatomy worked out in great detail by LeDoux and colleagues. Learning requires intact amygdala; reflex response unaffected by amygdalar lesions	Refs 14–16

Table 2. Continued

Procedure	Method	Neurobiology	Key references
Fear-potentiated startle	Rats jump if presented with a loud noise; jump height is greatly enhanced if the noise is presented with a CS which has previously been paired with a footshock UCS; response acquisition is very rapid and robust; control procedures have been worked out in detail	Physiology, pharmacology and neuroanatomy has been described in great detail by Davis and colleagues; learning requires intact amygdala; procedure is very useful for screening potential anxiolytics	Refs 17–20
Conditioned toxicosis (also known as taste-aversion learning or the Garcia Effect)	Pairing a novel taste with illness (even 24 h after novel food ingestion) produces one-trial learning in the rat which lasts for very long periods (months or more); learning is specific to novel taste, not peripherally but simultaneously presented cues	Physiology, pharmacology and neuroanatomy of the pathways involved have been worked out in reasonable detail by a variety of workers; essential structures for learning are the parabrachial nucleus, medial thalamus and the amygdala	Refs 21–27
Latent inhibition	Method described in *Table 1*	Underlying neuroanatomy, physiology and pharmacology have been worked out in reasonable detail; essential structures include the nucleus accumbens and hippocampus	Refs 28–31

Figure 1. The most studied temporal relationships between conditioned and unconditioned stimuli; time runs from left to right

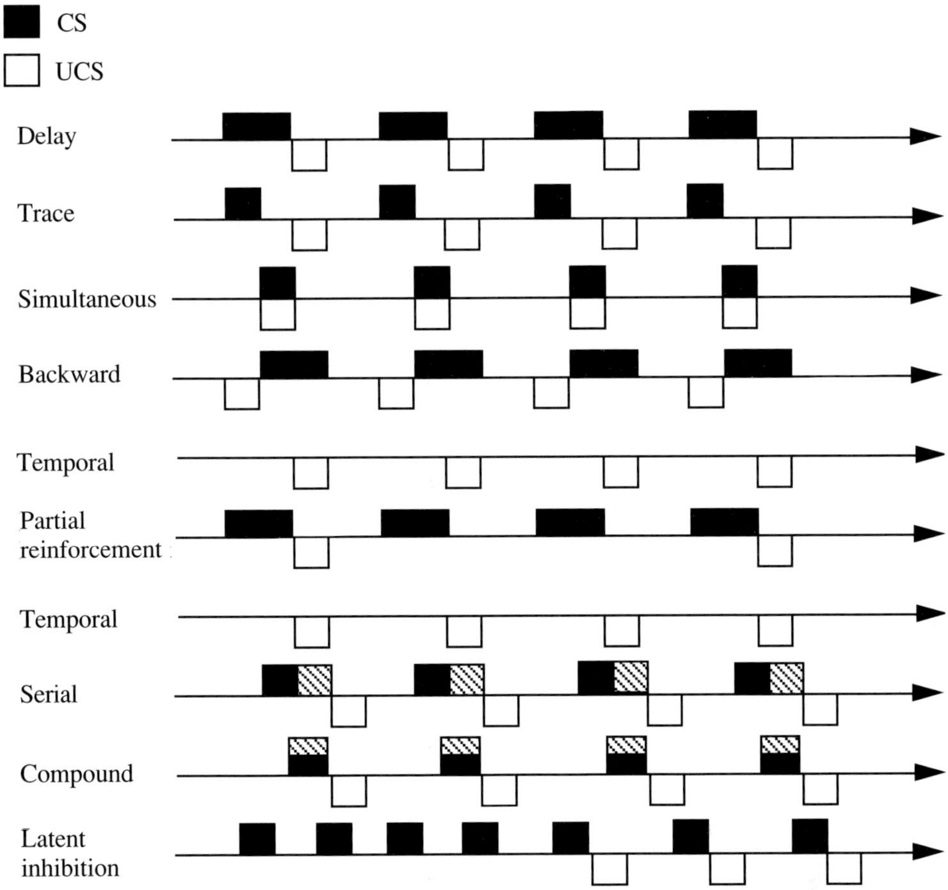

4. SPATIAL LEARNING AND MEMORY PROCEDURES

Rats have extremely good spatial learning abilities, as several decades of behavioral and neurophysiological research have demonstrated (see refs 5 and 6 for reviews). Two major paradigms – the water maze and the radial arm maze – have been used widely to assess spatial learning abilities of rats or mice. The water maze procedure was developed as a technique which would allow rats to draw inferences about the position of a stable platform in a swimming pool in the absence of local cues which would indicate the location of the platform (32). Both systems have their advantages and disadvantages, which are listed in *Table 3*.

Table 3. Advantages and disadvantages of two major paradigms

Water maze	Radial arm maze
For	
No motivational problems – problem cases (e.g. aged or uncooperative animals) are easily tested Olfactory cues are not present Animals learn readily and quickly	Distinguishes between motor impairment and spatial learning – if the animal has motor problems it can still choose an arm Different memory types can be tested: working memory, long-term (reference) memory, motor (egocentric) memory A tracking program is not essential Cheap, simple set-up, which can be dismantled and easily stored after use
Against	
Differences in platform finding latency are the product of learning *and* motor performance; it is not possible to subtract one from another; if the test group displays motor impairments it is impossible to separately assess spatial learning impairments Working memory cannot be tested independently A videotracking system is essential for evaluating the swim tracks of animals A fairly large pool has to be installed in a dedicated-use room	Animals must be well-handled and motivated to perform the tasks Appetitive task: the animals must be mildly food-deprived to be motivated to look for baited arms; may be a problem in some cases (e.g. in aged rats which need full food supplies to stay healthy)

4.1. Setting up a water maze

A small room should be set aside exclusively for water maze tests; it should contain obvious landmarks such as a large window, door, blackboard, posters etc. for orientation of animals. Light levels should be under artificial control; a water supply is also required.

Several MS-DOS and Macintosh tracking systems are available commercially (see ref. 33 for a cheap alternative). The program should measure the swim path length of the animal and provide percentages of swim path or time per pool quadrant. If the program does not include a graphics and statistics package it is useful to chose a program that produces data files compatible with the graphics and statistics packages used for data analysis.

4.2. Example of a water maze set-up

The pool for the maze is made of Fiberglas and is painted white. The dimensions are: diameter 120–200 cm, height 48 cm, depth of water 34 cm (26 cm with the platform exposed in the visual cue/discrimination task). The platform has a diameter of 10 cm and is 3 cm below the water surface in the water maze training. The water temperature is 22 ± 2°C and it is made opaque with titanium dioxide. The pool is situated in a room with visual cues such as a door, a curtain, a black cover over a window, and a water heater on the wall. The animals' movements are recorded with a video camera attached to the ceiling. The animals are marked with a black waterproof felt tip pen to make them visible to the tracking program. Other examples of water maze setups can be found in refs 32, 34 and 35. A transparent platform made of Perspex can also be used. Animals will then not be able to see the platform owing to water surface glare and there will be no need for the water to be opaque. If the pool is painted black, the rat will be visible with high contrast sufficient for the tracking program. Then animals do not have to be marked at all.

4.3 Examples of training protocols for spatial water maze tasks

Animals are put in the water from one of four different randomly alternated starting points. For the first three days each animal is given 6 trials per day with an intertrial interval of 5 min. A trial cutoff time of 120 s is used if the animal does not find the platform, after which the animal is moved manually onto the platform for 10 s.

Swim speed and path length must be measured. In the event of altered swim speed, little can be said about what causes a difference between groups in latencies; path length measurements can help. If path length is very different between groups, then the animals may be experiencing motor and learning impairments. If the path length is the same across groups, then latency differences are presumably due to motor impairments and not learning impairments.

Reversal task
In a reversal task, the location of the platform is moved from one quadrant to the next. Animals must learn the new location and 'override' the knowledge of the previous location. Reversal learning can be a sensitive test for animals that have no difficulties learning an initial task but cannot switch to a novel task (perseverance). For example, hippocampal lesions tend to produce perseverative behavior (36, 37).

Transfer test
To test for memory of the task, animals can be given a transfer task without the platform. Given 60 s to explore the pool, the ability to retrieve the stored information is tested. The animal's swim path is recorded and the pool surface is divided by the computer tracking program into four quadrants (see *Figure 2*). The amount of time spent, or the swim path swum by animals, in each quadrant is measured. The control group is expected to spend more time in the quadrant where the platform used to be. Only after some time animals tend to 'wander off' to look somewhere else.

4.4. Examples of training protocols for nonspatial water maze tasks

Visible platform test
The water level is lowered so that the platform is visible. To remove spatial elements from the task: (i) move the platform around randomly or (ii) use a curtain to occlude distal visual cues. This is a nonspatial visual discrimination task which has only a small learning component: it tests if the animals can see the target and move to it in a normal way. If the manipulation impairs sensory, motivational or motor ability, a difference in latency/distance between groups when performing the task is expected.

Visual discrimination tasks
To test the ability of animals to recognize and discriminate visual cues, a two-platform visual discrimination task can be used. A stable platform which supports the weight of the animal, and an unstable but anchored floating platform which does not support the animal's weight, are used. The animal has to learn which platform it can mount in order to escape the water. General motor skills, visual discrimination, and learning ability of a cue task are tested (38, 39). Cue learning is independent of the hippocampal formation and is a control in lesion studies if brain areas other than the hippocampus are affected, or to check if drug effects result in impaired hippocampal function (40).

Figure 2. Examples of swim paths of animals in the transfer test task. The small circle represents the previous location and surroundings of the platform. This circle (the 'annulus', about 20 cm in diameter) is produced by the tracking program. The times the animal crosses this circle is counted (annulus crossings) and gives an approximate idea of how precise the animal was in locating the (now absent) platform. Control experiments must be conducted for possible motor impairments, visual deficits and nonspecific nonspatial learning deficits.

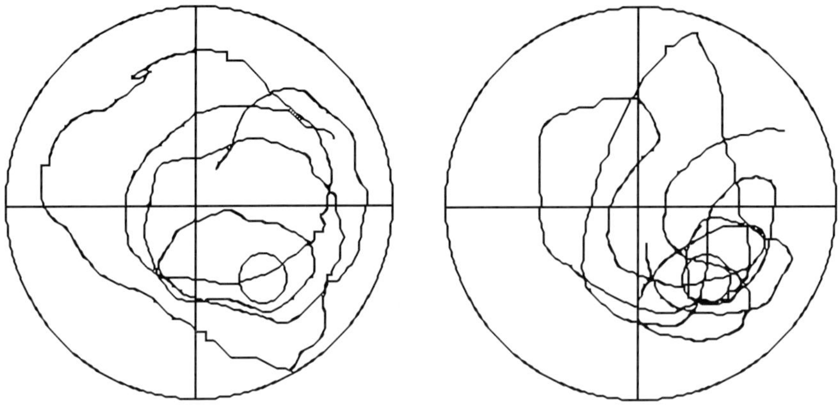

4.5 Example of a radial maze

The maze is made of 15 mm thick brown plastic-coated chipboard and consists of eight arms, each 19 cm wide, 60 cm long, and 30 cm high, and a central octagonal platform which is 46 cm in diameter. A white plastic weighing boat 5 cm wide is attached to the end of each arm in which a food pellet (45 mg) is placed if the arm is to be baited. The animals are not able to see the food pellet from the center platform. The animals are observed with a video camera from another room. The monitor is marked to allocate each arm a number from 1 to 8.

Detachable arms have the advantages of ease of storage and of reconfiguring the arms into a T-maze, or Y-maze, or a 4-arm plus maze. A fairly dark color for the maze is best as rats are somewhat photophobic. If a tracking program is used the maze should be dark brown or black to maximize contrast between animal and background. The advantage of a tracking program is that running speed can be measured to indicate if procedure has any nonspecific effect on motor activity. The height of the maze should not be too low, particularly as young rats sometimes try to jump out of the arms when the maze is still unfamiliar.

The designs of experiments vary widely in the literature and a selection of the most common designs are presented here. The 8-arm maze can be used to test for several different types of memory which are supposed to be controlled by different brain areas. Hence, a lesion or a drug might affect only some of a whole set of memory mechanisms.

Pretraining

It is important to handle the rats several days before the experiment. Sometimes it will be necessary to give animals pretraining sessions to familiarize themselves with the setup and the environment. These sessions can be used to assess the animals' general behavior.

Exploratory behavior
Healthy rats will investigate novel environments. There is a tendency to alternate between arms and not to visit arms twice (41). In a pretraining session, the animal is put into the center of the 8-arm maze and is allowed to investigate the maze freely for 5–10 min. Spatial orientation is made possible by extra maze cues such as doors and windows. One trial per animal is given. The arms are not baited.

The sequence of arm entries is recorded. Analysis of re-entry of arms will tell the experimenter if the animals are affected in a general way. If a rat only visits some of the arms but ignores others it shows a deficit in spatial awareness of the maze. Also, analyzing which angles between consecutively entered arms appear most often can give a general idea whether the animal chooses an 'egocentric' strategy, such as always entering the second arm to the left. Animals with impaired spatial perception of the environment can resort to egocentric strategies (see below).

Reinforced exploratory behavior
A reinforced version of the task described before can be used to analyze the animal's ability to orient itself in the maze. All arms are baited and the time needed to complete the task and number of arm entries required to complete the task are measured (42). As only one trial is given the task only tests working memory.

Spatial learning
Spatial memory can be tested by baiting one arm and giving several trials to the animal to learn which arm is baited. To make the task more complicated, two or three arms are baited in a simple spatial arrangement (e.g. arms 1, 3, 5) or more complex pattern (e.g. arms 1, 4, 5). Six trials are given per animal per day. Different types of memory can be analysed in this task.

Working memory. This memory can be tested using the results of the reference memory task. Arms which are visited twice within a trial are counted as a working memory error since the average rat usually does not visit an arm twice in a trial. Working memory is important to store all relevant memory within a trial. The animal forgot that it had visited the arm already and that it does not contain any bait any more (40, 43). Lesions that interfere with proper functioning of the hippocampal system affect this type of memory very much (44–46).

A delay introduced within the trial makes the task a more difficult working memory task (47). If three arms are baited in the experiment, an intratrial delay can be introduced by retaining the animal in one arm with a door made of plywood.

Reference memory (long-term memory). A reference memory error is defined as a visit to an arm that has not been baited before in previous trials. Some authors bait only one arm (48). This task is relatively simple and may be learned by animals that are only slightly impaired.

Latency. Measuring the time needed by animals to learn a spatial task is a general indicator of learning ability. Clearly, latencies of animals are dependent on motor activity just as in the water maze and can be influenced by effects of lesions or drugs on running speed. Therefore one should measure motor activity or running speed as well when using latency data as a measure for learning.

Reversal learning
After learning which arms are baited in the maze, different arms can be baited and animals have to ignore the learned information while learning the new task. Solving this task relies on

hippocampal function and impairments can produce perseverance for previously baited arms (41).

Egocentric memory
Spatial learning and orientation using external cues such as doors and windows (allocentric learning) is largely dependent on a functioning hippocampal system (36), whereas learning locations relative to body position (egocentric learning) is largely dependent on motor areas of the brain (49, 50). In the radial arm maze, animals learn egocentric tasks by remembering which arm to visit after leaving the starting arm (e.g. visit the second arm to the right, then the fourth arm to the right).

Rats are visually guided animals, preferring visual cues for orientation. To avoid this, the room should be dimly lit. Egocentric task learning usually takes longer than allocentric learning.

Example of an egocentric experiment. Three arms are baited. A fourth arm serves as a starting arm. Arms remain in the same position relative to each other (e.g. baited arms at 45°, 135°, and 225° turning right after leaving starting arm). This fixed pattern of baited arms and starting arm is rotated randomly after each trial relative to room coordinates. This ensures that the animal cannot rely on external cues (window, door) for orientation. Instead the egocentric pattern of body turns (45° to the right etc.) has to be remembered. See ref. 46 for details.

Cue learning
As a control task for possible impairments in nonspatial learning, visible discrimination, motor coordination, or motivation, a simple visual cue task can be used. One arm is marked with a checkerboard pattern or a similar dominant marking which makes it unmistakable. This arm is the only baited arm. Latency as well as arm visits are measured.

If this task is given to the animals after learning a spatial 8-arm task, it is possible to measure extinction of the previously learned task as well. This measure of extinction is comparable to the transfer task in the water maze and shows how well the animals learned or recall the task.

5. ACKNOWLEDGMENTS

Christian Hölscher is a Health Research Board Post-Doctoral Fellow. Shane O'Mara's research is supported under a European Union Human Capital and Mobility Research Network and by the Dean's Research Fund of Trinity College, Dublin.

6. REFERENCES

1. Rescorla, R.A. (1988) *Ann. Rev. Neurosci.*, **11**, 329.

2. Holland, P.C. (1993) *Curr. Opin. Neurobiol.*, **3**, 230.

3. Dickinson, A. (1980) *Contemporary Animal Learning Theory*, Cambridge University Press, Cambridge, p. 95.

4. Mackintosh, N.J. (1983) *Conditioning and Associative Learning*, Oxford University Press, Oxford.

5. O'Keefe, J. and Nadel, L. (1978) *The Hippocampus as a Cognitive Map*, Clarendon Press, Oxford, p. 102.

6. O'Mara, S.M. (1995) *Prog Neurobiol.*, **45**, 253.

7. Rolls, E.T. and O'Mara, S.M. (1993) in *Brain Mechanisms of Perception: From Neuron to Behavior* (L.R.S.T. Ono, M. Raichle, D. Perrett and M. Fukuda, eds), Oxford University Press, New York.

8. Rescorla, R.A. (1967) *Psychol. Rev.*, **74**, 71.

9. Rescorla, R.A. (1988) *Ann. Rev. Neurosci.*, **11**, 329.

10. Hearst, E. (1988) in *Stevens' Handbook of Experimental Psychology* (R.C. Atkinson *et al.*, eds), John Wiley, New York.

11. Thompson, R.F., Donegan, N.H. and Lavond, D.G. (1988) in *Stevens' Handbook of Experimental Psychology* (R.C. Atkinson *et al.*, eds), John Wiley, New York.

12. Gormezano, I. (1983) *Prog. Physiolog. Psychol. Psychobiol.*, **10**, 197.

13. Thompson, R.F. (1986) *Science*, **233**, 941.

14. Dickinson, A. (1980) *Contemporary Animal Learning Theory*, Cambridge University Press, Cambridge.

15. LeDoux, J.E. (1992) *Curr. Opin. Neurobiol.*, **2**, 191.

16. LeDoux, J.E. (1993) *Behav. Brain Res.*, **58**, 69.

17. Davis, M. (1992) in *The Amygdala: Neurobiological Aspects of Emotion, Memory and Mental Dysfunction* (J.P. Aggleton, ed.) Wiley: Chichester.

18. Davis, M. *et al.* (1993) *Behav. Brain Res.*, **58**, 175.

19. Campeau, S. and Davis, M. (1995) *J. Neurosci.*, **15**, 2301.

20. Campeau, S. and Davis, M. (1995) *J. Neurosci.*, **15**, 2312.

21. Okifuji, A. and Friedman, A.G. (1992) *Behav. Res. Ther.*, **30**, 23.

22. Chambers, K.C. (1990) *Ann. Rev. Neurosci.*, **13**, 375.

23. Yamamoto, T. *et al.* (1994) *Behav. Brain Res.*, **65**, 123.

24. Garcia, J., Hawkins, W.G. and Rusiniak, K. (1974) *Science*, **185**, 824.

25. Garcia, J. *et al.* (1968) *Science*, **160**, 794.

26. Hatfield, T., Graham, P.W. and Gallagher, M. (1992) *Behav. Neurosci.*, **106**, 286.

27. Willner, J. *et al.* (1992) *Behav. Neurosci.*, **106**, 315.

28. Purves, D., Bonardi, C. and Hall, G. (1995) *Behav. Neurosci.*, **109**, 366.

29. Solomon, P.R. (1987) in *Classical Conditioning*, 3rd edn (J. Garmezano, W.R. Prokasy and R.F. Thompson, eds), LEA, Hillsdale, NJ.

30. Kaye, H. *et al.* (1987) *Q. J. of Exper. Psychol.*, **39B**, 127.

31. Lubow, R.E. (1973) *Psychol. Bull.*, **79**, 398.

32. Morris, R.G.M. (1981) *Learn. Motiv.*, **12**, 239.

33. Wolfer, D.P. and Lipp, H.-P. (1992) *J. Neurosci. Meth.*, **41**, 65.

34. Frick, K.M., *et al.* (1995) *Neurobiol. Aging*, **16**, 149.

35. Spencer, R.L., O'Steen, W.K. and McEwen, B.S. (1995) *Behav. Brain Res.*, **68**, 139.

36. Morris, R.G.M., Hagan, J.J. and Rawlins, J.N.P. (1986) *Q. J. Exper. Psychol.*, **38B**, 257.

37. Rawlins, J.N.P. (1985) *Behav. Brain Sci.*, **8**, 479.

38. Bannerman, D.M. *et al.* (1994) *J. Neurosci.*, **14**, 7415.

39. Bannerman, D.M. *et al.* (1994) *J. Neurosci.*, **14**, 7404.

40. Olton, D.S., Becker, J.T. and Handelmann, G.E. (1979) *Behav. Brain Sci.*, **2**, 313.

41. Olton, D.S. and Samuelson, R.J. (1976) *J. Exper. Psychol.: Anim. Behav. Proc.*, **2**, 97.

42. Bohme, G.A. *et al.* (1991) *Europ. J. Pharmacol.*, **199**, 379.

43. Shapiro, L.M. and Caramanos, Z. (1990) *Psychobiology*, **12**, 231.

44. Tonkiss, J., Feldon, J. and Rawlins, J.N.P. (1990) *Behav. Brain Res.*, **36**, 113.

45. Olton, D.S. (1977) *Sci. Amer.*, **236**, 82.

46. Hölscher, C. and Schmidt, W.J. (1994) *Behav. Brain Res.*, **63**, 187.

47. Butelman, E.R. (1990) *Pharmacol. Biochem. Behav.*, **35**, 533.

48. Jarrard, L.E. (1979) *Behav. Brain Sci.*, **2**, 334.

49. Rasmussen, M., Barnes, C.A. and McNaughton, B.L. (1989) *Psychobiology*, **17**, 335.

50. Kesner, R.P., Farnsworth, G. and DiMattia, B.V. (1989) *Behav. Neurosci.*, **103**, 956.

CHAPTER 7
PREPARATION OF TISSUE FOR NEUROCHEMICAL ANALYSIS

M.P. Clements

1. INTRODUCTION

Isolation of the individual components of the peri-synaptic region has greatly advanced our understanding of the neurochemical mechanisms which underlie the functioning of the brain. This statement refers in particular to the mechanisms underlying the release of neurotransmitters, the site, expression and mode of action of receptors, and the role of second messenger cascades leading eventually to the gene expression and the synthesis of neural-specific proteins. This chapter describes selected methods for the preparation of tissue for neurochemistry from slices to synaptosomes, postsynaptic densities, synaptic plasma membranes, glia and synaptic vesicles. The choice of preparation depends on the particular analysis which is to be undertaken, so this chapter also provides a selection of recent publications indicating the uses of such material. The methods described are not exhaustive, so many of the references givc alternative methods for preparation of tissue for neurochemical analysis.

2. GENERAL POINTS FOR TISSUE PREPARATION

For the majority of neurochemical experiments, dissected brain areas can be used either immediately or stored frozen in 10% DMSO (1) without any detriment to the tissue and subsequent analyses. Generally, all tissue preparation, including dissection, is carried out on ice or at 4°C. The simplest, and quickest, tissue preparation is that of tissue slices or prisms; a brain region is dissected out and the tissue cross-chopped into prisms on a McIllwain tissue chopper (side-length varies from 250 µm to 400 µm). Slices can be used for neurochemical experiments where knowledge of the subcellular site of activity is not critical to the experimental protocol.

For a greater insight into the location of, for example, a particular enzyme or protein, or to reduce the contamination of extracellular neural components, it is necessary to prepare either synaptosomes or to further fractionate the tissue into its subcellular components. Subcellular fractionation requires homogenization of brain tissue using a pestle and tube with a clearance of 0.25 mm.

Choice of medium is critical for successful homogenization. Sucrose (0.32 M) is prepared in either Krebs buffer (NaCl 136 mM, KCl 2.54 mM, KH_2PO_4 1.18 mM, $MgSO_4.7H_2O$ 1.18 mM, $NaHCO_3$ 16 mM, $CaCl_2$ 1.3 mM, glucose 10 mM), or Krebs–Heinsleit buffer (NaCl 118.5 mM, KCl 4.7 mM, KH_2PO_4 1.18 mM, $MgSO_4.7H_2O$ 1.18 mM, $NaHCO_3$ 24.9 mM, $CaCl_2$ 2.5 mM, glucose 10 mM), or is buffered with bicarbonate (0.05 M), Tris–HCl (0.01 M), imidazol–HCl (0.01 M), or in the presence of ethylenediaminetetraacetic acid (EDTA). This is recommended to prevent agglutination or enzyme inactivation.

Following homogenization, centrifugation provides a pellet (P_1) and a supernatant (S_1), and each subsequent centrifugation step will provide a supernatant (S_2; S_3; etc) and a pellet (P_2;

P_3; etc.). Following lysis, centrifugation provides a pellet of lysed material (LP1) and a supernatant fraction (LS1).

The preparation of crude synaptosomes, pure synaptosomes, synaptoneurosomes, postsynaptic densites, synaptic plasma membranes, membranes enriched in glia and synaptic vesicles are shown by means of flow diagrams in *Figures 1–6*.

Table 1 gives a number of references indicating the use of the different tissue preparations, and *Table 2* gives information about viability of the preparations and problems which may be encountered during preparation.

Figure 1. Preparation of crude synaptosomes (2)

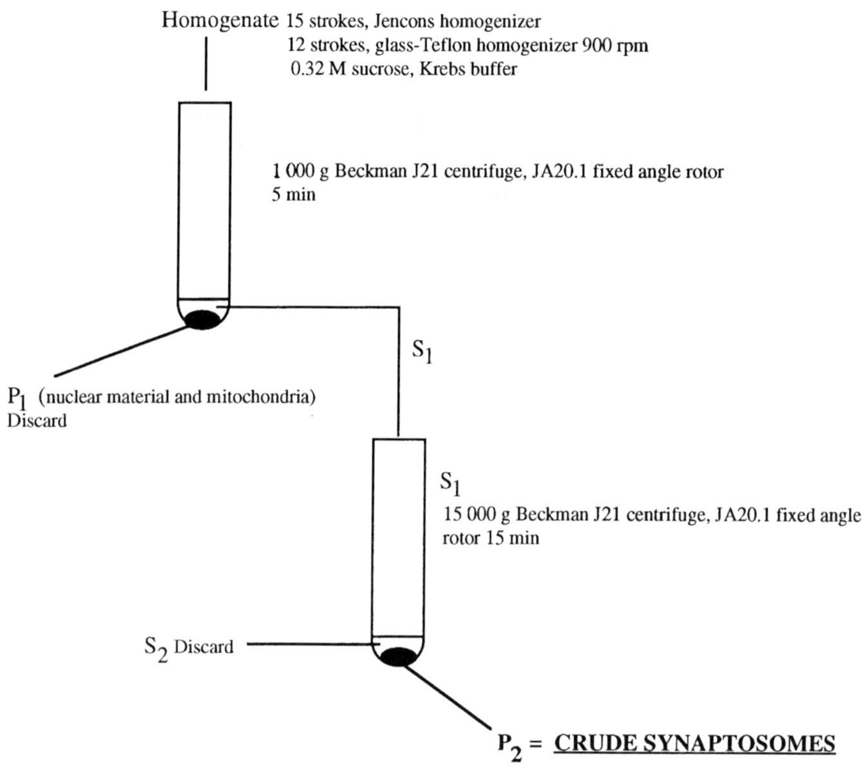

Figure 2. Preparation of pure synaptosomes (3)

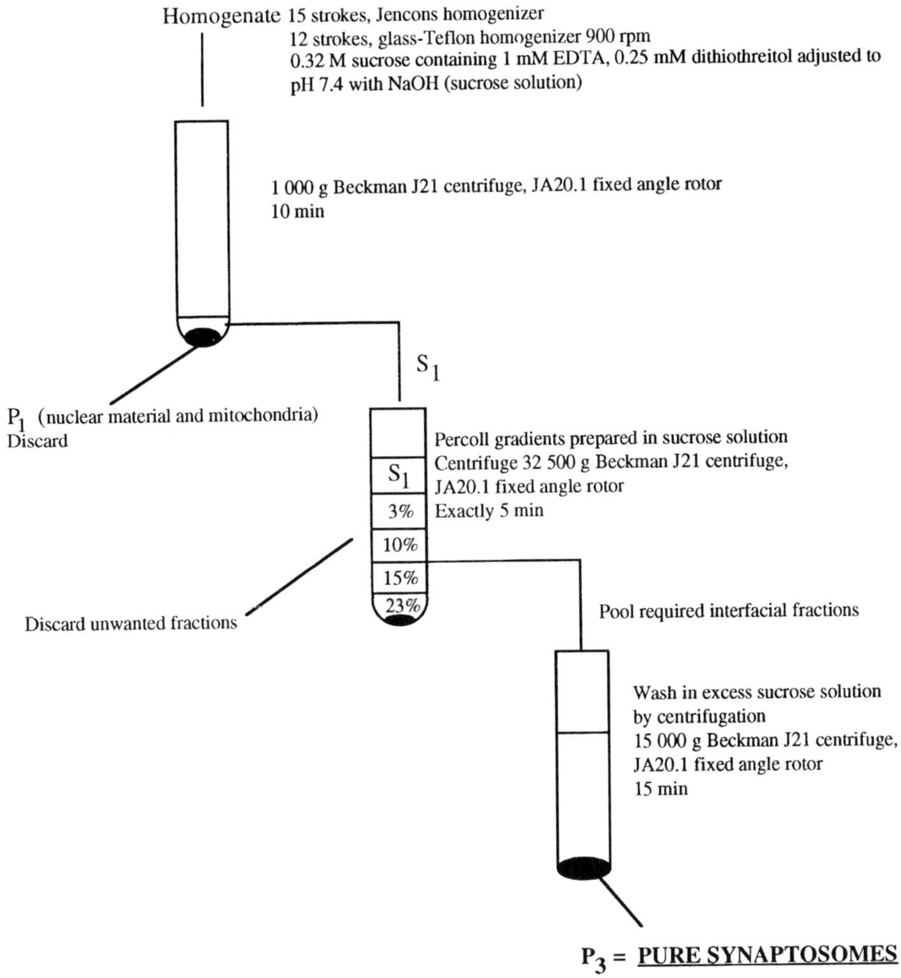

Neurochemical Analysis

Figure 3. Preparation of synaptoneurosomes (4)

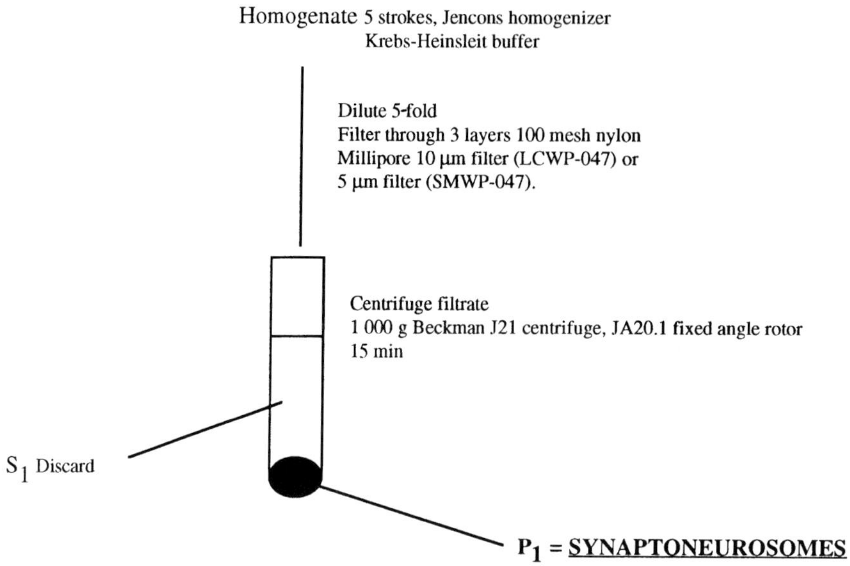

Figure 4. Preparation of postsynaptic densities and synaptic plasma membranes (5)

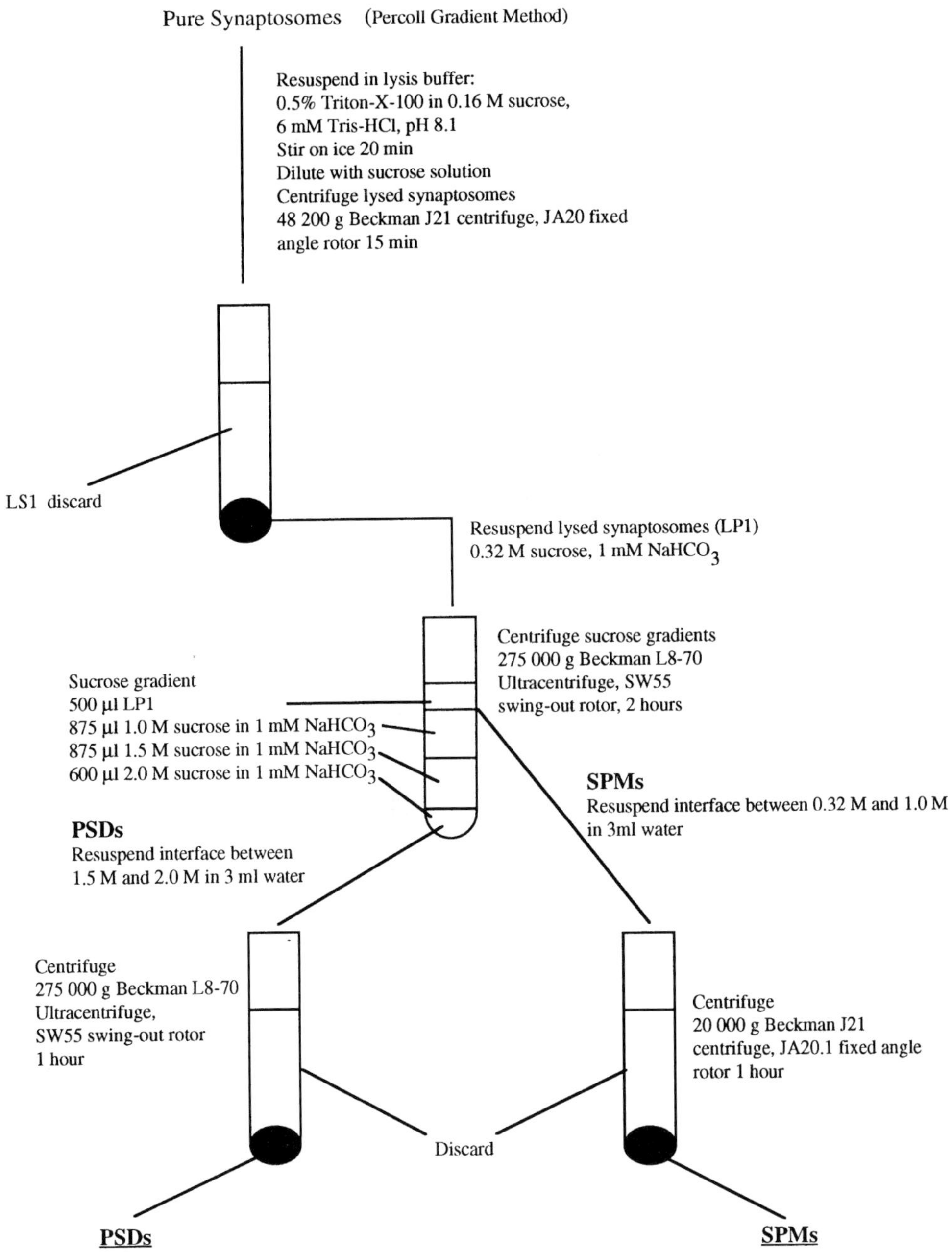

Figure 5. Preparation of glia-enriched membranes (6)

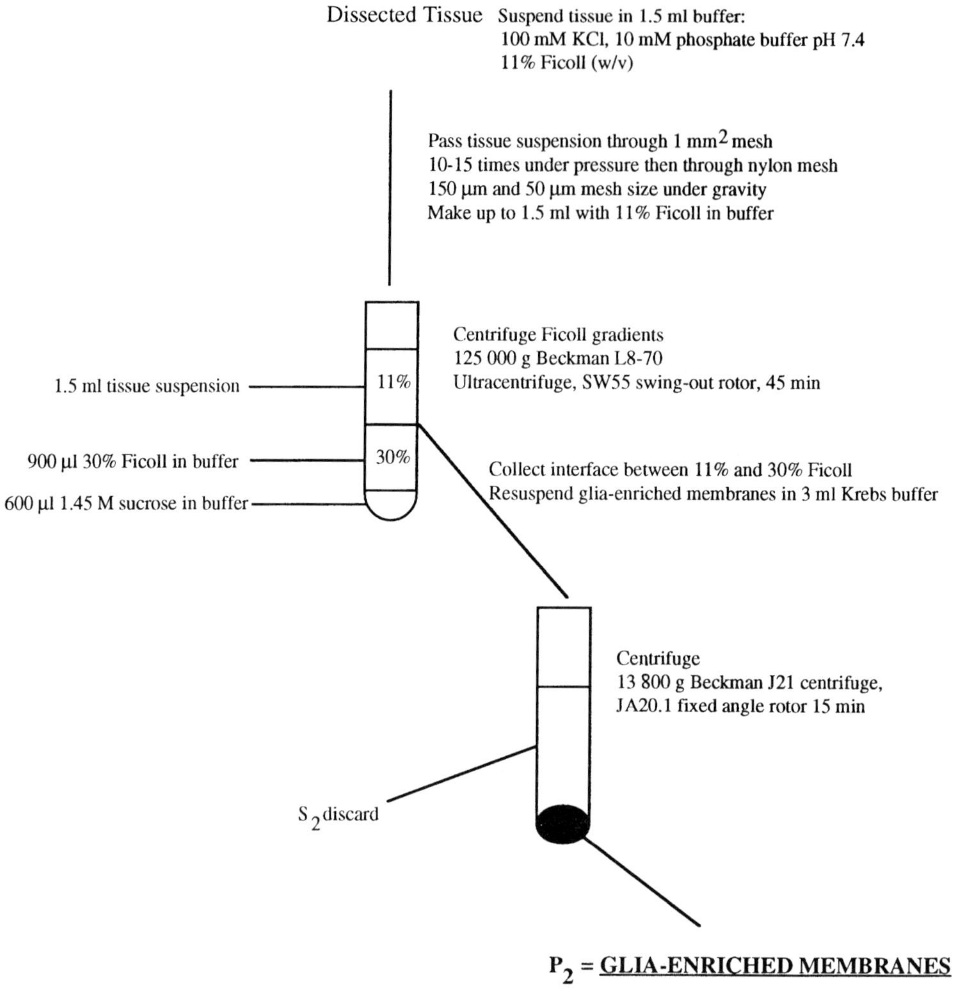

Figure 6. Preparation of synaptic vesicles (7)

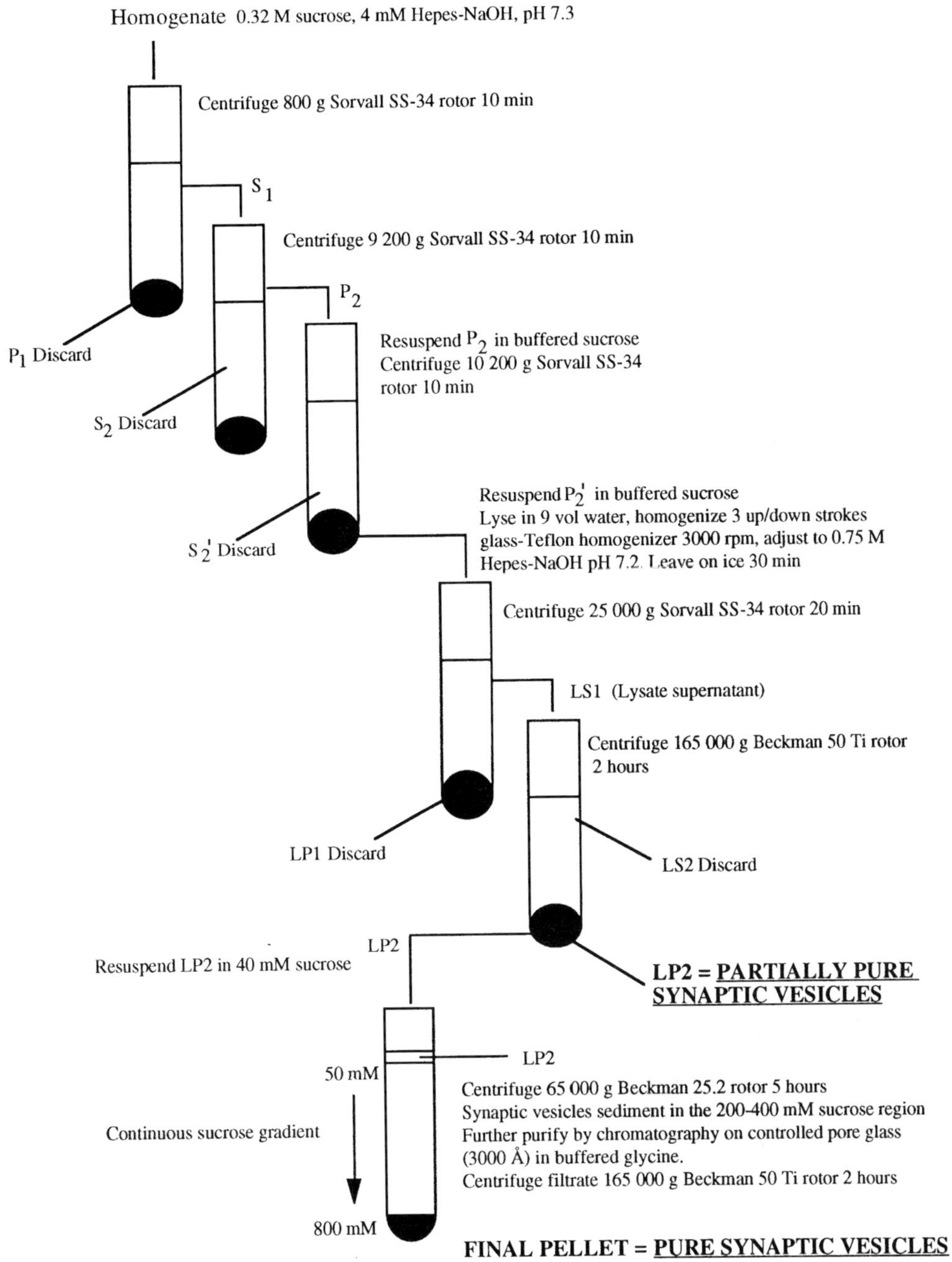

Table 1. Referenced examples of the use of different tissue preparations

Preparation	References
Tissue slices/prisms	
Uptake/release of neurotransmitters	8–10
Second messenger studies	11–18
Protein studies	19
Synaptosomes (crude and pure)	
Uptake/release of neurotransmitters	10, 14, 20–30
Calcium uptake studies	28, 31–39
Second messenger studies	23, 40–43
Protein studies	31, 44, 45
Synaptoneurosomes	
Uptake/release of neurotransmitters	46–51
Calcium uptake studies	51
Second messenger studies	4, 51–58
Ion channel studies	59–65
Protein studies	66, 67
Post-synaptic densities	
Second messenger studies	68
Protein studies	5, 6, 69–77
Synaptic plasma membranes	
Second messenger studies	68
Ion channel studies	62, 78
Protein studies	5, 6, 31, 75–77, 79–83
Glia-enriched membranes	
Second messenger studies	68
Protein studies	84
Synaptic vesicles	
Uptake/release of neurotransmitters	83
Protein studies	7, 45, 79–81

Table 2. Viability and problems of tissue preparations

Preparation	Viability	Problems
Slices/tissue prisms	Can be stored in 10% DMSO at –70°C for up to 1 month Fresh prisms viable up to 4 h with oxygenated buffer	Dead tissue at cut face Require preincubation in oxygenated buffer for up to 1 h with frequent changes of buffer prior to use
Crude synaptosomes	Can be prepared from frozen prisms Viable up to 3 h on ice	Contamination by extrasynaptosomal mitochondria is high Not a good preparation for studying cAMP accumulation

Table 2. Continued

Preparation	Viability	Problems
Pure synaptosomes	Can be prepared from frozen prisms Viable up to 3 h on ice	Remove Percoll by washing and centrifugation prior to experiment to prevent leaky synaptosomes Synaptosomal preparations from young animals sediment between the lower percentage Percoll fractions; e.g. synaptosomes from day-old chick sediment between 3% and 10% Percoll
Synaptoneurosomes	Must be prepared from fresh tissue Viable at room temperature for 2 h	Main contamination is resealed astrocytic processes with a small number of free mitochondria, endoplasmic reticulum and ribosome clusters Filtered preparation necessary for some second messenger and ion channel studies
Postsynaptic densities	Can be stored frozen for up to 3 months	Very small yield from brain regions Lysis buffer can be modified depending on use of PSDs Check purity by EM
Synaptic plasma membranes	Can be stored frozen for up to 3 months	Very small yield from brain regions Lysis buffer can be modified depending on use of SPMs Check purity by EM
Glia-enriched membranes	Can be stored frozen for up to 3 months	Very small yield from brain regions Check purity with glia-specific antibodies
Synaptic vesicles	Can be stored frozen for up to 3 months	Long preparation time (1.5 days) For a pure synaptic vesicle preparation, chromatography through controlled pore glass is necessary

3. REFERENCES

1. Haan, E.A. and Bowen, D.M. (1981) *J. Neurochem.*, **37**, 243.
2. Gray, E.G. and Whittaker, V.P. (1962) *J. Anat.*, **96**, 79.
3. Dunkley, P.R., Heath, J.W., Harrison, S.M., Jarvie, P.E., Glenfield, P.J. and Rostas, J.A.P. (1988) *Brain Res.*, **441**, 59.
4. Hollingsworth, E.B., McNeal, E.T., Burton, J.L., Williams, R.J., Daly, J.W. and Creveling, C.R. (1985) *J. Neurosci.*, **5**, 2240.
5. Cohen, R.S., Blomberg, F., Berzins, K. and Seikevitz, P. (1977) *J. Cell Biol.*, **74**, 181.
6. Hamberger, A., Eriksson, O. and Norrby, K. (1971) *Exp. Cell Res.*, **67**, 380.

7. Huttner, W.B., Schiebler, P., Greengard, P. and De Camilli, P. (1983) *J. Cell Biol.*, **96**, 1374.

8. Burke, S.P., Adams, M.E and Taylor, C.P. (1993) *Eur. J. Pharmacol.*, **238**, 383.

9. Gaur, S., Newcomb, R., Rivnay, B., Bell, J.R., Yamashiro, D., Ramachandran, J. and Miljanich, G.P. (1994) *Neuropharmacology*, **33**, 1211.

10. Raiteri, M., Costa, R. and Marchi, M. (1992) *Neurosci. Lett.*, **145**, 109.

11. Baird, J.G. and Nahorski, S.R. (1991) *J. Neurochem.* **57**: 629.

12. Brown, E., Kendall, D.A. and Nahorski, S.R. (1984) *J. Neurochem.*, **42**, 1379.

13. Li, X., Song, L. and Jope, R.S. (1990) *Neurochem. Res.*, **15**, 725.

14. Littman, L., Glatt, B.S. and Robinson, M.B. (1993) *J. Neurochem.*, **61**, 586.

15. Littman, L., Munir, M., Flagg, S.D. and Robinson, M.B. (1992) *J. Neurochem.*, **59**, 1893.

16. Littman, L. and Robinson, M.B. (1994) *J. Neurochem.*, **63**, 1291.

17. Nicoletti, F., Valerio, C., Pellegrino, C., Drago, F., Scapagnini, U. and Canonico, P.L. (1988) *J. Neurochem.*, **51**, 725.

18. Palmer, E., Nangel-Taylor, K., Krause, J.D., Roxas, A. and Cotman, C.W. (1990) *Devel. Brain Res.*, **51**, 132.

19. Boss, V. and Conn, P.J. (1992) *J. Neurochem.*, **59**, 2340.

20. Blasi, J., Chapman, E.R., Link, E., Binz, T., Yamasaki, S., De Camilli, P., Südhof, T.C., Niemann, H. and Reinhard, J. (1993) *Nature*, **365**, 160.

21. Cahill, C.M., White, T.D. and Sawynok, J. (1993) *J. Neurochem.*, **60**, 886.

22. Chéramy, A., Desce, J.M., Godeheu, G. and Glowinski, J. (1994) *Neurochem. Int.*, **25**, 145.

23. Coffey, E.T., Herrero, I., Sihra, T.A., Sánchez-Prieto, J. and Nicholls, D.G. (1994) *J. Neurochem.*, **63**, 1303.

24. Freeman, E.J., Terrian, D.M. and Dorman, R.V. (1990) *Neurochem. Res.*, **15**, 743.

25. Gehlert, D.R., Schober, D.A., Hemrick-Luecke, S.K., Krushinski, J., Howbert, J.J., Robertson, D.W., Fuller, R.W. and Wong, D.T. (1995) *Neurochem. Int.*, **26**, 47.

26. Hens, J.J.H., Ghijsen, W.E.J.M, Dimjati, W., Wiegant, V.M., Oestreicher, A.B., Gispen, W.H. and De Graan, P.N.E. (1993) *J. Neurochem.*, **61**, 602.

27. Herrero, I., Miras-Portugal, M.T. and Sánchez-Prieto, J. (1992) *Nature*, **360**, 163.

28. McMahon, H.T., Foran, P., Dolly, J.O., Verhage, M., Wiegant, V.M. and Nicholls, D.G. (1992) *J. Biol. Chem.*, **267**, 21338.

29. Richter-Levin, G., Canevari, L. and Bliss, T.V.P. (1995) *Behav. Brain Res.*, **66**, 37.

30. Sabrià, J., Pastor, C., Clos, M.V., Garcia, A. and Badia, A. (1995) *J. Neurochem.*, **64**, 2567.

31. Abe-Dohmae, S., Suzuki, T. and Tanaka, R. (1992) *Brain Res.*, **598**, 225.

32. Azimi-Zonooz, A. and Litzinger, M.J. (1992) *Int. J. Devl. Neurosci.*, **10**, 447.

33. Fontana, G. and Blaustein, M.P. (1993) *J. Neurochem.*, **60**, 843.

34. Huidobro, A., Blanco, P., Villalba, M., Gómez-Puertas, P., Villa, A., Pereira, R., Bogónez, E., Martínez-Serrano, A., Aparicio, J.J. and Satrústegui, J. (1993) *Neurobiol. Aging*, **14**, 479.

35. Lundy, P.M., Frew, R. and Hamilton, M.G. (1990) *Biochem. Pharmacol.*, **40**, 651.

36. Maubecin, V.A., Sanchez, V.N., Rosato Siri, M.D., Cherksey, B.D., Sugimori, M., Llinás, R. and Uchitel, O.D. (1995) *J. Neurochem.*, **64**, 2544.

37. Pocock, J.M., Venema, V.J. and Adams, M.E. (1992) *Neurochem. Int.*, **20**, 263.

38. Scheer, H.W. (1990) *Can. J. Physiol. Pharmacol.*, **68**, 1049.

39. Suszkiw, J.B., Murawsky, M.M. and Fortner, R.C. (1987) *Biochem. Biophys. Res. Comm.*, **145**, 1283.

40. Clements, M.P., Errington, M.L., Lynch, M.A. and Bliss, T.V.P. (1990) *Eur. J. Neurosci.*, **2**, 383.

41. Lynch, M.A., Errington, M.L., Clements, M.P., Bliss, T.V.P., Redini-Del Negro, C. and Laroche, S. (1990) *Prog. Brain Res.*, **83**, 251.

42. Lynch, M.A. and Voss, K.L. (1991) *J. Neurochem.*, **56**, 113.

43. Strosznajder, J., Samochocki, M. and Duran, M. (1994) *J. Neurochem.*, **62**, 1048.

44. Edelmann, L., Hanson, P.I., Chapman, E.R. and Jahn, R. (1995) *EMBO J.*, **14**, 224.

45. Schiavo, G., Shone, C.C., Rossetto, O., Alexander, F.C.G. and Montecucco, C. (1993) *J. Biol. Chem.*, **268**, 11516.

46. Bernard, J., Chabot, C., Gagné, J., Baudry, M. and Masicotte, G. (1995) *Brain Res.*, **671**, 195.

47. Bernard, J., Lahasaïni, A., Baudry, M. and Masicotte, G. (1993) *Brain Res.*, **628**, 340.

48. Ebstein, R.P. and Daly, J.W. (1982) *Cell Mol. Neurobiol.*, **2**, 193.

49. Ebstein, R.P. and Daly, J.W. (1982) *Cell Mol. Neurobiol.*, **2:** 205.

50. Ebstein, R.P., Seamon, K., Creveling, C.R. and Daly, J.W. (1982) *Cell Mol. Neurobiol.*, **2**, 179.

51. Iredale, P.A., Martin, K.F., Hill, S.J. and Kendall, D.A. (1993) *Biochem. Pharmacol.*, **45**, 407.

52. Dudek, S.M., Bowen, W.D. and Bear, M.F. (1989) *Devel. Brain Res.*, **47**, 123.

53. Mayat, E., Lebrun, F., Sassetti, I. and Récasens, M. (1994) *Int. J. Dev. Neurosci.*, **126**, 1.

54. Mayat, E., Lerner-Natoli, M., Rondouin, G., Lebrun, F., Sasseti, I. and Récasens, M. (1994) *Brain Res.*, **645**, 186.

55. Musgrave, M.A., Madigan, M.A., Bennett, B.M. and Goh, J.W. (1994) *J. Neurochem.*, **62**, 2316.

56. Récasens, M., Sassetti, I., Nourigat, A., Sladeczek, F. and Bockaert, J. (1987) *Eur. J. Pharmacol.*, **141**, 87.

57. Strosznajder, J. and Samochocki, M. (1992) *Adv. Exp. Med. Biol.*, **318**, 251.

58. Strosznajder, J., Samochocki, M. and Duran, M. (1994) *Neurochem. Int.*, **25**, 193.

59. Engblom, A.C. and Akerman, K.E. (1993) *Biochem. Biophys. Acta.*, **1153**, 262.

60. Ito, Y., Segawa, K. and Fukuda, H. (1995) *Synapse*, **19**, 188.

61. Morrow, A.L. and Paul, S.M. (1988) *J. Neurochem.*, **50**, 302.

62. Samochocki, M. and Strosznajder, J. (1993) *Neurochem. Int.*, **23**, 261.

63. Samochocki, M. and Strosznajder, J. (1994) *J. Neurochem.*, **63**, 1522.

64. Schwartz, R.D., Wagner, J.P., Yu, X. and Martin, D. (1994) *J. Neurochem.*, **62**, 916.

65. Titulaer, M.N., Ghijsen, W.E., Kamphuis, W., De Rijk, T.C. and Lopes Da Silva, F.H. (1995) *J. Neurochem.*, **64**, 2615.

66. Giambalvo, C.T. and Wagner, R.L. (1994) *J. Neurochem.*, **63**, 169.

67. Weiler, I.J., Hawrylak, N. and Greenough, W.T. (1995) *Behav. Brain Res.*, **66**, 1.

68. Clements, M.P., Bliss, T.V.P. and Lynch, M.A. (1991) *Neuroscience*, **45**, 379.

69. Beesley, P.W., Mummery, R. and Tibaldi, J. (1995) *J. Neurochem.*, **64**, 2288.

70. Carlin, R.K., Grab, D.J. and Siekevitz, P. (1981) *J. Cell. Biol.*, **89**, 449.

71. Dosemaci, A. and Reese, T.S. (1995) *Synapse*, **20**, 91.

72. Grab, D.J., Carlin, R.K. and Siekevitz, P. (1981) *J. Cell. Biol.*, **89**, 433.

73. Grab, D.J., Carlin, R.K. and Siekevitz, P. (1981) *J. Cell. Biol.*, **89**, 440.

74. McNeill, R.B. and Colbran, R.J. (1995) *J. Biol. Chem.*, **270**, 10043.

75. Rostas, J.A.P., Brent, V.A., Heat, J.W., Neame, R.L.B., Powis, D.A., Weinberger, R.P. and Dunkley, P.R. (1986) *Neurochem. Res.*, **11**, 253.

76. Valez-Faircloth, M., Guadano-Ferrez, A., Henzi, V.A. and Fremeau, R.T. Jr (1995) *J. Biol. Chem.*, **270**, 15755.

77. Willmott, T., Skitsa, I., Mummery, R. and Beesley, P.W. (1992) *J. Neurochem.*, **58**, 2037.

78. Abe, T., Koyano, K., Saisu, H., Nishiuchi, Y. and Sakakibara, S. (1986) *Neurosci. Lett.*, **71**, 203.

79. Babitch, J.A. and Benavides, L.A. (1979) *Neuroscience*, **4**, 603.

80. Mayanil, C.S.K. and Knepper, P.A. (1993) *Dev. Neurosci.*, **15**, 133.

81. Schiavo, G., Rossetto, O., Catsicas, S., Polverino de Laureto, P., DasGupta, B.R., Benfenati, F. and Montecucco, C. (1993) *J. Biol. Chem.*, **268**, 23784.

82. Sower, A.C., Bird, E.D. and Perrone-Bizzozero, N.I. (1995) *Mol. Chem. Neuropath.*, **24**, 1.

83. Winter, H.C. and Ueda, T. (1993) *Neurochem. Res.*, **18**, 79.

84. Rose, S.P.R. (1967) *Biochem. J.*, **102**, 33.

85. Barnekow, A., Jahn, R. and Schartl, M. (1990) *Oncogene*, **5**, 1019.

86. Benfenati, F., Valtorta, F., Chieregatti, E. and Greengard, P. (1992) *Neuron*, **8**, 377.

87. Brose, N., Petrenko, A.G., Südhof, T.C. and Jahn, R. (1992) *Science*, **256**, 1021.

88. Fischer von Mollard, G., Südhof, T.C. and Jahn, R. (1991) *Nature*, **349**, 79.

89. Jahn, R. and Südhof, T.C. (1994) *Ann. Rev. Neurosci.*, **17**, 219.

90. Jahn, R., Schiebler, W., Ouimet, C. and Greengard, P. (1985) *Proc. Natl. Acad. Sci.*, **82**, 4137.

91. Washborne, P., Schiavo, G. and Montecucco, C. (1995) *Biochem. J.*, **305**, 721.

CHAPTER 8

CENTRAL NEUROTRANSMITTERS

K.M. O'Boyle

1. HISTORICAL BACKGROUND

The concept of chemical neurotransmission was first tentatively proposed by Du Bois Reymond in 1877, but was only explicitly stated in 1905 when Elliott suggested that a specific chemical substance mediates the effect of nerve stimulation. By the 1930s, contributions from Loewi, Dale and others had provided sufficient proof for the theory of chemical neurotransmission to gain general acceptance (for a review, see ref. 1). Since then, the number of candidate neurotransmitters has increased dramatically, so much so that the question of whether there are 'superfluous transmitters' has now being posed (2).

This chapter summarizes the properties of well-defined neurotransmitter substances found in the central nervous system (CNS) and gives an overview of the steps involved in neurotransmission. Important techniques used to study aspects of neurotransmitters and neurotransmission are summarized at the end of the chapter. Where possible, the reader is directed to reviews or textbooks for further information.

2. IDENTIFICATION OF CENTRAL NEUROTRANSMITTERS

It is now well established that communication between neurons in the CNS occurs through chemical synapses in the vast majority of cases. The chemicals involved in these events are classed as either neurotransmitters, neuromodulators or neurohormones depending on their characteristics.

Neurotransmitters are released from presynaptic nerves and act at nearby postsynaptic sites. Their duration of action is usually brief. The term neuromodulator is a vaguer term, often referring to chemicals which modulate the response of neurons or other cell types and act at a site distant to the site of release. Neurohormones are characterized by the fact that they are released into the bloodstream rather than into a synaptic space. They have effects on neurons following re-entry from the blood into the extraneuronal space.

There are several strict criteria which should be fulfilled for a substance to be considered as a neurotransmitter and these are listed in *Table 1*. Establishing that a chemical substance is a neurotransmitter is far more difficult for central synapses than for peripheral synapses, partly because of the anatomical complexity of the CNS, but also due to the difficulty in accessing a single set of synaptic terminals. *Table 2* lists well-defined, small molecule neurotransmitters, candidate agents which meet some of the criteria for neurotransmitters, and neuropeptides for which there is generally insufficient information to make a proper classification. The structures of many of these agents are depicted in *Figure 1*.

Table 1. Criteria for classifying a neurotransmitter

(i)	The substance is present in neurons together with the enzymes necessary for its synthesis
(ii)	Depolarization of the nerve terminal of the neuron causes the release of the substance by a Ca^{2+} dependent process
(iii)	Receptors for the substance exist on the postsynaptic cell
(iv)	Exogenous administration of the substance mimics the effects of the endogenous substance released by stimulation of the presynaptic nerve. The actions of both endogenously released and exogenously applied substance are blocked by the same specific antagonist
(v)	A mechanism exists in the neuron or the synaptic cleft for the removal and/or deactivation of the substance

Table 2. Selected neurotransmitters, candidate neurotransmitters and neuropeptides in the CNS

Small molecule neurotransmitters
Acetylcholine, adrenaline, γ-aminobutyric acid (GABA), aspartate, dopamine, glutamate, glycine, histamine, 5-hydroxytryptamine (5-HT, serotonin), noradrenaline

Candidate neurotransmitters
Adenosine, adenosine triphosphate, cholecystokinin, endogenous opioids (dynorphin, β-enkephalin, leucine enkephalin), neurotensin, nitric oxide (NO), prostaglandins, somatostatin, substance P, vasopressin

Neuropeptides
Adrenocorticotropic hormone, androgens, angiotensin I, II and III, bombesin, bradykinin, calcitonin, carnosine, cortisol, estrogens, gastrin, gastrin inhibiting peptide, leutinizing hormone, melanocyte-inhibiting factor, melanocyte-stimulating hormone, melatonin, neural growth factor, neuronal polypeptide, neuropeptide Y, oxytocin, secretin, substance K, thyroid stimulating hormone, vasoactive intestinal peptide

3. BASIC STEPS IN NEUROCHEMICAL TRANSMISSION

The main processes that occur in a classical chemically transmitting synapse are summarized in *Table 3*. These events are well characterized for many neurotransmitters (e.g. acetylcholine and the monoamine neurotransmitters except histamine) but may well be different for some other neurotransmitters (e.g. amino acids, peptides). All of the steps listed in *Table 3* are potential targets for drugs, although not all of the 10 steps indicated in the table occur at all synapses (3). This scheme provides a useful basis for understanding the mechanism of action of drugs which act by facilitating or inhibiting neurotransmission, many of which have very useful therapeutic applications (4–6).

3.1. Uptake of precursor substances and synthesis of neurotransmitters

The process of neurotransmission begins with the transport of precursor molecules from the bloodstream into the neuron and the conversion of these into the neurotransmitter by biosynthetic enzymes. Peptide neurotransmitters are synthesized in the cell body only, whereas the site of synthesis of the other neurotransmitters will depend on the location of the biosynthetic enzymes. Details of precursor molecules and synthetic enzymes are specific for individual neurotransmitters, and it is the presence of the appropriate transporter molecules and enzymes which determines where a neurotransmitter is synthesized. Originally, it was

Figure 1. Structures of central neurotransmitters. Abbreviations: Arg, arginine; Asn, asparagine; Asp, aspartic acid; Gln, glutamine; Glu, glutamic acid; Gly, glycine; His, histidine; Ile, isoleucine; Leu, leucine; Lys, lysine; Met, methionine; Phe, phenylalanine; Pro, proline; Trp, tryptophan; Tyr, tyrosine; Val, valine

MONOAMINE NEUROTRANSMITTERS

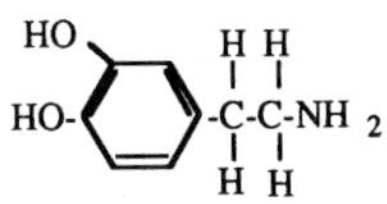

Dopamine

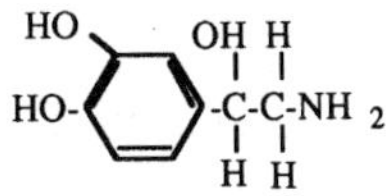

Noradrenaline

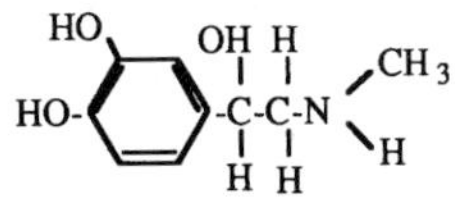

Adrenaline

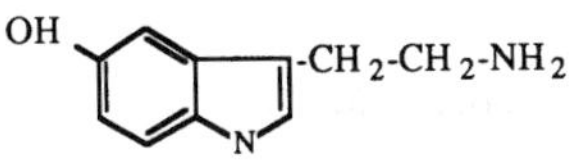

5-Hydroxytryptamine
(serotonin)

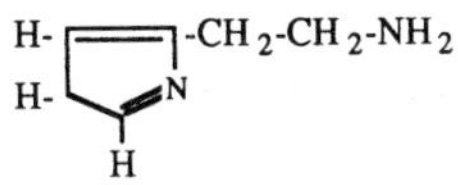

Histamine

ACETYLCHOLINE

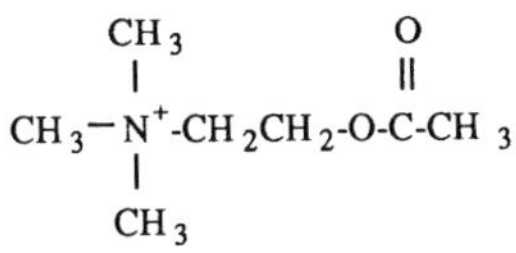

AMINO ACID NEUROTRANSMITTERS

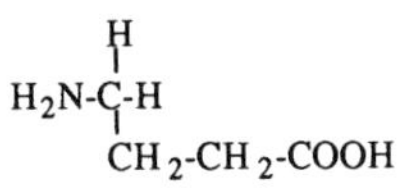

γ-Aminobutyric Acid
(GABA)

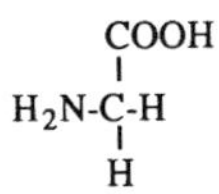

Glycine

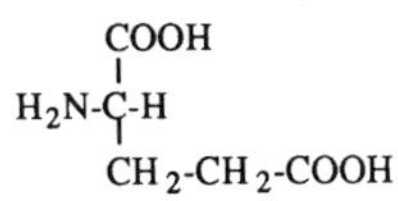

Glutamate

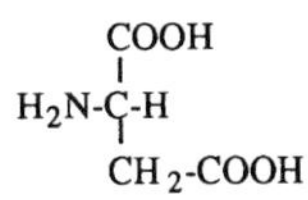

Aspartate

PEPTIDE NEUROTRANSMITTERS

Substance P Arg-Pro-Lys-Pro-Gln-Gln-Phe-Phe-Gly-Leu-Met
Neurokinin B Asp-Met-His-Asp-Phe-Phe-Val-Gly-Leu-Met

Neurotensin Glu-Leu-Tyr-Glu-Asn-Lys-Pro-Arg-Pro-Arg-Tyr-Ile-Leu

Leu-enkephalin Tyr-Gly-Gly-Phe-Leu
Dynorphin A Tyr-Gly-Gly-Phe-Leu-Arg-Arg-Ile-Arg-Pro-Lys-Leu-Lys-Trp-Asn-Asn-Gln

Met-enkephalin Tyr-Gly-Gly-Phe-Met
ß-endorphin Tyr-Gly-Gly-Phe-Met- +26 residues

Cholecystokinin octapeptide (CCK-8) Asp-Tyr-Met-Gly-Trp-Met-Asp-Phe

Table 3. Key stages in neurochemical transmission

Uptake of precursor substances
Synthesis of transmitter in neuron
Storage of transmitter in vesicles
Depolarization of nerve terminal by propagated action potential
Influx of Ca^{2+} in response to depolarization
Release of neurotransmitter by exocytosis
Diffusion to postsynaptic membrane and interaction with postsynaptic receptors
Diffusion to presynaptic membrane and interaction with presynaptic receptors
Re-uptake of neurotransmitter into presynaptic neuron (not all neurotransmitters)
Inactivation of neurotransmitter by enzymes (in neuron or in synaptic cleft)

thought that a single neuron synthesized a single neurotransmitter, but it now seems clear that some neurons may synthesize and release more than one transmitter substance (7). Co-transmission of neurotransmitters is discussed in more detail below. This discrete localization of the necessary machinery for neurotransmitter synthesis confers a chemical specificity on neurons such that a given neuron releases the same transmitter substance (or set of neurotransmitters) from every one of its synaptic terminals (originally stated by Dale (8)). All the well-defined CNS transmitters have different, although often overlapping, anatomical distributions. Details of the distribution, precursor molecules and synthetic enzymes for the well-defined neurotransmitters are summarized in *Table 4*.

3.2. Storage of neurotransmitters in vesicles

Once synthesized, the neurotransmitter is stored in membrane bound vesicles. The properties of vesicles differ somewhat for different transmitters (see *Table 5* and ref. 9). Two main types of vesicles have been identified: large dense-core vesicles (LDCVs) and small synaptic vesicles (SSVs). These have different membrane and membrane-associated proteins and are different sizes.

LDCVs have large diameters, >70 nm, originate from the trans-Golgi network, and appear to contain a dense core under the electron microscope. LDCVs are thought to be manufactured in the nerve cell body and then transported down the axon to the nerve terminal. Neuropeptides and some small molecule neurotransmitters such as the catecholamines (dopamine, noradrenaline, adrenaline) and 5-hydroxytryptamine are stored in LDCVs.

SSVs are approximately 50 nm in diameter, are clustered in the terminal region where they are synthesized *de novo* and can be recycled and refilled after transmitter release.

Catecholamine LDCVs and SSVs can both exist in the same neuron. Neurotransmitters are concentrated within these vesicles because of the action of transporter molecules located in the vesicle membrane. Some transporters are highly selective for a single neurotransmitter (e.g. the glutamate vesicular transporter) whereas others show less selectivity (e.g. the monoamine transporter or the GABA/glycine vesicular carrier).

3.3. Nerve depolarization, Ca^{2+} entry and transmitter release

Neurons are electrically excitable cells; i.e. a depolarization of the membrane can be sustained and propagated along the nerve as an action potential. At rest, the membrane is polarized (–60 mV, inside negative) because of the unequal intracellular and extracellular concentrations of ions. Typically, nerve cells have high intracellular concentrations of K^+,

Table 4. Overview of the properties of well-defined neurotransmitter systems in the CNS

Neurotransmitter	Anatomy/cytology	Precursors	Synthetic enzymes	Termination of action
Amino acids				
Inhibitory (neutral)				
Glycine	Spinal interneurons; some interneurons in brain stem	Serine	Serine hydroxy methyltransferase	Uptake
GABA*	supraspinal interneurons; spinal interneurons mediating presynaptic inhibition; abundant in nigrostriatal system	Glutamate	Glutamic acid decarboxylase*	Uptake; GABA transaminase*
Excitatory (acidic)				
Glutamate*	Interneurons at all levels	Glutamine; Krebs cycle intermediates	Glutaminase*; Krebs cycle transaminases*	Uptake
Aspartate*	Interneurons at all levels	Krebs cycle intermediates	Krebs cycle transaminases*	Uptake
Acetylcholine (ACh*)	Cell bodies at all levels; short and long connections. Important pathways: forebrain, septohippocampal, interneurons in striatum and nucleur accumbens, inhibitory pathway from spinal motoneurons	Choline, acetate	Choline acetyl-transferase*	Acetylcholine esterase*

Table 4. Continued

Neurotransmitter	Anatomy/cytology	Precursors	Synthetic enzymes	Termination of action
Monoamines				
Adrenaline*	Cell bodies in midbrain and medulla project to diencephalon	Tyrosine	Tyrosine hydroxylase*, DOPA decarboxylase, dopamine β-hydroxylase*, phenylethanolamine *N*-methyl transferase*	Uptake; monoamine oxidase, catechol *o*-methyl transferase
Noradrenaline*	Cell bodies clustered in pons and medulla project to cortex, hippocampus, hypothalamus, cerebellum and spinal cord	Tyrosine	Tyrosine hydroxylase*, DOPA decarboxylase, dopamine β-hydroxylase*	Uptake; monoamine oxidase, catechol *o*-methyl transferase
Dopamine*	Cell bodies at all levels; short, medium and long connections. Three main pathways: nigrostriatal, mesolimbic and tuberoinfundibular	Tyrosine	Tyrosine hydroxylase*, DOPA decarboxylase	Uptake; monoamine oxidase, catechol *o*-methyl transferase
5-hydroxytryptamine* (5-HT, serotonin)	Cell bodies in midbrain and pons project to cortex, limbic system, spinal cord and hypothalamus	Tryptophan	Tryptophan hydroxylase*, DOPA decarboxylase	Uptake; monoamine oxidase, aldehyde dehydrogenase
Histamine*	Cell bodies in posterior hypothalamus project to cortex and midbrain	Histidine	Histidine decarboxylase*	Histamine methyl transferase, monoamine oxidase, aldehyde dehydrogenase

Peptides				
Substance P*	Some unmyelinated primary sensory neurons in spinal cord and medulla	Substance P gene	Enzymes of protein synthesis	Neuropeptidases
Opioid peptides*	Throughout CNS	Opioid peptide genes	Enzymes of protein synthesis	Neuropeptidases

* indicates components to which antibodies have been raised and which may be used in immunocytochemistry

Table 5. Vesicular storage of neurotransmitters

Neurotransmitter	Vesicle	ATP	Protein	Electron dense	Vesicular transporter
Dopamine, adrenaline, noradrenaline, serotonin	SSV, LDCV	Yes	Yes	Yes[a]	DA, NA, ADR, 5-HT and Hist share common vesicular transporter
Acetylcholine	Small vesicles	Yes	Yes	No	ACh probably accumulated in exchange for 2 H^+
Neuropeptides	LDCV	?	Little	Yes[b]	?
Glutamate	Glutamatergic synaptic vesicles	Yes	Yes	?	Transporter highly selective for Glu versus Asp

ACh, acetylcholine; ADR, adrenaline; DA, dopamine; 5-HT, 5-hydroxytryptamine; LDCV, large dense-core vesicles; NA, noradrenaline; SSV, small synaptic vesicle
[a]High density due to presence of neurotransmitter; can be depleted by reserpine
[b]High density due to chromogranin; not depleted by reserpine
Table adapted from Nicholls (9)

and high extracellular concentrations of Na^+ and Cl^-. Depolarization is achieved by opening voltage-sensitive sodium channels which results in an inward current. The depolarization also spreads passively along the axonal membrane. If the depolarization is above a threshold value, then more sodium channels open and the action potential is propagated along the axon. The membrane repolarizes either by a delayed opening of voltage-sensitive potassium channels or by leakage of potassium ions out of the cell. A Na^+/K^+ ATPase pump restores the ions to their original resting values. A full account of this process can be found in refs 10 and 11.

Arrival of the action potential at the nerve terminal triggers the opening of voltage-sensitive Ca^{2+} channels. The subsequent rise in cytosolic free calcium $[Ca^{2+}]_i$ provides the signal for fusion of the secretory vesicles with the nerve terminal membrane and the release of the vesicle contents. This process is known as 'exocytosis' and it takes place at specialized sites in the nerve terminal called 'active zones' (12). At the nerve terminal, synaptic vesicles are linked together within a cytoskeletal meshwork consisting of actin filaments, or fodrin molecules. The vesicle protein synapsin I is important for anchoring the vesicles in this meshwork. While the process of exocytosis is not fully understood, it is thought that vesicle-associated membrane proteins act in concert with cytosolic and plasma membrane proteins to coordinate the docking and fusion of vesicles with the presynaptic membrane. Proteins which are thought to be important for exocytosis include synapsin I, synaptophysin, synaptotagmin, and synaptobrevin (6, 9, 13, 14).

The mechanism of release from SSVs and LDCVs appears to be different (9). SSVs cluster in a region in the nerve terminal which contains a high concentration of Ca^{2+} channels. They release their contents directly into the synaptic space. In contrast to SSVs, LDCVs do not cluster at these active zones in the nerve terminal region but rather LDCV exocytosis is most often seen at non-synaptic sites. Hence, the contents of LDCVs are not released immediately on to postsynaptic receptors but have first to diffuse some distance before interacting with their receptors. This is consistent with a neuromodulatory role for neuropeptides which are typically stored in LDCVs. A consequence of the vesicular storage of neurotransmitters is that quanta (multimolecular packets of preset size) of transmitter are released.

The exocytosis of different neurotransmitters may be coupled to distinct subclasses of Ca^{2+} channels. Thus, the Ca^{2+} channel associated with monoamine release can be inhibited, at least in part, by the N-type Ca^{2+} channel inhibitor ω-conotoxin-GVIA, whereas the Ca^{2+} channels involved in amino acid release are resistant to this agent. There is also evidence that the Ca^{2+} channel involved in release of a particular transmitter may differ in different brain regions (9).

3.4 Regulation of release by presynaptic receptors

The release of many neurotransmitters can be regulated by the action of presynaptic receptors located on the nerve terminal. These receptors may be activated by the neurotransmitter released from that nerve (autoreceptors) (15) or, alternatively, they may be activated by transmitters released from adjacent neurons (heteroreceptors) (16). Autoreceptors are usually inhibitory, thereby providing a self-regulating mechanism to control release. Furthermore, activation of autoreceptors may be achieved by concentrations of transmitter lower than those required to activate postsynaptic receptors. In contrast, activation of presynaptic heteroreceptors may result in the inhibition *or* the stimulation of the release of another transmitter. For example, the release of acetylcholine can be inhibited by dopamine acting at receptors located on the cholinergic nerve terminal, whereas the release of dopamine can be stimulated by acetylcholine acting at receptors located on the dopaminergic nerve terminal (16).

3.5. Postsynaptic neurotransmitter receptors

This topic is dealt with in more detail in Chapters 12 and 15 of this volume, and see ref. 17. Briefly, neurotransmitter receptors are viewed as important regulators of synaptic action. They are proteins which contain a binding site for the neurotransmitter and a means for signaling across the nerve cell membrane to elicit a change within the cell. Most neurotransmitters can interact with more than one type of receptor; receptor subtypes are distinguished by their amino acid sequences, the different drugs which selectively stimulate or inhibit them, and by the different intracellular changes which they produce. Thus, the exact postsynaptic effect of a neurotransmitter depends on the nature of the receptor at that particular site.

Neurotransmitter receptors can be divided into two broad classes according to the speed of the response produced. Fast neurotransmitter receptors give responses within milliseconds and are usually associated with an ion channel; examples are nicotinic cholinergic receptors and many amino acid receptors (18). Slow neurotransmitter receptors take hundreds of milliseconds to minutes to respond and are usually coupled to guanine nucleotide binding (G) proteins (19); examples for this class include $GABA_B$, metabotropic glutamate receptors and all the peptide and monoamine receptors except 5-hydroxytryptamine$_3$.

3.6. Termination of action of neurotransmitters

If chemicals are to provide a rapid signaling system at synapses then it is essential that they are rapidly inactivated. In general, there are two ways by which this is achieved: enzymatic degradation of the neurotransmitter in the synapse, or uptake of the neurotransmitter into surrounding cells. The steps involved in the termination of the effects of the well-defined neurotransmitters are known, and are summarized in *Table 4*.

4. CO-LOCALIZATION OF NEUROTRANSMITTERS

Originally it was assumed that a single neuron contained only one neurotransmitter. However, since the discovery of the peptide neurotransmitters it has become increasingly apparent that this is not always the case. Most, if not all peptide neurotransmitters coexist with other small molecule neurotransmitters; both are stored in separate sub-populations of vesicles. Furthermore, many vesicles for small molecule neurotransmitters also contain adenine nucleotides such as ATP which will also be released and may have pre- or postsynaptic effects. Examples of neurotransmitter coexistence are listed in *Table 6*.

Table 6. Examples of co-localization of neurotransmitters in neurons

Small molecule transmitter	Co-localized with:	
	peptide	**other**
Acetylcholine	Vasoactive intestinal peptide Substance P	ATP
Dopamine	Cholecystokinin Neurotensin	ATP
GABA	Somatostatin Cholecystokinin Enkephalin Substance P Dynorphin	Nitric oxide
Noradrenaline	Somatostatin Enkephalin Neuropeptide Y Neurotensin	ATP
5-hydroxytryptamine	Substance P Enkephalin	ATP

Data from refs 20–22.

It must be noted, however, that while the co-localization of neurotransmitters may be a widespread phenomenon, it does not occur in an all-or-none fashion. Thus, studies on striatal interneurons show that two spiny projection cell types can be distinguished on the basis of cytochemical differences: both cell types are GABAergic, but one set contains substance P and dynorphin as co-transmitters while the other uses enkephalin (20). There is also evidence that different proportions of the various co-transmitters are mobilized or replenished at different rates and that the ratio of released co-transmitters may vary according to duration or frequency of stimulation (22).

5. TECHNIQUES FOR STUDYING NEUROTRANSMITTERS

The number of techniques available to neuroscientists to study central nervous system transmitters and the stages of neurotransmission has exploded in the last 10–15 years. Some important techniques are summarized in *Table 7*.

Table 7. Practical techniques for studying neurotransmitters and neurotransmission

Technique	Procedure	Notes	References
Immunocytochemistry *measures antibody binding to specific neurochemical*	(i) Raise antibody against a specific neurochemical, e.g. neurotransmitter or enzyme related to the neurotransmitter (ii) Incubate antibody with tissue containing target antigen (iii) Localize bound antibody either by (a) use of a radio- or fluorescence-labeled antibody, or (b) incubating antibody-bound tissue with second antibody directed against first antibody. Second antibody can then be visualized by conjugation with an appropriate marker, e.g. a fluorescent molecule, gold particles, horseradish peroxidase, biotin or immunological markers (iv) Examine stained tissue by light or electron microscopy Tissue may be stained pre- or post-embedding for light or electron microscopy	Use to study neurotransmitter distributions and to characterize morphologically those neurons which express the transmitter (or enzyme) See *Table 4* for neurochemicals against which antibodies have been raised	23–25
Histochemistry *measures enzyme activity in brain sections*	(i) Perfuse animal with appropriate fixative (ii) Cut unfrozen brain sections on a vibrating microtome, or frozen sections on a cryostat (protocols for (i) and (ii) depend on the stability of the enzyme to be examined) (iii) Incubate sections in a cocktail of solutions with a suitable substrate such that action of enzyme results in formation of a colored product	Use to determine the exact location of specific enzymes related to transmitters in brain sections and, hence, to map neurochemically distinct cell types Has been applied to the study of acetylcholinesterase, monoamine oxidase and GABA-transaminase	26

Table 7. Continued

Technique	Procedure	Notes	References
***In situ* hybridization** *measures localization of mRNA*	(i) Cut fresh-frozen brains in a cryostat and mount on glass slides (ii) Prepare suitable oligonucleotide or ribonucleotide probe (probe is often radiolabeled or labeled with alkaline phosphatase) (iii) Hybridize probe to mRNA under optimal conditions of salt and temperature (iv) Visualize hybridized probes	Use to study expression of peptide neurotransmitters or to study transmitter receptor expression and second messenger pathways	27
Radioligand binding *measures receptors in membranes*	(i) Homogenize and centrifuge tissue to make a membrane preparation (ii) Incubate membranes with radioligand which has high affinity and selectivity for target receptor site. Parallel tubes contain a displacing agent to define non-specific binding (iii) Separate membrane-bound ligand from free ligand by filtration or centrifugation (iv) Quantify the amount of bound ligand by liquid scintillation or gamma counting	Use to study regional distributions, densities and pharmacological profiles of neurotransmitter receptors Also use to study neurotransmitter transporters	17, 28
Receptor autoradiography *measures receptors in brain slices*	(i) Cut sections of brain tissue (ii) Incubate sections with radioligand which has high affinity and selectivity for target receptor site. Parallel incubations contain a displacing agent to define non-specific binding (iii) Wash sections to remove unbound ligand (iv) Expose sections to appropriate photographic film	Use to study regional distributions, densities and pharmacological profiles of neurotransmitter receptors in brain slices or tissue sections	17, 29

	(v) Quantify amount of binding by e.g. densitometric analysis		
Synaptosomes *vesicles formed from resealed neuronal plasma membranes*	(i) Homogenize neural tissue (ii) Carry out differential and density gradient centrifugation in sucrose, Ficoll or Percoll	Use to study physiological and pharmacological aspects of synaptic function Pre-load synaptosomes with radiolabeled transmitter or precursor and study release	30
^{45}Ca and fura-2 *measure neuronal Ca^{2+}*	(i) Isolate synaptosomes (see above) (ii) Measure ^{45}Ca uptake *or* load synaptosomes with calcium sensitive fluorescent probe such as fura-2 and measure $[Ca]_i$	Use to study neuronal Ca^{2+} channels and role of Ca^{2+} in neurotransmitter release	31
***In vivo* microdialysis** *measures release of transmitter and/or metabolites*	(i) Anesthetize animal (ii) Implant dialysis cannula (iii) Insert dialysis probe (iv) Perfuse with buffer and collect dialysis samples for subsequent analysis of transmitter/metabolite content by e.g. HPLC (v) Confirm location of dialysis probe	Use to study the *in vivo* release of neurotransmitters or their metabolites	32

6. REFERENCES

1. Bacq, Z.M. (1975) *Chemical Transmission of Nerve Impulses: a Historical Sketch*, Pergamon Press, Oxford.

2. Bowers, C.W. (1994) *Trends Neurosci.*, **17**, 315.

3. Bloom. F.E. (1990) in *Goodman and Gillman's The Pharmacological Basis of Therapeutics*, 8th edn (A.G. Gilman, T.W. Rall, A.S. Nies and P. Taylor, eds), Pergamon Press, Oxford, p. 244.

4. Rang, H.P., Dale, M.M. and Ritter, J.M. (1995) *Pharmacology,* 3rd edn, Churchill Livingstone, Edinburgh.

5. Leonard, B.E. (1992) *Fundamentals of Psychopharmacology*, John Wiley, Chichester.

6. Strange, P.G. (1992) *Brain Biochemistry and Brain Disorders*, Oxford University Press, Oxford.

7. Hökfelt, T., Fuxe, K. and Oernow, B. (1987) *Prog. Brain Res.*, **68**, 1.

8. Dale, H.H. (1935) *Proc. R. Soc. Med.*, **28**, 319.

9. Nicholls, D.G. (1994) *Proteins, Transmitters and Synapses*, Blackwell Scientific, London.

10. Hodgkin, A.L. and Huxley, A.F. (1952) *J. Physiol.*, **117**, 500.

11. Aidley, D.J. (1989) *The Physiology of Excitable Cells*, 3rd edn, Cambridge University Press, Cambridge.

12. Smith, S.J. and Augustine, G.J. (1988) *Trends Neurosci.*, **11**, 458.

13. Burgoyne, R.D. and Cheek, T.R. (1995) in *Neurotransmitter Release and its Modulation* (D.A. Powis and S.J. Bunn, eds), Cambridge University Press, Cambridge, p. 5.

14. Sollner, T. and Rothman, J.E. (1994) *Trends Neurosci.*, **17**, 344.

15. Brock, J.A. (1995) in *Neurotransmitter Release and its Modulation* (D.A. Powis and S.J. Bunn, eds), Cambridge University Press, Cambridge, p. 81.

16. Fredholm, B.B. (1995) in *Neurotransmitter Release and its Modulation* (D.A. Powis and S.J. Bunn, eds), Cambridge University Press, Cambridge, p. 104.

17. Watson, S. and Girdlestone, D. (1995) *Receptor and Ion Channel Receptor Supplement*, Elsevier, Amsterdam, p. 1.

18. Conley, E. (1994) *The Ion Channel Facts Book*, Academic Press, London.

19. Watson, S. and Arkinstall, S.A. (1994) *The G-Protein Linked Receptor Facts Book*, Academic Press, London.

20. Kawaguchi, Y., Wilson, C.J., Augood, S.J. and Emson, P.C. (1995) *Trends Neurosci.*, **18**, 527.

21. Kaplan, H.I., Sadock, B.J. and Grebb, J.A. (1994) *Kaplan and Sadock's Synopsis of Psychiatry*, 7th edn, William & Wilkins, Baltimore, p. 144.

22. Sneddon, P. (1995) in *Neurotransmitter Release and its Modulation* (D.A. Powis and S.J. Bunn, eds), Cambridge University Press, Cambridge, p. 22.

23. Totterdell, S., Ingham, C.A. and Bolam, J.P. (1992) in *Experimental Neuroanatomy: A Practical Approach* (J.P. Bolam, ed.), IRL Press, Oxford, p. 103.

24. Ingham, C.A. (1992) in *Experimental Neuroanatomy: A Practical Approach* (J.P. Bolam, ed.), IRL Press, Oxford, p. 129.

25. Priestly, J.V. (1987) in *Neurochemistry: A Practical Approach* (A.J. Turner and H.S. Bachelard, eds). IRL Press, Oxford, p. 65.

26. Vincent, S.R. (1992) in *Experimental Neuroanatomy: A Practical Approach* (J.P. Bolam, ed.), IRL Press, Oxford, p. 153.

27. Gerfen, C.R., Young, W.S. and Emson, P. (1992) in *Experimental Neuroanatomy: A Practical Approach* (J.P. Bolam, ed.). IRL Press, Oxford, p. 173.

28. Yamamura, H.I., Enna, S.J. and Kuhar, M.J. (1985) *Neurotransmitter Receptor Binding*, Raven Press, New York.

29. Wharton, J. and Polak, J.M. (1993) *Receptor Autoradiography: Principles and Practice*, Oxford Science Publications, Oxford.

30. Gordon-Weeks, P.R. (1987) in *Neurochemistry: A Practical Approach* (A.J. Turner and H.S. Bachelard, eds), IRL Press, Oxford, p. 1.

31. Grynkiewicz, G., Poenie, M. and Tsien, R.Y. (1985) *J. Biol. Chem.*, **260**, 3440.

32. Robinson, T.E. and Justice, J.B. (1991) *Microdialysis in the Neurosciences*, Elsevier, Amsterdam.

CHAPTER 9
NEURONAL CELL CULTURES

L.M. Haynes

1. INTRODUCTION

Studies of the nervous system in tissue culture since 1907 (1) have provided invaluable insights into the embryological origins of nerve cells, the development of their cytological and biochemical identities, their functional connectivity and their interactions with glial and other cell types within nervous systems. In spite of the limitations of all culture systems, in the past decade neuronal tissue culture has increasingly become a mainstream technique in use in multidisciplinary laboratories. With the improvement of cell-selection techniques for culture from mixed cell suspensions in the 1980s, it became possible to grow neurons from the central nervous system of the chick embryo (2) with greater than 90% purity. At the same time, techniques for culture of nerve cells from mature vertebrates (3) and from invertebrates (4) were reported.

The generation of monolayer cell cultures of neurons requires expertise in microdissection, a knowledge of cell-selection principles, and the development of an ability to diagnose the health of nerve cells during the process of adaptation to culture conditions. The laboratory for primary culture work must provide an environment of adequate sterility and basic items of equipment (CO_2 incubator, dissecting and universal compound microscopes, refrigerated centrifuge). For work involving the manipulation of viruses and DNA transfection, laboratory operational standards must be reviewed. Full details are given in ref. 5.

2. BASIC STRATEGIES FOR NEURONAL CULTURE FROM PRIMARY NERVOUS TISSUES

Nervous tissues are made up of a variety of cell types. Thus cell suspensions from vertebrate brain, even when prepared from accurately-dissected central nervous system (CNS) tissue fragments, can contain astroglia, oligodendroglia, microglia, endothelial, mesenchymal and ependymal cells. Likewise, peripheral nervous system (PNS) cell suspensions may contain fibroblasts, Schwann cells, satellite cells and smooth muscle cells as well as neurons.

The yield of neurons which survive in monolayer culture from vertebrate nervous tissue depends for the most part on the stage of development. Neurons from any brain region are most capable of autonomous survival when cultured on or shortly after their birth date in embryogenesis. Neurons cultured after that time will require enhanced media and substratum supplementation if they are to survive at all. *Table 1* gives the optimum dates for cell culture of neuronal types of common application.

Mature neurons rely on their capacity for autonomous regeneration to adapt to culture conditions. Those which do not readily regenerate do not grow. Mature neurons from invertebrate CNS or segmental ganglia adapt readily to culture conditions for this reason. The valuable lifetime of most primary neuronal cell cultures is between 3 and 14 days. The critical

Table 1. Optimum date ranges for transfer of neurons of developing rodent and avian nervous system for differentiation in monolayer cell culture

Cell source	Age range	Comments	
		Rat/mouse[a]	Chick/quail
Cerebral cortex (telencephalon)	E12–E15[b]	E8	
Corpus striatum	E15–E16[b]		
Forebrain basal ganglia	E16–E18		
Hippocampus	E17–E21		
Hypothalamus (diencephalon)	E15–E18		
Midbrain/colliculi	E16–E17		
Cerebellar cortex	P1–P8		Earler times for Purkinje cells
Brainstem			
Spinal cord	E12–E14	E6	
Dorsal root ganglion	E14–adult	E8	Growth factor-dependence increases with age in fetal cultures
Superior cervical ganglion	E15–P7		
Enteric ganglion	Postnatal		
Ciliary ganglion		E6–E9	

[a]Developmental stages in mid-gestation mouse brain are 24–40 h ahead of those in rat from time of conception, depending on the stage of development.
[b]Later dates are possible with greater difficulty.
E = days post-conception; P = days post-partum.

period during which adaptation to culture conditions takes place is between 2 and 4 days. Synaptogenesis where it occurs is normally evident by the end of the first week in culture (6).

2.1. Dissection

Dissection of material for primary cultures takes place in two stages: gross dissection of the animal and microdissection of the tissue fragment. The former procedure should not normally be carried out in the laminar flow cabinet but in a designated sterile preparation area. An exception is the delivery of embryos from avian eggs. A neural tissue culture/dissecting guide (e.g. ref. 7) and atlases of mammalian embryonic central nervous systems (8, 9) are helpful aids in the laboratory.

For microdissection the stereomicroscope should provide a broad magnification range from 1.5× upwards with a powerful but cool source of *cis*-illumination. Operating microscopes with a long working distance (>200 mm) objective are the best choice for the culture laboratory since they enable permanent remote mounting (e.g. to a wall or the rear of the laminar flow cabinet), reducing the number of surfaces which require sterilization before work. Dissection may be done in a clean-room or in a laminar flow cabinet with full front access or hinged front panel. The microscope arrangement for dissection in a laminar flow cabinet must take into account possible disruption of sterile air currents caused either by bulky equipment or alteration of the cabinet itself. The best dissecting instruments are durable for long life and repeated repair. Home-made microneedles and scalpels may be needed for some applications. Apart from good basic sterile technique, the following generally-applicable rules improve sterility and yield of high-quality cultures:

(i) Lay out all equipment, plasticware, freshly-prepared culture medium and reagents by the work station to avoid movement once dissection has commenced. Adhere to a timed protocol for dissection, pooling of tissue, cell dissociation, cell population enrichment, if applicable, and plating.
(ii) Work on a manageable scale. Large dissections should be carried out by 2–3 people. Keep dissected tissues in small groups in a covered vessel (e.g. petri dish) to limit the number of times the contents of collecting dishes are exposed to the atmosphere. Wash dissected tissue by transfer between dishes at least twice prior to collection.
(iii) Keep dissected material for as short a time as possible at ambient temperature. Dissected tissue should be collected in stages in a complete, dextrose-supplemented saline (e.g. Puck's G balanced salt solution (BSS)) and kept at 4°C until tissue dissociation.

2.2. Cell dissociation

Cell dissociation for preparing neuronal monolayer cultures is a critical step. The nervous system is almost unique in comprising mainly process-bearing cells. It therefore follows that for all but stem cell cultures from embryonic forebrain and mitotic cells from other germinal brain regions, the plasma membranes of the majority of developing neurons will have undergone extensive but, in viable cells, temporary rupturing during dissociation. The problem is analogous to that for cells dissociated from tissues which are joined by *zonula occludens* or *nexus*-type junctions. Dissociation by whatever means can lead to the death prior to plating of cell numbers greater than those for other tissues, and cell swelling after plating, due to fluid exchange between medium and cytoplasm before resealing of the plasma membrane. The problem is exacerbated after plating by the fact that viable neurons can be very sensitive to free radicals, proteases and other toxic proteins released from dead cells in culture, and it is frequently undesirable to perform a medium change once seeding is complete to remove these dead cells. It must also be borne in mind that in those neurons which survive in culture, the axon and all or most of the dendrites have undergone *de novo* regeneration. Many if not all of these problems are avoided by the choice of organotypic rather than monolayer culture techniques. However, damaging consequences of tissue disaggregation can be reduced by:

(i) where possible using methods employing minimum shear forces for cell dissociation (e.g. Nitex bag methods)
(ii) performing the dissociation step in calcium/magnesium-free dextrose-supplemented BSS, reducing the accumulation of extracellular calcium in the cytoplasm
(iii) if enzymic dissociation is employed, inactivating or washing away the dissociating enzyme prior to the application of mechanical force.

It should be remembered that tissue disaggregation and plating, which are often rapid processes at the end of a laborious dissection, are the most critical steps in preparing primary neural cell cultures. For speed, simplicity and improved asepsis, it is common to use the shear forces generated by trituration by Pasteur pipette or syringe and hypodermic needle to achieve substantial cell dissociation. The following general principles should be observed:

(i) Employ siliconized pipettes with tips fire-polished to the appropriate diameter. Choice of tip diameter will be dictated by the size and fragility of the cells in suspension.
(ii) During cell dissociation by trituration, recover and pool dissociated cell supernatants from tissue fragments frequently to avoid mechanical damage to cells through repetitive pipetting. Avoid frothing.
(iii) Plate dissociated and washed cells as soon as possible into prewarmed and gassed culture vessels.

Mechanically-disaggregated neural tissues, particularly where no enzyme has been used, frequently yield numerous microscopic cell clumps. For buoyant cells which adhere relatively poorly, the cells in these clumps will die rapidly and remain suspended in the culture media to damage surviving cells on the substratum. This phenomenon should not be confused with clumping which occurs 24 h or more after seeding. Tissue disaggregation should be performed in sterile tubes with conical bases. Allow 1–2 min after dissociation for clumps to settle to the bottom of the tube and pipette the supernatant into a fresh tube before plating. The cells may be transferred through a 40 μm pore diameter disposable cell strainer (Becton Dickinson) to remove clumps larger than about 10 cells.

Disaggregation media should provide the mildest environment for cell separation in the shortest possible time. *Table 2* lists standard disaggregation approaches for the commonest types of neuronal monolayer culture. Enzymes are normally required for the preparation of neurons from all regions of the nevous system other than the cerebral and cerebellar cortices. The bacterial enzyme clostripeptidase A (commonly termed 'collagenase') is required to disaggregate smooth muscle and connective tissues such as perineurium. The proteolytic activities regarded as impurities in commercially-available enzyme preparations are normally desirable to expedite dissociation of complex connective tissues. However, collagenase purified free of other protease activites is sometimes chosen to avoid irreversible damage to cell surface receptors.

Table 2. Disaggregation procedures for the most commonly-used mammalian neuronal monolayer culture models

Tissue source	Age	Procedure
Forebrain	E12–E16	Mechanical alone
	E16–E19 (mouse)	Mechanical alone or 0.05% trypsin
	E18–E19 (rat)	0.05% trypsin (+ hyaluronidase and DNase if clumps persist)
	E20–P7	Mechanical (avoid trituration)
	E16–E19 basal forebrain	0.05% trypsin/collagenase
Cerebellum	P7–P8	Versene or 0.25% trypsin
Sensory ganglia	E14–E16	0.25% trypsin
	Postnatal	Trypsin/collagenase or papain

E = embryonic; P = postnatal.

2.3. Plating

Gross experimental inaccuracies frequently result from non-uniform plating of cells either within or between culture wells or flasks. *Table 3* lists plating procedures essential for the production of high-quality neuronal cultures. The biochemical properties of mass-produced postmitotic neuronal monolayers vary substantially with plating density. Poor viability and slow differentiation characterize low-density cultures, whilst glial overgrowth and overcrowding result from high plating densities. Thus intermediate densities give best performance with the optimum around $2–2.5 \times 10^6$ cells per 1000 mm^2. Neurons from some tissues (e.g. rodent embryo forebrain) show a pronounced tendency to reaggregate from a monolayer within hours. The adhesive quality of the substratum must be sufficient to overcome this tendency. Incomplete solubilization of the substratum material, non-uniformity of the surface and substratum concentration can all contribute to cell reaggregation. The choice of surface area size, plasticware source and culture vessel design may all be factors which improve the uniformity and quality of neuronal monolayers. Finally, cell mixing artefacts may result in variations between culture wells and consequently magnify data variation.

Table 3. General rules for achieving reproducible monolayers of neurons

Evaluate the appropriate cell density for each neuronal culture
Evaluate optimum substratum concentration for good spreading and minimum cell reaggregation
Mix cell suspension after plating every 2–4 wells
Avoid plating undissociated cell clumps
Ensure even spreading of cell suspension by shaking
Use appropriate well-size and medium volume to avoid regional variations in cell density and viability
Inspect culture wells after 2 and 4 days and reject those which do not meet quality criteria (see section 3)

2.4. Media

Neuronal culture is carried out where possible in a minimal defined medium supplemented with a serum-free cocktail of constituents designed to support differentiation and protein synthesis. Stem cells, however, benefit from the broad spectrum of growth factors found in serum. Neuronal types derived from late fetal and early postnatal brain of mammals may not adapt to serum-free medium unless provided with a biomatrix substratum.

Table 4 lists media and supplements which are employed for most neural culture applications. Eagle's Minimal Essential Medium (MEM) and Dulbecco's Modification (DMEM) are in very common use. The latter has frequently been employed in a 1:1 mixture with Ham's F12 Nutrient Mixture. The mixture contains little of benefit to primary neuronal cultures that cannot be supplied by DMEM alone and contains over 40 μM glutamate, which enhances growth in neuronal cultures but may be expected to desensitize or downregulate (10) excitatory amino acid receptors and influence synaptic transmission. Defined media formulations such as Neurobasal (Life Technologies International) are designed to overcome this problem.

Table 4. Defined media for popular mammalian neuronal monolayer cultures

Medium	Typical uses
MEM	Superior cervical ganglion neurons
DMEM	Fetal rat brain, sensory ganglia
F12/F14	Adult sensory ganglia, retinal ganglion neurons, basal forebrain, fetal mouse brain (1:1 with DMEM)
L15	Superior cervical ganglion neurons, spinal motor neurons

Defined media supplements for many applications of primary neuronal culture are very similar to those with universal suitability originally formulated by Faivre-Bauman *et al.* (11). These are based on the principle that serum can be replaced by carrier proteins (transferrin and albumin) and hormones which promote growth and metabolism (insulin, thyroxine, steroids) providing the cell gains adequate purchase and support on the substratum (see section 2.5). However, some proprietary supplements include tissue-specific growth factors and cytokines. Moreover, neuronal types from both central and peripheral nervous systems display partial or absolute requirements for neurokines, including neurotrophins, ciliary neurotrophic factor, glial-derived neurotrophic factor, heregulins and other neuronal differentiation-inducing factors. Neurons in the central nervous system receive copious excitatory synaptic inputs from an early stage in embryonic development. They are therefore in some cases dependent not only on target-derived trophic influences but on afferent-induced depolarization for healthy development. Synaptic depolarization can be replaced by chemically-induced depolarization with high (15–25 mM) extracellular potassium chloride when cell contacts are interrupted during disaggregation and monolayer culture. Glucose supplementation to up to 6 g/l has been employed in neuronal cell cultures. There is some evidence that high glucose promotes growth in serum-free culture. However, modifications to the medium which raise the osmolarity of the environment to more than 350 mOsm per liter should be avoided. Moreover, particular attention should be paid to the concentration of liquid media by evaporation during culture. Periods of culture of 7 days or more without medium change, use of culture plates offering high medium–gas equilibration and incomplete humidification of the CO_2 incubator can all contribute to cell death by shrinkage.

2.5. Substrata

Adhesive substrata are an absolute requirement for all types of primary neuronal cell in culture. For viability, animal cells require to gain purchase on the substratum within about 30 minutes and to spread the plasma membrane to form focal adhesion contacts within the next few hours. Neurons do not adopt the classical mode of pseudopodial spreading but rather extend one or more thin processes, reducing the number of adhesion contacts which are initially formed with the surface.

Whilst serum assists in both anchorage and spreading in non-neuronal cells (through the deposition of vitronectin, fibronectin and other glycoprotein spreading factors on the culture surface), adherence and process formation in cultured neurons are aided by different conditions; the former by the presence and the latter by the absence of serum from the medium. Thus serum-free media promote the differentiation of neurons but are counterproductive for anchorage and therefore viability. Improvements to the environment have been made by coating the surface with polymers of basic amino acids or other positively-charged moiety (see *Table 5*). These provide an adhesive surface for cell surface-associated glycoproteins as well as those secreted *de novo* in culture to form the extracellular matrix. They thus assist in both anchorage and process-formation.

Biomolecular substrata are used to actively promote process-formation and differentiation in culture (see *Table 5*). The extracellular matrix of the CNS is dominated by laminin, which

Table 5. Culture substrata suitable for neurons growing on a plastic or chemically-clean glass surface

Coating	Concentration range (μg ml^{-1})
Poly-L-lysine[a]	5–20
Poly-D-lysine[a]	5–50
Poly-ethyleneimine[a]	100–200
Collagen types I/III (dermal or tendon)	0.3–1.0
Laminin	5–10
Matrigel (regular of growth factor-reduced)	1:20–1:80
Midkine	0.001–0.01
Pleiotropin	0.001–0.01
Fetal bovine serum	1:10

[a]Polyamino-acid and poly-ethyleneimine substrata are washed 2–3 times prior to application of cells. Biological substrata may be applied over a surface precoated with polyamino-acid to improve evenness of spreading.

promotes receptor-mediated neurite outgrowth. Similar responses can be obtained with proteoglycans, collagens and fibronectin. For serum-free culture, the surface may previously be treated with dilute serum, which provides a source of surface-bound glycoproteins. Such a treatment does not expose neurons to those proteins which inhibit differentiation when serum is added continuously to the culture medium. Recently the role in neuronal differentiation of minor extracellular matrix factors, including midkine and pleiotropin/HBNF, has come to light (12). Such factors can be used at 1000 × lower concentrations for coating substrata than laminin and may replace a requirement for neurokines in some types of neuron. Monolayer cultures can be grown on a surface precoated with a thin layer of Matrigel (Becton Dickinson), applied at a dilution of 1/20–1/80. This preparation, which is an extract of murine tumors induced by *in vivo* inoculation of Engelbreth–Holm Swarm (EHS) sarcoma cells, contains laminin, nidogen, proteoglycans, cellular fibronectin and a variety of growth factors (13). Ammonium sulphate precipitation yields a growth factor-reduced preparation, which nevertheless still contains a number of growth factors and macromolecules in undefined concentrations. Both preparations support neuronal growth and accelerate differentation, reducing the need for added growth factors.

3. DIAGNOSIS OF NEURONAL HEALTH IN CULTURE

Researchers who are introduced to primary neuronal culture after working with cell lines are normally unprepared for the increased labour-intensity of the technique and the inability of normal neurons to 'look after themselves'. Neuronal monolayer culture protocols suffer from two types of imperfection in inexperienced hands, both of which can be observed with relatively little training with the inverted microscope: firstly, poor appearance of cultures shortly after plating, and secondly, deterioration within two or three days of cultures which were apparently in good condition on day 1. The former problem normally arises through errors in disaggregation and plating technique and the suitability of the substratum. The latter arises through suboptimal environmental conditions. It reflects the fact that viable cells are slowly deteriorating rather than showing full adaptation to the culture conditions leading to *de novo* gene expression, protein and lipid biosynthesis. Optimization of environmental conditions must take into account the temperature, gas content, pH, osmolarity and supply of energy-giving metabolites in the medium adjacent to the cells, presence of correct growth and differentiation-promoting macromolecules in the media and on the growth surface and, in some circumstances, the presence of other supporting cells and a depolarizing influence. Furthermore, active steps to prevent necrosis and/or apoptosis may be required.

Table 6 provides a troubleshooting guide with some suggested remedies. The best solution to persistent problems, however, is to learn directly from a researcher with first-hand experience of the chosen model.

4. NEURONAL CULTURE PROTOCOLS

The most recent practical textbooks on cell culture applied to neuroscience (e.g. 14–16) do not provide comprehensive guides to the culture of neurons. However, since 1990 the number of new culture models for neurons, and applications of these models, has increased substantially. It is beyond the scope of this chapter to give detailed protocols for specific neuronal culture models. However, *Table 7* lists 20 references exemplifying well-cited and trouble-free methods for monolayer culture of CNS and PNS neurons at different ages for mammals and birds. A complete guide to cell and organized cultures for study of living neurons will form part of a forthcoming book of which the author is editor (17).

Table 6. Common causes of poor culture quality or neuronal degeneration after plating

Evidence	Diagnosis	Solution
Poor quality soon after plating		
Excessive debris	Presence of myelin or other cell fragments	Use 1–3 preplating stages or centrifuge through 3.5–5% bovine serum albumin
Excessive dead cells/ poor yield	Cell damage during dissociation	Review dissociation method
Medium too pink and poor cell health	pH over 7.3, cells not respiring	Add pyruvate or stronger buffer (e.g. 5–10 mM Hepes); use plates with increased gas exchange
Poor neurite extension, floating cells	Poor anchorage	Increase substratum-coating concentration
Subsequent deterioration after 1–3 days in culture		
Cell death/shrinkage	Poor energy supply (hypoxia/hypoglycemia)	Add glucose, check glutamine has not expired[a]
	Trace chemical contaminants	Use 1X pre-prepared medium and or proprietary growth supplement
	Free radical damage	Add vitamin C or reduced glutathione to both dissociation and culture media
	Plating density too low?	Evaluate
	Critical growth factors/ biomatrix substratum support needed	Evaluate
	Inappropriate presence of/ contact with glia	Enrich cell suspension for neurons, use serum-free medium
Excessive apoptosis	Inappropriate soluble trophic support	Evaluate trophic factors, consider using serum
'Centre-well/meniscus artefact' in small wells	Local degeneration due to uneven gas exchange	Use 24-well plates or larger

[a]Use medium well before expiry date (normally very optimistic). Preferably buy without glutamine and add from frozen stock or use Glutamax-1™. Do not store complete medium. Use once only.

Decisions about experimental precautions, such as the preparation of tissue cultures from living anesthetized or recently-sacrificed animals, the use of antibiotics in cultures and the scale of culture experiments are made by individual experimenters and are unlikely critically to affect the quality of the product.

Several regular international practical courses exist which cover cell culture of neurons, including one in the author's laboratory.

Table 7. Protocols for popular monolayer culture preparations

Species	Preparation	Stage	Reference
Rat	Superior cervical ganglion	Embryonic and postnatal	18
	Sensory ganglion	Embryonic	19
		Neonatal	20
		Adult	21
	Cerebellar granule cells	Neonatal	22
	Hippocampal pyramidal cells	Embryonic	23
		Perinatal	24
	Cerebral cortex	Embryonic	25
	Mesencephalic (dopaminergic)	Embryonic	26
	Diencephalic (hypothalamic, peptidergic)	Embryonic	27
	Hypothalamic	Neonatal	28
	Spinal cord (isolated dorsal and ventral)	Embryonic	29
	Motor neurons	Embryonic	30
Chick	Sensory ganglion/placode	Embryonic	31
	Ciliary ganglion	Embryonic	32
	Cerebral cortex	Embryonic	33
	Retinal photoreceptors	Embryonic	34
	Motor neurons	Embryonic	35
Amphibian			
Xenopus	Neural tube	Larval	36
Molluscan			
Helisoma	Giant neuron	Adult	37

Cell Cultures

5. REFERENCES

1. Harrison, R.G. (1907) *Anat Rec.*, **1**, 116.

2. Pettman, B., Louis, J.C. and Sensenbrenner, M. (1979) *Nature*, **281**, 378.

3. Goldberg, S.S.S. and De Boni, U. (1988) *J. Neurobiol.*, **14**, 195.

4. Schacher, S. and Proshansky, E. (1983) *J. Neurosci.*, **3**, 2403.

5. Various authors (1993) Chapter 1 in *Cell and Tissue Culture: Laboratory Procedures* (A. Doyle, J.B. Griffiths and D.G. Newell, eds), Wiley, Chichester.

6. Weiss, S., Pin, J.-P., Sebben, M., Kemp, D.E., Sladeczek, F., Gabrion, J. and Bockaert, J. (1986) *Proc. Natl. Acad. Sci. USA*, **83**, 2238.

7. Shahar, A., DeVellis, J., Vernadakis, A. and Haber, B. (eds) (1989) *A Dissection and Tissue Culture Manual of the Nervous System*, Wiley–Liss, Chichester.

8. Schambra, U.B., Lauder, J.M. and Silver, J. (1992) *Atlas of Prenatal Mouse Brain*, Academic Press, San Diego.

9. Altman, J. and Bayer, S.A. (1995) *Atlas of Prenatal Rat Brain Development*, CRC Press, Boca Raton.

10. Resink, A., Villa, M., Boer, G.J., Mohler, H. and Balazs, R. (1995) *Eur. J. Neurosci.*, **7**, 1700.

11. Faivre-Bauman, A., Puymirati, J., Loudes, C. and Tixier-Vidal, A. (1984) in *Methods for Serum-Free Culture of Neuronal and Lymphoid Cells* (D.W. Barnes, D.A. Sirbasku and G.H. Sato, eds), Liss, New York, p. 37.

12. Muramatsu, T. (1994) *Dev. Growth Diff.*, **36**, 1.

13. Kleinman, H.K., McGarvey, M.L., Hassell, J.R., Star, V.L., Cannon, F.B., Laurie, G.W. and Martin, G.R. (1986) *Biochemistry*, **25**, 312.

14. Federoff, S. and Richardson, A. (eds) (1992)

Protocols for Neural Cell Culture, Second Edition, Humana, New Jersey.

15. Cohen, J. and Wilkin, G. (1996) *Neural Cell Culture: A Practical Approach*, Oxford University Press, Oxford.

16. DeVellis, J., Haber, B. and Shahar, A. (eds) (1996) *Cell Culture and Transplantation Manual of the Nervous System*, 2nd edn, Wiley–Liss, in press.

17. Haynes, L. (ed.) The Neuron in Tissue Culture *IBRO Handbook Series: Methods in the Neurosciences*, Wiley, Chichester, in preparation.

18. Johnson, M.L. and Argyro, V. (1983) *Meth. Enzymol.*, **103**, 334.

19. Wood, P.M. (1980) *Brain Res.*, **196**, 247.

20. Burgess, G.M., Mullaney, I., McNeil, M., Dunn, P.M. and Rang, H.P. (1989) *J. Neurosci.*, **9**, 3314.

21. Lindsay, R.M. (1989) *J. Neurosci.*, **8**, 2394.

22. Thangnipon, W., Kingsbury, A., Webb, M. and Balazs, A.R. (1983) *Brain Res.*, **11**, 179.

23. Mattson, M.P., Dou, P. and Kater, S.B. (1988) *J. Neurosci.*, **8**, 2087.

24. Banker, G.A. and Cowan, W.M. (1987) *J. Comp. Neurol.*, **187**, 469.

25. Rosenberg, P.A. and Dichter, M.A. (1994) *J. Neurosci.*, **9**, 2654.

26. Reisert, I., Engele, J. and Pilgrim, C. (1989) *Cell Tissue Res.*, **255**, 411.

27. Loudes, C., Faivre-Bauman, A. and Tixier-Vidal, A. (1983) *Meth. Enzymol.*, **103**, 313.

28. Yang, Z., Lee, D., Huang, W., Copolov, D.L. and Lim, A.T. (1994) *Brain Res.*, **648**, 99.

29. Guthrie, P.B., Brenneman, D.E. and Neale, E.A. (1987) *Brain Res.*, **420**, 313.

30. Martinou, J.C., Thai, A.L.V., Casser, G., Roubinet, F. and Weber, M.J. (1989) *J. Neurosci.*, **9**, 3645.

31. Lindsay, R.M., Thoenen, H. and Barde, Y.-A. (1985) *Dev. Biol.*, **112**, 319.

32. Ebendal, T., Stromberg, I., Olson, L. and Beiger, A. (1985) *Neuroscience*, **16**, 181.

33. Daval, J.L., Louis, J.C., Gerard, M.J. and Vincendon, G. (1983) *Neurosci. Lett.*, **36**, 299.

34. Adler, R. and Hatlee, M. (1989) *Science*, **243**, 391.

35. Dohrmann, U., Edgar, D. and Thoenen, H. (1986) *Dev. Biol.*, **118**, 209.

36. Peng, H.B., Baker, L.P. and Chen, Q. (1991) in *Methods in Cell Biology*, vol. 36, (B.K. Kay and H.B. Peng, eds), Academic Press, San Diego, p. 511.

37. Ridgway, R.L., Syed, N.I., Bulloch, A.G.M. and Lukowiak, K. (1991) *J. Neurobiol.*, **22**, 377.

CHAPTER 10
PROTEIN PHOSPHORYLATION IN NEURONAL TISSUE: APPROACHES, ISSUES AND INTERPRETATION

J.A.P. Rostas and A.T.R. Sim

1. INTRODUCTION

Phosphorylation of proteins (by protein kinases) and their dephosphorylation (by protein phosphatases) is the addition or removal of a phosphate group at one or more sites on the surface of a protein (*Figure 1*). The three major amino acids that become phosphorylated are serine, threonine and tyrosine and the phosphate group comes from ATP which donates its γ-phosphate. The vast majority of phosphate incorporation in protein occurs at serine and threonine residues and is controlled by a diverse group of kinases and phosphatases. Tyrosine phosphorylation accounts for less than 10% of total phosphorylation and is controlled by another diverse group of kinases and phosphatases. The addition or removal of a phosphate group produces a change in shape of the protein which in turn alters one or more of its properties and hence regulates some aspect of its function. However, the exact functional consequence of phosphorylation at a particular site is known only for a few proteins and sites. Protein phosphorylation occurs in all cells but nervous tissue is particularly rich in protein kinases and phosphatases. This reflects the importance of protein phosphorylation and dephosphorylation in the regulation of neuronal function.

The aim of this chapter is to provide the reader with a broad understanding of the principles and practical issues involved in the analysis of protein phosphorylation changes in nervous tissue and a general framework for designing and interpreting phosphorylation experiments.

Figure 1. Phosphorylation and dephosphorylation by protein kinases and protein phosphatases.

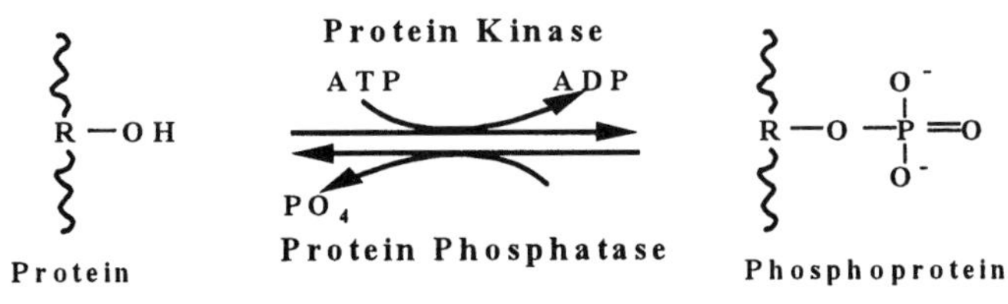

2. EXPERIMENTAL APPROACHES

Figure 2 shows the general, perhaps ideal, sequence of events when investigating the role of protein phosphorylation in neuronal function: from the original identification of the change in phosphorylation to the evidence that the change is not only associated with a physiological response but necessary for it to occur. Two kinds of evidence may be obtained to show that a physiological response produces a change in protein phosphorylation.

Figure 2. Overview of the process of investigating the role of protein phosphorylation in a physiological response.

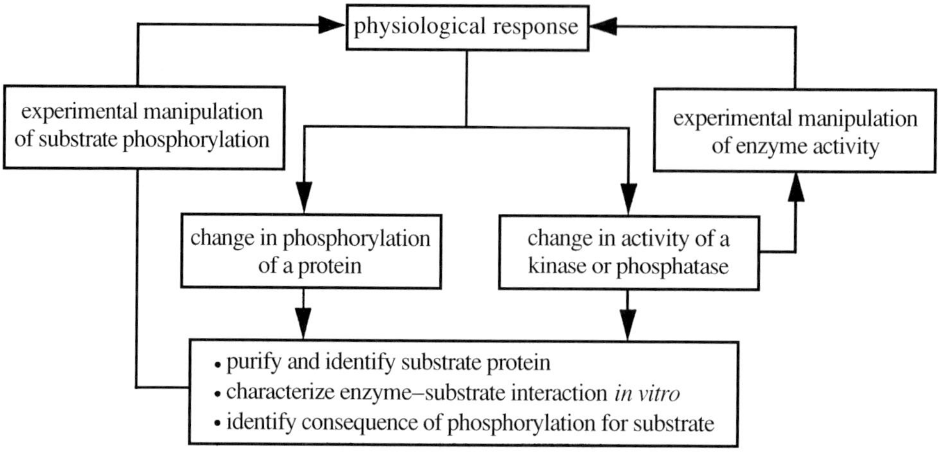

(i) If the enzyme involved is known, the activity of that enzyme can be modified *in vivo* (e.g. by injection of specific inhibitors or activators, by overexpression or gene knockout experiments) to test whether the physiological response is blocked or enhanced.

(ii) If one or more proteins have been identified whose phosphorylation is altered after the response, the following need to be established:

- the identity of the protein(s)
- the methods necessary to purify it and measure its phosphorylation in a mixture (e.g. production of a specific antibody, identification of characteristic phosphopeptides)
- the identity of the protein kinase or protein phosphatase which acts on it *in vivo* (this is usually done by initially characterizing the enzyme substrate interactions *in vitro*)
- the effect of phosphorylation on the function of the protein.

Once the above are known, the protein's phosphorylation state can be manipulated *in vivo* (e.g. by molecular biological techniques, by injecting phosphorylated or dephosphorylated purified protein or antibodies to the protein) to test whether this enhances or blocks the physiological response.

The experiments involved in such studies can be broadly categorized into five approaches based on the question being asked: these approaches are described below. This chapter focuses mainly on experimental approaches 2, 3 and 4.

Approach 1: 'Show involvement'
Aim: To establish that protein phosphorylation is involved in regulating a functional response *in vivo*

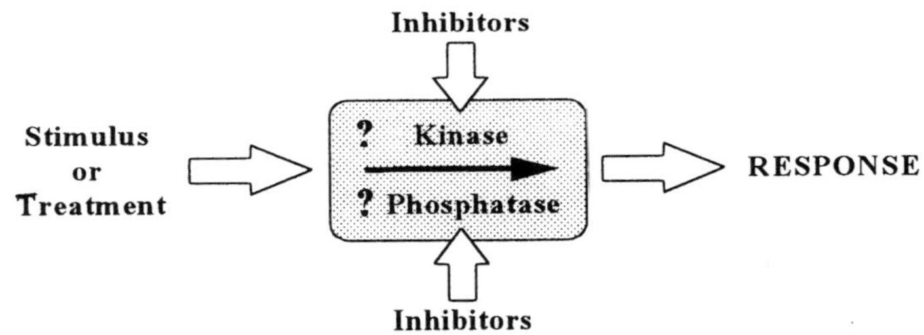

Procedure	Method	Advantages	Limitations
Elicit the physiological response in the presence and absence of inhibitors of protein kinases or protein phosphatases	1. Apply cell permeable inhibitors to cell medium or inject *in vivo* OR 2. Inject membrane impermeable inhibitors into cells	1. Does not require measurement of protein phosphorylation 2. Implicates protein phosphorylation in a process 3. Can be done at different levels of analysis (single cells, tissue slices, cell culture, whole animal)	1. Toxicity or side-effects of some inhibitors will limit their use in animals 2. Lack of enzyme specificity of some inhibitors (*Tables 1 and 2*) 3. Lack of cellular specificity with cell permeable inhibitors

Approach 2: 'Identify change'
Aim: To identify endogenous changes in kinase or phosphatase activity, or substrate phosphorylations which correlate with a physiological response *in vivo*

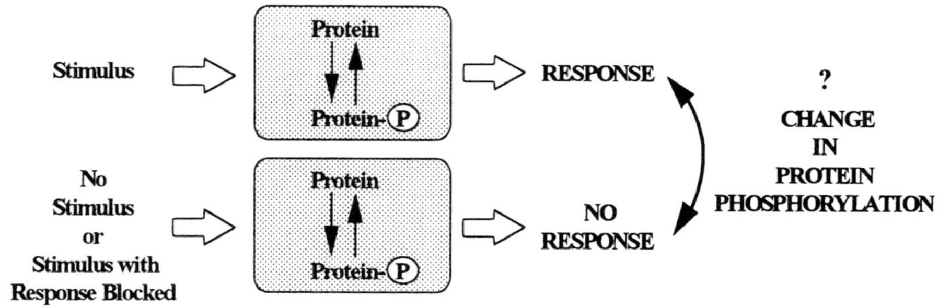

Procedure	Method	Advantages	Limitations
Compare stimulated cells with unstimulated cells or stimulated cells in which the response to the stimulus has been blocked, and measure the difference in endogenous protein phosphorylation or the activity of kinases or phosphatases	1. Measure *in vivo* phosphorylation of endogenous proteins using: – non-isotopic methods (section 10); – isotopic labeling of intact cells (section 8); – *post hoc* phosphorylation assays of lysed tissue *in vitro* (section 7) 2. Measure kinase or phosphatase activity using *in vitro* assays (section 6)	1. May identify the functional target of a phosphorylation or dephosphorylation response 2. Can correlate the degree of change in phosphorylation with the degree of change in the physiological response of the cell	1. Usually cannot be done at a single cell level except with antibodies to phosphorylated epitopes (section 10.4) 2. Phosphorylation state of endogenous proteins must be preserved (section 9)

Approach 3: 'Identify enzyme'
Aim: To identify the kinase or phosphatase that acts on a known phosphoprotein *in vivo* to regulate a physiological response

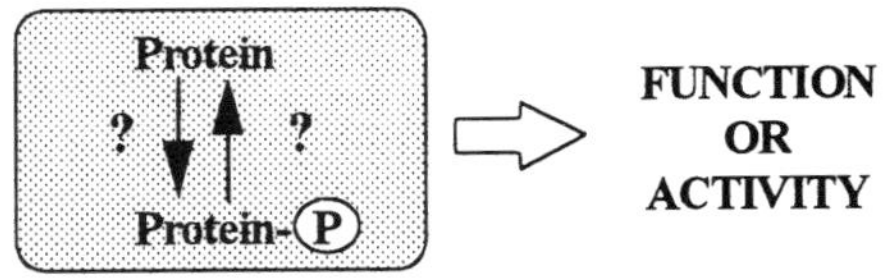

Procedure	Method	Advantages	Limitations
Modulate protein phosphorylation in intact cells and measure changes in the functional response of the cells and phosphorylation state of a particular endogenous protein	1. Modulate kinase or phosphatase activities *in vivo* using: – stimuli which produce intracellular responses that alter kinase or phosphatase activity (e.g. rise in cytosolic Ca^{2+}, inhibition of phosphodiesterase); – membrane permeable enzyme inhibitors or activators; – intracellular injection of membrane impermeable enzyme inhibitors or activators – intracellular injection or viral transfection to introduce enzymes directly 2. Establish that the phosphorylation state of the target protein has changed 3. Identify the enzyme responsible for the change using Approaches 1 and 5	1. Can be done in intact cells 2. Can be done in fragments of tissue (e.g. membrane patches) if the function of the protein can be measured in these fragments 3. Can correlate degree of change in phosphorylation of a substrate protein with the degree of change in the functional property of the protein	1. Assumes that a protein suspected of being regulated by phosphorylation has been identified (e.g. ion channel or enzyme) and methods for measuring its function are available (e.g. patch clamp recording or enzyme activity) 2. If the identified protein has no measurable functional activity, the change in phosphorylation can only be correlated with the cellular response and the causal link between the two is only implied 3. If the identified protein is of low abundance, changes in phosphorylation state will be difficult to measure unless the protein can be purified and concentrated from the cells

Approach 4 : 'Identify substrate'
Aim : To identify the target substrate for an enzyme *in vivo* that has been strongly implicated in, or is known to be responsible for, the regulation of a physiological response

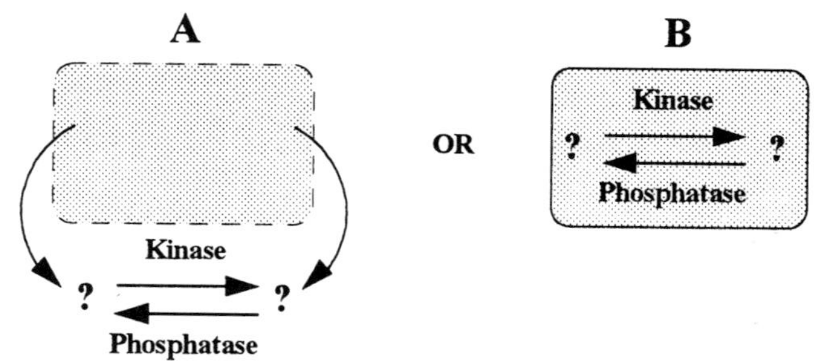

Procedure	Method	Advantages	Limitations
A. In lysed tissue			
Identify potential substrates for a specific enzyme known to be involved in a response *in vivo*	1. Lyse cells with or without fractionation of the proteins 2. Phosphorylate *in vitro* (section 6.1) using either: – the activity of the endogenous kinase by selectively activating it using specific combinations of activators and inhibitors – purified exogenous kinases in combination with specific inhibitors to inactivate endogenous enzymes	1. Identifies a range of potential substrates 2. Can identify substrates for low abundance enzymes (if exogenous enzymes are used for *in vitro* labeling) 3. This approach has a high sensitivity when combined with fractionation of proteins before labeling	1. Need to have identified the kinase or phosphatase involved in the response 2. Need purified enzymes or ways of ensuring the specific activation of endogenous enzymes 3. Need to confirm that a potential substrate identified in this way is phosphorylated *in vivo*

	3. For phosphatases take proteins phosphorylated as in 2 and measure dephosphorylation using: – endogenous phosphatases using specific combinations of activators and inhibitors – purified exogenous phosphatases in combination with specific inhibitors to inactivate endogenous enzymes 4. SDS-PAGE and autoradiography or other detection method		
B. In intact tissue Identify endogenous proteins whose phosphorylation is changed following the increase or decrease in the activity of a specific enzyme *in vivo*	1. Alter the activity of a specific enzyme *in vivo* by: – intracellular injection of purified enzyme or specific inhibitors or activators – cell permeable agents that lead to specific increases or decreases in the activity of the target enzyme 2. Measure changes in phosphorylation of endogenous proteins by SDS-PAGE and autoradiography or other detection method	1. Can identify substrates *in vivo* for enzymes known to be involved in a response 2. Can show that potential substrates identified using lysed tissue are phosphorylated *in vivo*	1. Need to have identified the kinase or phosphatase involved in the response 2. Substrate must be abundant enough to be detected in whole cells 3. Need to show that the change in phosphorylation correlates with the change in cellular response (Approach 2)

Approach 5: 'Characterize *in vitro*'

Aim : To determine the specific sites at which a particular enzyme acts on a particular protein and how phosphorylation at that site affects the function of the protein

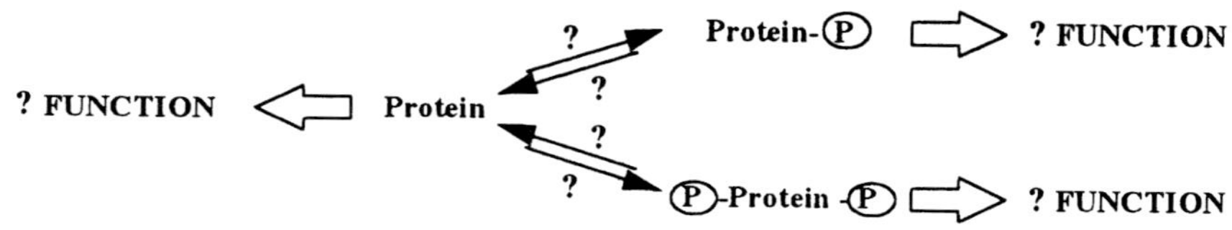

Procedure	Method	Advantages	Limitations
In vitro analysis of enzyme–substrate interactions	1. Using purified substrates and enzymes reaction conditions are extended to ensure phosphorylation by kinases to high stoichiometry (see section 3.1) for quantitative removal of phosphate by phosphatases 2. Site specificity of individual enzymes is determined by site analysis (section 3.1) 3. The functional effect of phosphorylation at a particular site is determined	1. Provides unambiguous proof that a particular enzyme implicated in the phosphorylation of a particular protein *in vivo* can act on that protein 2. Stoichiometry and site analysis can be determined for individual enzymes 3. Functional effect of phosphorylation at individual sites can be measured	1. Requires purification of substrates and enzymes 2. Often requires sophisticated and special techniques for peptide analysis, protein sequencing or enzyme kinetics

3. CHANGES IN PROTEIN PHOSPHORYLATION: GENERAL CONSIDERATIONS

3.1. Changes in endogenous substrate phosphorylation

When interpreting the effects of phosphorylation on an endogenous protein, three issues need to be addressed:

(i) the stoichiometry of phosphorylation (the number of moles of phosphate bound per mole of protein);
(ii) the sites (specific amino acid in the protein) on which the phosphorylation is occurring; and
(iii) the consequences of the phosphorylation in terms of the function of the protein being phosphorylated.

Stoichiometry of phosphorylation

- This can never be more than the number of phosphorylation sites in a protein and is usually less because of phosphatase activity in tissues and because, in multiple phosphorylated proteins, phosphorylation at one site often inhibits phosphorylation at another site.
- In test tube experiments with pure substrates and kinases, phosphorylation of a site at low stoichiometry (much less than 1 mole phosphate per mole protein) is generally regarded as having little functional significance.
- In intact tissues low stoichiometry phosphorylation may be functionally significant because:
  - biological regulation often involves subtle changes in the relative activities of competing or antagonistic pathways and therefore requires changes in only a small proportion of proteins;
  - tissue heterogeneity, which is particularly high in nervous tissue, can make a high stoichiometry phosphorylation that occurs in a small proportion of cells in a tissue appear like a low stoichiometry phosphorylation overall (e.g.. a substrate found in many cells may be phosphorylated to high stoichiometry at a particular site in only a few cells because the kinase is activated or present only in those cells).

Site of phosphorylation – significance of site analysis

- In general, phosphorylation of a protein at a different site has a different functional consequence and hence a different biological significance; i.e. for proteins that can be phosphorylated at multiple sites, it is biologically important to determine which site is phosphorylated under various conditions.
- In many cases kinases and phosphatases are selective for certain sites on a protein.
- Low stoichiometry changes are often impossible to detect without site analysis. Experiments that limit their analysis to the total level of phosphorylation of a particular protein may miss significant biological changes unless the protein is known to have only one phosphorylation site.

Methods for site analysis

- This is usually done by purifying the labeled protein (e.g. as a band on an SDS polyacrylamide gel or as an immunoprecipitate), partially digesting with specific proteases, separating the peptides containing the different sites and measuring the level of radioactivity in each. The two most widely used methods involve:
  - HPLC separation of the peptides coupled with on line detection of radioactivity;
  - SDS-PAGE of the peptides coupled with autoradiography. The latter is particularly versatile because it can be used as part of a two dimensional technique to investigate changes in the phosphorylation of many proteins at once (1, 8). Descriptions of these techniques are given in ref. 2.

Functional consequences of phosphorylation

- This can only be determined once the identity and normal function of the phosphorylated protein is known.

3.2. Changes in kinase or phosphatase activity

When the induction of a physiological change or application of a stimulus produces a change in protein kinase or protein phosphatase activity, three types of change could occur:

(i) transient change in activity;
(ii) stable change in activity; or
(iii) change in enzyme concentration.

Transient change in activity

- Can be caused by a transient increase in the concentration of an intracellular activator (e.g. cyclic nucleotides or Ca^{2+}).
- The activation may be so brief that enzyme activity is back to normal levels by the time the treated tissue is taken for analysis. Even if the tissue can be disrupted while the activity *in vivo* is elevated, such changes in kinase or phosphatase activity will not be detected in *post hoc* assays *in vitro* because the intracellular activator concentration will not be maintained in disrupted tissue.
- Changes in the phosphorylation level of endogenous substrates usually provide the only way of measuring transient enzyme activation.

Stable change in activity

- Many kinases and phosphatases are regulated by molecular modifications which can be rapid in onset but relatively stable in duration (e.g. autophosphorylation, phosphorylation by other kinases, dephosphorylation, translocation or proteolysis).
- Changes of this type generally survive tissue disruption, provided appropriate precautions are taken (see section 9), and the consequent changes in activity can be measured in disrupted tissue *in vitro* (*post hoc* assays).

Change in concentration

- A change in enzyme concentration can be confirmed in disrupted tissue by measuring either the levels of the protein using immunoassays or enzyme activity under maximally activated conditions (see section 6.1).
  Increases:
  - involve protein synthesis and generally require hours to occur;
  - are only of potential importance in experiments where the treatment (e.g. chronic exposure to a drug) or the induced change (e.g. long-term potentiation) is sufficiently long lasting.

  Decreases:
  - may occur quickly if a specific protease is activated;
  - in practice, however, brief treatments that reduce enzyme activity usually do so by inhibition rather than degradation.

3.3. Deciphering changes occurring in cells

When a change in protein phosphorylation is known to occur in a cell, the change could be due to an alteration in one or more of the following:

(i) the activity or concentration of one or more protein kinase;
(ii) the activity or concentration of one or more protein phosphatase;

(iii) the concentration of one or more endogenous substrate protein;
(iv) the sites being phosphorylated on the same endogenous substrate protein.

Distinguishing between these mechanisms usually requires a number of different approaches involving studies in intact tissue as well as *post hoc* assays on disrupted tissue.

4. USE OF INHIBITORS

In vivo roles of various kinases and phosphatases can be investigated using specific cell permeable inhibitors to treat intact cells (see *Tables 1 and 2*) and measure the functional consequences. However, because of some overlap in specificity of most if not all of these inhibitors, results need to be interpreted with some caution. The majority of kinase inhibitors target the ATP binding sites which are considerably homologous between different kinases. Similarly, all known protein phosphatases have considerable homology such that no inhibitor is absolutely specific. Thus, while the specificity of a particular inhibitor can be controlled *in vitro* by using it at different concentrations, this is more difficult to achieve *in vivo*. Subcellular localization of the kinase/phosphatase of interest as well as variable tissue permeability of the inhibitors must also be taken into consideration when interpreting such data. Non-permeable inhibitors can also be used in permeabilized cell systems or if they are micro-injected into individual cells; in the latter approach, the effects of the inhibitor in a tissue can be confined to the injected cell.

In vitro these inhibitors are useful in situations where substrates or conditions are not absolutely selective for the kinase/phosphatase being investigated. Examples are the use of the PKA inhibitor (PKI) to inhibit PKA when activating PKG with cGMP(because cGMP also activates PKA) and the use of inhibitor-1 to inhibit protein phosphatase 1 when measuring PP2A activity towards glycogen phosphorylase (because glycogen phosphorylase can also be dephosphorylated by PP1).

5. USE OF ISOTOPES

Measurement of phosphate incorporation into proteins is most routinely done using isotopically labeled phosphate incorporated at the γ position in ATP. The most favored isotope is ^{32}P. ^{32}P is a high-energy β-emitter, such that incorporation of very small amounts of labeled phosphate can be readily detected by autoradiography or scintillation counting. Indeed the nature of this isotope is such that photon emission can be detected in the absence of added scintillant (Cerenkov counting) using the ^{3}H-channel, although the efficiency of counting is reduced. The disadvantages of using ^{32}P are primarily its short half-life (14.29 days), requiring careful planning of experiments, and the potential hazards from using such a high-energy β-emitter. It should be noted, however, that once appropriate handling and isolation precautions are exercised, dangers from exposure to ^{32}P are minimal. Furthermore, these high-energy isotopes can now be supplied with an inert colored marker such that distribution and contamination can be monitored visually.

An alternative isotope often used is ^{35}S. In this case one of the oxygen atoms on the γ-phosphate of ATP is replaced by a sulfur atom, so called adenosine trithiophosphate (ATPγS). The use of the ^{35}S isotope has the advantages of safety, being a considerably lower energy emitter, and longevity, having a half-life of 87.4 days. The primary disadvantage of the use of ^{35}S is in autoradiography where exposure times are necessarily much longer. It must also be noted that ATPγS is not an ideal phosphate donor in protein phosphorylation reactions and the kinetics of thiophosphorylation are considerably different from those using ATP. Typically, phosphorylation reactions are much slower using ATPγS. Furthermore,

Table 1. Some useful protein kinase inhibitors

Inhibitor	Enzyme specificity	Ki/IC50	Membrane permeable	Comments	Reference
K252a	CaMPK-II>PKA>PKG>PKC	10 nM	Yes	Potent, general kinase inhibitor	11
KN62	CaMPK-II	1 μM	Yes	Calmodulin binding site antagonist; use KN04 as inactive analog	12
KN93	CaMPK-II	0.4 μM	Yes	As KN62; use KN92 as inactive analog; note KN62 and KN93 also block calcium channels	13
CaMPK-$II_{281-302}$	CaMPK-II>PKC	2 μM	No	Pseudosubstrate inhibitor; degree of selectivity is controversial (48); Ala^{286} substituted autophosphorylation site prevents inhibitor being phosphorylated	14
PKI_{5-22}	PKA	2 nM	No	Walsh inhibitor peptide (Wiptide)	15
RpCAMPS	PKA	11 μM	Yes		16
H7	PKA=PKG=PKC>CaMPK-II	3 μM	Yes		17
H8/H9	PKG=PKA=PKC	0.5 μM	Yes		18
H89	PKA≫PKG>CaMPK-II>PKC	50 nM	Yes	Use H85 as inactive analog	19
KT5823	PKG≫PKC>PKA	200 nM	Yes		20
Gö6976	PKC	8 nM	Yes	Selective for α,β1 isoforms; use Gö6352 as inactive analog	21
Calphostin C	PKC>PKG>PKA	50 nM	Yes	Regulatory domain inhibitor	22
Bisindolylmaleimide (I–IV)	PKC>PKA	5–70 nM	Yes	Isoform specific variants; use bisoindolmaleimide V as inactive analog	23
PKC_{19-36}	PKC	147 μM	No	Pseudosubstrate inhibitor	24
Staurosporine	PKC=PKA=PKG=CaMPK-II	10 nM	Yes	Non-specific kinase inhibitor	25

Notes:
(i) Ki/IC50 given is for first named enzyme.
(ii) Non-specific actions of inhibitors must be considered.
(iii) Specificity of response must be determined using a range of inhibitors with different structures or mechanism of inhibition.
(iv) Wherever possible, use inactive analogs as controls.

Table 2. Some useful protein phosphatase inhibitors

Inhibitor	Enzyme specificity	Ki/IC50	Membrane permeable	Comments	Reference
Okadaic acid	PP2A>PP1≫PP2B	1 nM	Yes	Use Norokadone as analog	26
Calyculin A	PP2A=PP1≫PP2B	1 nM	Yes		27
Tautomycin	PP1>PP2A≫PP2B	1 nM	Yes		28
Microcystin LR	PP2A=PP1≫PP2B	0.1 nM	No	Algal toxin; >50 structural variants	29
Nodularin	PP2A>PP1≫PP2B	0.1 nM	No	Algal toxin	30
Cantharidic acid	PP2A	50 nM	Yes	Pesticide; specificity for phosphatases doubtful	31
Inhibitor-I/DARPP-32	PP1	4 nM	No	PP1 inhibitor protein; activated by PKA phosphorylation	32
Phospho I-$1_{9–41}$	PP1	4 nM	No	PP1 inhibitor peptide	33
Inhibitor-2	PP1	12 nM	No	PP1 inhibitor protein	32
Cyclosporin A	PP2B	33 nM	Yes	Immunosuppressant; requires endogenous cyclophilins	34
FK506	PP2B	32 nM	Yes	Requires endogenous FK binding protein; use rapamycin as inactive analog	34
CaN $A_{467–492}$	PP2B	25 µM	No	Autoinhibitory peptide	35
Sodium fluoride	All	20 mM	No	Non-specific; reversible by manganese	36
Sodium molybdate	All	20 mM	No	Non-specific	36
Sodium pyrophosphate	All	5 mM	No	Non-specific	36

Notes: (i) Ki/IC50 given is for first named enzyme.
(ii) Non-specific actions of inhibitors must be considered.
(iii) Specificity of response must be determined using a range of inhibitors with different structures or mechanism of inhibition.
(iv) IC50 values given for inorganic phosphatase inhibitors represent appropriate concentrations to use rather than actual IC50.
(v) Wherever possible, use inactive analogs as controls.
(vi) Ki/IC50 for phosphatase inhibitors varies according to phosphatase concentration.

protein-thiophosphate ester bonds formed during phosphorylation reactions are very resistant to protein phosphatase attack such that the turnover of such thiophosphate is minimal. This characteristic is nevertheless often advantageous in situations where maintenance of the phosphorylation state of a given protein is required.

Recently an alternative phosphorus isotope, ^{33}P, has become available. This isotope has considerably lower energy than ^{32}P and a half-life of 25.4 days. The principal advantage of this isotope over ^{32}P is in the area of safety and longevity. However, because of the lower energy, higher specific radioactivity is necessary to achieve the same sensitivity in phosphorylation reactions.

6. *IN VITRO* PHOSPHORYLATION/DEPHOSPHORYLATION EXPERIMENTS

Phosphorylation in lysed tissue with radiolabeled ATP *in vitro* can be used to:

(i) identify endogenous substrates for an endogenous kinase or phosphatase;
(ii) identify the endogenous kinases or phosphatases which label each site on a protein;
(iii) measure the activity of protein kinases or phosphatases in a tissue;
(iv) measure the direct (i.e. not intact cell mediated) effects that agents such as drugs can have on protein kinase or phosphatase activity.

6.1. Substrate identification

The use of tissue extracts to identify substrates (Approach 4, section 2) for a particular protein kinase or protein phosphatase requires several conditions to be met. If not, failure to detect substrates using this method does not mean absence of a substrate.

Basic method

- Protein kinase substrates are identified by an increase in radiolabel.
- Protein phosphatase substrates are identified by prior phosphorylation and monitoring the decline in radioactivity.

Typical approach

- Use specific activators or inhibitors (see section 4) to favor the enzyme of interest and compare with results in the absence of exogenous activator/inhibitor. (It should be noted that positive results obtained using this approach are not absolute proof and purification of the relevant substrate will ultimately be required. This is because complete inhibition of all endogenous kinases cannot be achieved and it is possible that activation of a particular kinase does not lead to the direct phosphorylation of the substrate by that kinase but facilitates the phosphorylation of the substrate by another endogenous kinase. For example activation of PKA by elevation of cAMP in hepatocytes leads to the phosphorylation of acetyl CoA carboxylase by the AMP-activated protein kinase, itself regulated by protein phosphorylation.)
- Ultimately the 'characterization *in vitro*' approach (section 2) must be used.

Essential criteria

- The kinase or phosphatase must be present at sufficient concentration and activity to be detected; if not, purified exogenous enzymes must be used.
- Substrate proteins must be present at a high enough copy number to be detected by the assay method used.
- Substrate proteins must have sufficiently low levels of *in vivo* phosphorylation so that vacant sites are available for labeling *in vitro*.

Limitations

- Subcellular fractionation may lead to artefactual translocation of kinases, phosphatases or substrates such that detection of phosphorylation in the disrupted tissue does not necessarily imply that the protein and the kinase or phosphatase are co-localized in the intact cell.
- Many substrates are phosphorylated or dephosphorylated by more than one enzyme; thus in this approach, it is essential to maximize conditions for the enzymes of interest and minimize it for others.
- Measurement of protein phosphatase activity in this way requires prior phosphorylation in such a way that all the radioactive ATP is used up. This can be done by either designing phosphorylation reactions to completely utilize the ATP or to quench the radiolabeled ATP by adding a vast excess of non-radioactive ATP. In this case the turnover of phosphate continues but the loss of radioactivity can be detected and used as a measure of phosphatase activity.

6.2. Measurement of protein kinase activity

Once a particular protein has been identified as a substrate for a given protein kinase, this protein can be used to measure the activity of that protein kinase in a tissue extract or subcellular fraction. Either the endogenous protein or exogenously added purified protein can be used. Synthetic peptides based on the phosphorylation sites (if known) may be used. A number of options and considerations exist. Detailed interpretation of results is given in *Table 3*.

Endogenous substrates (see Protocol 13.2)

- Least preferred option because concentration of substrate is unknown and usually not in great excess compared to kinase.
- Most useful when specific to a particular kinase.
- Results confounded by multiple phosphorylation sites.

Exogenous protein substrates (see Protocol 13.1)

- Results confounded by multiple phosphorylation sites.
- Large proteins may have accessibility problems in some subcellular fractions, particularly membrane fractions.

Exogenous peptide substrates (see Protocol 13.1 and *Table 4*)

- Concentration can be controlled and saturating concentration can be guaranteed.
- Can be custom designed for kinase of interest.
- Can be designed to contain one or more phosphorylation sites.
- Greater susceptibility to proteolysis must be considered.

Activity vs concentration

- Under fully optimized conditions, the activity of an enzyme may be used as a measure of its concentration.
- Using activity as a measure of concentration does not take into account pools of inactivated or inhibited enzyme.
- Can only use activity to measure concentration by using an exogenous substrate because substrate concentration must be in excess.

Optimization

- Conditions cannot be optimized for all kinases in one assay using endogenous substrates:
  - use separate assays for each kinase;
  - use compromised set of conditions – see section 13.2.

Table 3. Interpretation of *post hoc* phosphorylation results[a]

Protein phosphorylation change *in vivo*				***Post hoc* phosphorylation change**	
Endogenous substrate	**Endogenous kinase**	**Endogenous phosphatase**	**Substrate stoichiometry**	**Endogenous substrate**[c]	**Exogenous substrate**
Changes in kinase or phosphatase activity					
N	Increase	N	Increase	Decrease	N or increase[b]
N	N	Increase	Decrease	Increase	N
N	Decrease	N	Decrease	Increase	N or decrease[b]
N	N	Decrease	Increase	Decrease	N
Decrease in substrate or kinase concentration					
Decrease	N	N	Increase	Decrease	N
N	Decrease	N	Decrease	Increase	Decrease
Increase in substrate or kinase concentration					
Increase	N	N	Decrease	Increase	N
N	Increase	N	Increase	Decrease	Increase

[a]The table describes an endogenous substrate with one phosphorylation site acted on by one kinase and one phosphatase and shows the types of *in vivo* change which can alter the stoichiometry of phosphorylation of this substrate following some treatment of intact tissue. The *post hoc* phosphorylation changes *in vitro* describe the differences obtained between pre- and post-treatment tissue in the ability to incorporate radioactivity into either the endogenous substrate or a suitable exogenous substrate. N = no change

[b]If the change in kinase activity is due to a change in the enzyme which is stable enough to survive the tissue lysis (e.g. due to autophosphorylation or translocation) the change can be measured *post hoc* using exogenous substrate. If the change in activity is due to a transient intracellular change (e.g. due to a change in the concentration of some intracellular second messenger which will be diluted during tissue lysis) *post hoc* assays using exogenous substrate will show no change.

[c]It may be necessary to add exogenous kinase or to extend the period of labeling in order to see this change if the substrate is present at saturating concentrations in the tissue compared with the kinase activity.

Endogenous activators

- Fully optimized conditions may mask the presence of endogenous activators:
  - use sub-optimal conditions.

Time course

- Linear reaction rates are essential for accurate activity measurements.
- Linearity is influenced by:
  - substrate concentration (must be in excess);
  - ATP concentration (must be in excess throughout experiment);
  - reaction temperature (must be constant);
  - endogenous inactivating factors such as proteolysis or dephosphorylation particularly with extended incubation times (use an inhibitor cocktail, section 9);
  - heat denaturation even at mild temperatures such as 37°C (avoid pre-incubation);
  - the time taken for tissue to warm up to incubation temperature (needs >5 s for most samples).

6.3 Measurement of protein phosphatase activity

Once a particular protein has been identified as a substrate, this protein can be used to measure activity of that protein phosphatase in a given tissue extract or subcellular fraction.

Table 4. Useful peptide substrates for serine/threonine protein kinases and phosphatases

Peptide sequence	Name	Enzyme specificity	Comments	Reference
LRRASLG	Kemptide	PKA>PKG		37
TRSAIRRASTIEMP	PL_{8-21}	PKA/PKG/PKC/CaMPK-II	$Phospholamban_{8-21}$	38
QKRPSQRSKYL	MBP_{4-14}	PKC	Myelin basic $protein_{4-14}$	39
PLSRTLSVAAKK	GS_{1-12}	PKC (also PP2A)	Glycogen $synthase_{1-12}$	40
AAKIQASFRGHMARKK	NG_{28-43}	PKC	$Neurogranin_{28-43}$	41
FKKSFKL	$MARCKS_{159-165}$	PKC	MARCKS $protein_{159-165}$	42
PLARTLSVAGLPGKK	Syntide 2	CaMPK-II	Glycogen synthase phosphorylation site 2	43
KKALRRQETVDAL	Autocamtide 2	CaMPK-II	Based on CaMPK-II autophosphorylation site	44
RRATVA		PP2A>PP2C	Phosphorylate with PKA	45
RRREEETEEEAA		PP2A	Phosphorylate with PKA	46
DLDVPIPGRFDRRVSVAAE	RII_{81-99}	PP2B>PP2A>PP2C	Phosphorylate with PKA	47

Notes: (i) Concentration of peptide used depends on enzyme source and must be experimentally determined to be in excess.
(ii) Protein phosphatase substrates must be prior phosphorylated to maximal stoichiometry using long incubation times and excess kinase. ATP must also be removed prior to assay.

The protein is first phosphorylated by the appropriate kinase using ATP (usually radiolabeled) and the phosphatase activity determined by measuring the phosphate released – see Protocol 13.4. A number of options and considerations exist.

Choice of substrates

- There are few unique substrates for individual phosphatases:
  - use inhibitors.
- Phosphatase activity is dependent upon the substrate used. Thus, measurement of high or low activity with one substrate does not necessarily mean the same for other substrates:
  - use substrate appropriate to function being investigated.
- Native proteins are better substrates than peptides. See *Table 4* for some useful peptides.

Chelators and reducing agents

- Phosphatase activity is particularly sensitive to the buffer conditions:
  - standardize for all procedures.
- Inclusion of BSA to chelate non-specific inhibitors is essential.
- Inclusion of reducing agents to prevent disulfide bond formation is essential.

Sample dilution

- Increasing sample dilution actually increases phosphatase activity/mg protein but decreases the total dephosphorylation measured:
  - choose compromise conditions which reliably and reproducibly measure the maximal possible activity/mg protein.

Time course

- Protein phosphatase activity is only linear up to 30% dephosphorylation.
- Linearity is influenced by:
  - substrate concentration (must be in excess);
  - reaction temperature (must be constant);
  - endogenous inactivating factors such as proteolysis (use an inhibitor cocktail, section 9);
  - heat denaturation even at mild temperatures such as 37°C (avoid pre-incubation);
  - the time taken for tissue to warm up to incubation temperature (needs >5 s for most samples).

Proteolysis of substrate

- May contribute to false positive measurements using the TCA precipitation method by failure to precipitate small peptide fragments containing the phosphorylated residue:
  - use acid/molybdate extraction (section 13.4).

Freezing

- Protein phosphatase activity can be measured after freezing samples.
- Freezing does however lead to changes in PP1/PP2A ratio.
- Activity may become dependent upon Mn^{2+}.
- For accurate measurement use fresh tissue.

7. *POST HOC* PHOSPHORYLATION: DECIPHERING CHANGES OCCURRING IN CELLS

The aim of *post hoc* phosphorylation experiments is to infer what protein phosphorylation change occurred *in vivo* from what change in protein phosphorylation can be demonstrated in lysed tissue *in vitro*. *Post hoc* phosphorylation studies involve experiments where radiolabeled ATP is added to a mixture of substrate proteins and kinases *in vitro*. Because the radioactive phosphate can only be incorporated at vacant phosphorylation sites, the incorporation obtained in these *post hoc* studies is dependent upon the initial phosphorylation state of the protein and the prevalent protein phosphatase activity. In the absence of any protein phosphatase activity, *post hoc* protein phosphorylation is inversely proportional to the phosphorylation state *in vivo*.

The consequences of giving a cell a stimulus which activates kinases or phosphatases are shown in *Figure 3* using as an example an endogenous protein which can be phosphorylated at three different sites. The effects of the stimulus can be deduced by comparing the results of *post hoc* protein phosphorylation experiments on cells in the resting and post-stimulus states (Identify change or Identify substrate approaches, section 2). If the tissue is lysed under appropriate conditions (see section 9) the molecules of the endogenous substrate which were phosphorylated *in vivo* will retain their bound phosphate. Unless the experiment involves preincubation of the tissue with radioactive phosphate (see section 8) this phosphate attached *in vivo* is not radioactive. *Figure 3* shows the results of *post hoc* phosphorylation experiments in disrupted tissue from the resting, dephosphorylated- and phosphorylated-states.

If only a single kinase is activated during the *in vitro* assay (e.g. by addition of a selective activator such as cAMP or a purified active kinase) there will be an increase in the phosphorylation of the protein at the site for which the kinase is selective (in *Figure 3* the

Figure 3. Changes in phosphorylation of a protein during a *post hoc* assay *in vitro* where a single kinase has been activated. The diagram shows eight molecules of a protein which contains some non-radioactive phosphate (white circles). During the assay both radioactive (black circles) and non-radioactive phosphate can be added.

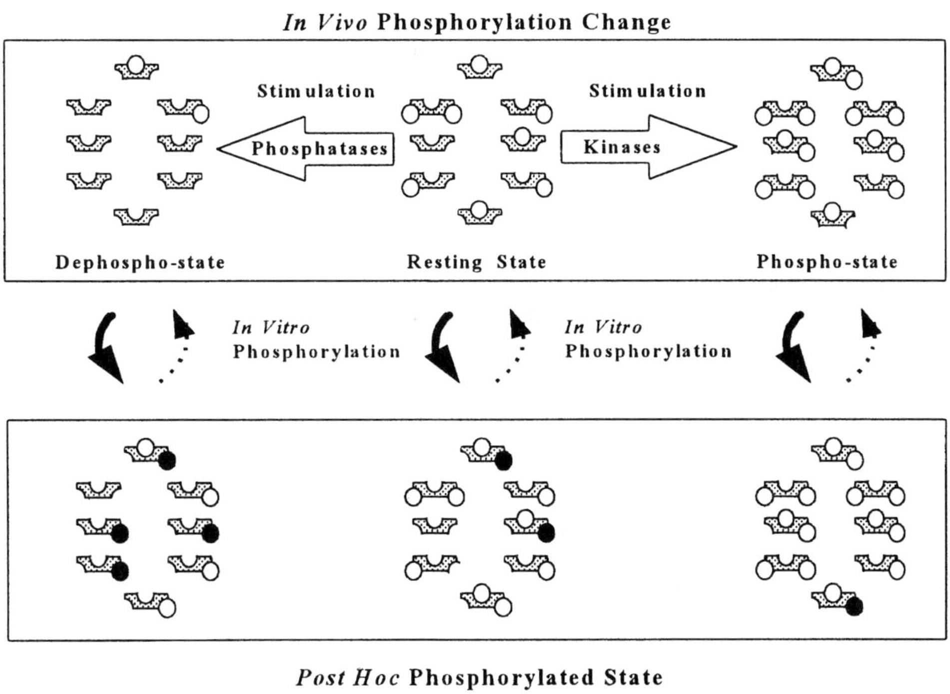

kinase is selective for the site on the right). Such selective kinase activation can seldom be achieved without extraction of the substrate or inactivation of some or all endogenous kinases (see below). If many kinases are activated during the *in vitro* assay radioactivity will become incorporated into more than one site and the interpretation of the results will be more difficult. The difficulty of providing specific activation in the *in vitro* labeling part of the *post hoc* assay means that few attempts to decipher *in vivo* changes from *post hoc* labeling experiments have succeeded when the whole tissue lysate is used. Phosphatase activities play a very minor role in such assays because incubation times are generally short (less than 1 minute) providing little opportunity for them to remove radioactive phosphate that has been added by the kinases. Phosphatases may remove some of the non-radioactive phosphate that was present *in vivo* but this would only be noticed in the assay as an opportunity to label another site which had previously been occupied by non-radioactive phosphate.

To illustrate how it is possible to infer what happened to the endogenous substrate during the treatment of the cell from the difference between the *post hoc* labeling patterns in the pre- and post-treatment states, let us first consider the stimulus which activated phosphatases. In this case, because the increased phosphatase activity *in vivo* has made more sites available for labeling *in vitro*, the substrate from the post-treatment tissue will show increased *post hoc* labeling under conditions which stimulate any of the kinases which phosphorylate this protein. Increased *post hoc* labeling by more than one kinase would suggest that the dephosphorylation *in vivo* occurred at more than one site and site analysis would reveal that dephosphorylation had occurred at all three sites to a similar extent. The stimulus which activated kinases *in vivo*, resulted in increased phosphorylation at all sites – and the right hand site in particular – so that there is a greatly decreased incorporation of radioactive phosphate in the *in vitro post hoc* assay.

Table 3 shows how different changes in protein phosphorylation *in vivo* would be reflected in the results of a radioactivity-based *post hoc* assay. In practice, experiments involving short-term changes lasting seconds to minutes (e.g. depolarization, receptor activation) produce transient or stable activation or inhibition of kinases and/or phosphatases. In some cases a short stimulation may activate proteases that could degrade substrates or enzymes leading to a rapid drop in their concentration. Increases in substrate and enzyme concentrations are only potentially important in experiments where the treatment or the induced change is sufficiently long lasting to allow protein synthesis to occur.

If the changes in enzyme activity are transient, distinguishing between an activation of a kinase and an inhibition of a phosphatase or inhibition of a kinase and activation of a phosphatase (*Table 3*) requires additional strategies such as the use of inhibitors in the intact cell. If changes in the kinase activity are stable, *post hoc* assays using exogenous substrates can distinguish between kinase and phosphatase effects. Similarly, if stable changes in phosphatase activity are produced these can be measured in *post hoc* assays using exogenous labeled substrates (endogenous substrates cannot be used unless the tissue has been prelabeled with radioactive phosphate).

There is a specialized version of the *post hoc* assay called 'back-phosphorylation' which was popularized for measuring the phosphorylation of the acid-soluble synaptic vesicle protein synapsin-1 (3). In this approach, intact tissue is disrupted taking care to inhibit or abolish all phosphatase activity, and the endogenous substrate protein of interest is purified from the treated tissue. This purified protein is then incubated *in vitro* with excess quantities of exogenously added, purified protein kinase specific for one site on the substrate. The assay works just like that illustrated in *Figure 3* except that all vacant sites can be labeled during extensive incubations because of the excess amount of kinase and the absence of phosphatase activity. In this way one can make quantitative assessments of the degree of change in the phosphorylation of a particular protein. If the added kinase is specific for one site on the protein, this obviates the need for site analysis. This technique has found limited application because:

(i) it can only be used if the phosphorylation sites on the substrate and their kinase specificity are known;
(ii) if the substrate can be purified from tissue quickly in a way that ensures no contamination with endogenous phosphatases or kinases; and
(iii) if sufficient purified kinase(s) specific for one or more of the sites on the substrate is available.

8. PHOSPHORYLATION IN INTACT TISSUE

The alternative to the indirect *post hoc* assay is to analyse directly changes in protein phosphorylation in intact tissue. For studies of tyrosine phosphorylation, the most common method is the use of antibodies to phosphotyrosine (see section 10), but for studies of the phosphorylation of serine and threonine sites, experiments usually involve radioactivity. The principles are indicated in *Figures 4 and 5*.

8.1. Labeling intact tissues

The most common method involves incubating functional, intact tissue with exogenous radioactive phosphate which is taken up by the cells using the normal phosphate transport

Figure 4. Incorporation of radioactive phosphate into proteins by intact cells. Radioactive phosphate (black circles) is taken up by cells and mixed with their own non-radioactive phosphate (white circles). Intracellular mitochondria synthesize a mixture of radioactive and non-radioactive ATP which is used by the cell to phosphorylate proteins. Phosphatase activity can recycle the radioactive phosphate.

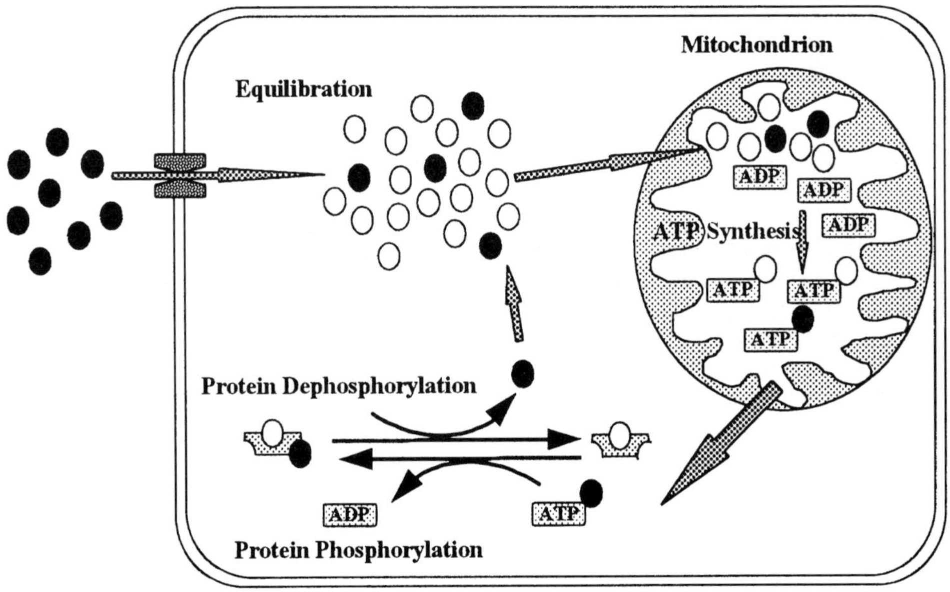

Phosphorylation

systems. Once inside it equilibrates with the intracellular pool of phosphate and becomes incorporated into radioactive ATP (*Figure 4*) as well as other phosphate containing molecules such as phospholipids. Because of the high turnover rate of ATP, the efficiency of labeling of this compound is relatively high. As the intracellular pool of ATP becomes radioactive, endogenous phosphoproteins become labeled with radioactive phosphate owing to the basal activity of protein kinases and phosphatases. Once steady-state labeling (or something approaching it) has been achieved, the intact cells can be given a stimulus (e.g. depolarization, drugs, synaptic stimulation) or their function can be manipulated (e.g. by the intracellular injection of specific enzyme inhibitors or activators or the active enzymes themselves) and the effects of these manipulations on the phosphorylation of specific proteins can be measured. In practice true steady-state labeling can only be achieved in culture where the tissue can be maintained in a viable state for many hours. Fortunately, if the experiment is designed to study the effects of a brief stimulation, it is not necessary to have reached a true steady state. To maximize the incorporation of radioactivity, experimenters often use an extracellular medium that has no added non-radioactive phosphate. However, this results in depletion of the intracellular phosphate and compromises long-term tissue viability. If necessary, a low amount (<0.5 mM) of phosphate can be included in the medium provided that a greater amount of radioactive phosphate can be used to maintain the degree of labeling. A protocol for labeling synaptosomes is given in section 13.3.

8.2. Analyzing protein phosphorylation in intact cells

Advantages

- Only substrates normally available to kinases *in vivo* are labeled – fewer false positives.
- The anatomical specificity of a response in a tissue can be demonstrated where the precise location (cellular or subcellular) of a particular substrate is already known.
- Specifically localized substrates (cellular or subcellular) may be used to measure localized changes in widely distributed kinases or phosphatases.
- Effects of both brief (e.g. synaptic activity) and prolonged (e.g. hormone action) stimulation can be studied.
- Both protein kinase and phosphatase activities can be demonstrated simultaneously.

Disadvantages

- Most treaments activate more than one pathway:
  - individual site analysis is necessary.
- Enzyme specificity of phosphorylation/dephosphorylation at a particular site must be established experimentally and not assumed from *in vitro* analysis.
- High background levels of labeling – a greater number of proteins become labeled in intact cells necessitating the use of higher resolution methods.

8.3. Deciphering changes occurring in intact cells

When a change in phosphorylation is observed after a stimulus or treatment what does this mean? Changes in radioactivity incorporation in intact tissue generally correspond to changes in kinase or phosphatase activity *in vivo*, in contrast to the *post hoc* assays *in vitro*, where the relationship is inverse (see section 7). However, when interpreting the results of phosphorylation in intact systems one must be aware of factors which might change the amount of radioactivity incorporated into substrates without having any functional effect. Two major factors need to be considered: effects on specific activity and stoichiometry.

Changes in specific activity

As illustrated in *Figure 4*, the radioactive phosphate labels only a proportion of the ATP molecules synthesized by the mitochondria. Therefore, any treatment which alters the specific activity of the cellular phosphate pool or the ATP pool will produce a change in the level of radioactivity incorporated into proteins even if the levels of protein phosphorylation do not change. Thus, it is important to control experiments for confounding effects on phosphate transport and pool size, energy metabolism and mitochondrial function.

Stoichiometry and turnover

In order to have a functional effect, the stoichiometry of phosphorylation must change. Since most sites that are phosphorylated *in vivo* undergo dephosphorylation and rephosphorylation at some rate, and the cell contains a mixture of radioactive and non-radioactive ATP, it is possible for the amount of radioactivity incorporated at a particular site to change without a change in the stoichiometry of phosphorylation at that site and hence without any functional consequence. This is called turnover or exchange.

Figure 5 illustrates examples of some of the phosphorylation changes that can occur in intact tissue following a stimulus. The tissue labeled to a steady state is shown in the middle panel and, as in *Figure 4*, only a single substrate which can be phosphorylated on three different sites is considered. If the tissue is given a brief stimulation (seconds to minutes) which predominantly activates a kinase that phosphorylates the right hand side (R) site but also gives a slight stimulation to the phosphatases that act on all the sites (*Figure 5*, right side), the overall incorporation of radioactivity into the protein increases from 3/8 to 4/8 but analyzing the protein as a whole, hides a much more complex change which would only be revealed by site analysis. The strong activation of the kinase has resulted in a large increase in radioactivity at the R site (1/8 to 3/8) and the stimulation of the phosphatases has resulted in the removal of a phosphate from one of each of the sites including one that was already radioactive. The overall stoichiometry of phosphorylation of the protein has not changed but the distribution of the phosphates is different so that the functional state of the protein is likely to have changed significantly to favour the effects of R site phosphorylation at which the stoichiometry has increased from 3/8 to 5/8. If a brief stimulation is given which predominantly activates the phosphatases that act on the left hand side (L) and top (T) phosphorylation sites, the overall incorporation of radioactivity decreases from 3/8 to 2/8 and site analysis would reveal that the effect occurred specifically at the L site. In fact, phosphatase activity removed two phosphates from the T site as well but the activation of the kinase that acts on the T site resulted in a radioactive phosphate being added to one T site.

Figure 5. Changes in phosphorylation of a protein in a radioactively labeled intact cell following application of a stimulus which either stimulates phosphorylation or dephosphorylation predominantly. In the center panel the diagram shows eight molecules of a protein which has been labeled to a steady state in the cell with a mixture of radioactive (black circles) phosphate and non-radioactive (white circles) phosphate.

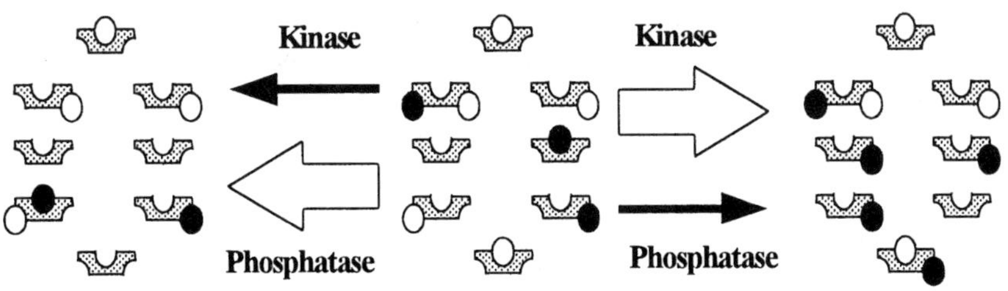

The overall stoichiometry decreased from 8/8 to 6/8 and the functional change in the protein will reflect the decreased effect of L and T site phosphorylation.

If the tissue treatment is prolonged (lasting hours), the same general principles apply but the likelihood of multiple pathway involvement increases as does the contribution of phosphate turnover to the pattern of incorporation of radioactivity.

Endogenous substrates whose phosphorylation can be measured in such a way must be abundant (or must be capable of being labeled to a very high specific activity and purified from cells). Often, therefore, they will not be the proteins whose regulation is the cause of the functional change being measured. Nevertheless, these proteins can serve as endogenous markers for the changes in activity of the kinases and phosphatases whose activity has been changed by the tissue treatment.

8.4. Hybrid techniques

Though used less frequently, there are some hybrid techniques for studying phosphorylation in intact cells which combine various aspects of the *post hoc* assays with the intact cell approach.

Phosphorylation of intact cells with ^{32}P-ATP

The plasma membranes of cells are impermeable to ATP. Therefore if intact cells are used in a *post hoc* assay in which radioactive ATP is added outside the cell for a brief period, the only source of protein phosphorylation should be kinases and endogenous substrates which are exposed to the extracellular medium. This approach has been used to identify exokinases and also to estimate what proportion of CaMPK-II is located pre- and post-synaptically at CNS synapses (4).

Phosphorylation of permeabilized cells with ^{32}P-ATP

The plasma membrane can be permeabilized chemically (5) or electrically (6) so that radioactive ATP can gain access to the inside of the cell. A *post hoc* assay involving radioactive ATP in permeabilized cells has several advantages:

- kinases and substrates are still in relatively normally location and relationship;
- non-membrane permeable activators or inhibitors may be used to control the activity of specific enzyme;
- ATP-γ-S may be used to provide phosphorylation resistant to dephosphorylation;
- functional correlates of phosphorylation changes can be measured simultaneously.

9. PRESERVING PHOSPHORYLATION STATE

Most phosphorylated groups are labile to varying degrees such that attention to preserving phosphorylation states must be given. The most important considerations are time, temperature and the use of inhibitor cocktails to minimize dephosphorylation and proteolysis.

Considerations

- Minimize delay in removing tissue from animals – ATP levels drop rapidly altering endogenous balance between kinase and phosphatase activity.
- Rapid cooling in excess cold buffer has minimal artefactual consequences and is the preferred method. If a whole rat brain is dropped into 500 ml of rapidly stirring ice cold isotonic sucrose, a deep structure such as the hippocampus reaches 10°C within 15 seconds and 4°C within 30 seconds (Rostas and Dhanjal, unpublished).
- Decapitating a rat and dropping the head immediately into liquid nitrogen, has been advocated as an 'instant' freezing approach, but this method has some drawbacks:

- Because of the insulating effect of the boiling nitrogen gas around the warm head, the freezing is nowhere near instant since deeper structures within the brain (e.g. hippocampus) may take at least 15 seconds to freeze (Rostas and Dhanjal, unpublished).
- Freezing and thawing is known to affect the activities of many kinases and phosphatases which may complicate the interpretation of *post hoc* activity assays.
- Once frozen tissue is thawed, lysosomal enzymes are released which increase the likelihood of both dephosphorylation and proteolysis; thawing and homogenization of the tissue in a cocktail of protease and phosphatase inhibitors is one solution to this problem.

- Use of inhibitor cocktails is essential but not 100% effective – see *Table 1*.
  - Microcystin-LR is highly specific for some but not all phosphatases
  - Non-specific, inorganic inhibitors effect most phosphatases but attention must be paid to non-specific effects caused by the high concentrations necessary.
  - Chelating agents are also useful but non-specific effects are possible.
  - Protease inhibitors are essential but never achieve complete prevention of proteolysis.
  - Where phosphatase inhibitors are used, tissue cannot be subsequently used to measure phosphatase activity.
  - In practice, a satisfactory general purpose cocktail can be produced from microcystin, one or more of the inorganic inhibitors, PMSF plus another protease inhibitor with or without a chelator of divalent cations.

10. NON-ISOTOPIC METHODS

While the use of radiolabeled ATP is the most widely used approach, a number of non-isotopic methods are becoming available. These have the advantage that they measure all phosphate and are therefore not affected by isotopic dilution or turnover. In addition the safety factor of not requiring radiolabel may be attractive to some investigators. However, these assays are largely only available in kit form and are thus expensive. More importantly, these assays are limited in their ability to see small changes in response to a stimulus, particularly if the background level of phosphorylation is high.

Post hoc *assays for serine/threonine kinase activity*

- A colorimetric assay based on eletrophoretic or affinity separation of phospho and dephospho forms of peptides tagged with coloured dye is commercially available with a range of substrates for different kinases. Although such methods do not allow for the analysis of large numbers of assays, they do allow for an alternative to radioactive assays for small numbers of sample.

Para-nitrophenol assay for protein phosphatase activity

- Non-specific.
- Low sensitivity.
- Useful for broad screen of activity (e.g. column fractions).

Malachite Green assay for protein phosphatase activity

- Colorimetrically detects free phosphate.
- High background.
- Requires phosphate-free labware.
- Low sensitivity.

Chemical derivatization of phosphorylated residues on purified substrates

- Converts phosphorylated sites to unique, stable amino acid derivatives.

- Requires detailed knowledge of substrates and kinases involved.
- Requires expensive instrumentation for detection.
- Requires specialized knowledge.

Antibodies to phosphoaminoacids
- Highly sensitive and specific for tyrosine phosphorylated residues.
- Commercially available kits for tyrosine kinase detection.
- Similar antibodies for serine/threonine kinases not widely used.
- Serine/threonine phosphorylated residues not uniformly available for antibody recognition.

Antibodies to phosphorylated epitopes
- Highly specific phospho- and dephospho- antibodies available for a number of proteins.
- Can be used to quantitate phosphorylation changes *in vivo*.
- Expensive and labour intensive to produce.
- Requires detailed knowledge of substrate, its phosphorylation site and the specificity of the sites for particular kinase *in vitro*.
- Assumes accurate correlation of *in vitro* knowledge with *in vivo* response.

11. LOCATING KINASES AND PHOSPHATASES WITH ANTIBODIES

The wide range of antibodies to protein kinases and protein phophatases that are now available provide a number of ways in which to investigate the localization of these enzymes following short or prolonged stimulation.

Advantages
- Determines cellular location of enzyme whose effect is being measured.
- Provides spatial information.
- Determines nature and distribution of the isoforms present.
- Determines changes in subcellular location of enzymes following cellular stimulation.
- Determines concentration of enzyme in the tissue.
- Wide range of applications – e.g. western blotting, immunohistochemistry, immunoprecipitation, ELISA, RIA.

Disadvantages
- Specificity of antibody must be determined.
- Uniform accessibility of antigen is assumed.
- May not detect all isoforms of a particular enzyme.
- Antigen recognition may be affected by fixation techniques used.
- Recognition may be affected by post-translation modification of the enzyme, e.g. carboxymethylation of PP2A inhibits C-terminal antibody recognition.
- Only semiquantitative in histochemical applications.

12. CORRELATING BIOCHEMICAL RESULTS WITH ELECTROPHYSIOLOGY AND ANATOMY

The ultimate aim of experiments of the type outlined in this chapter is to correlate the molecular changes in protein phosphorylation with the physiological changes occurring at specific cellular locations so that the existence of a causal link between the two can be tested. However, there is a problem because, in general, the biochemical techniques on the one hand and the electrophysiological and anatomical techniques on the other hand work on a different scale. Electrophysiological techniques measure changes in single cells or groups of cells within a piece of tissue. Single cell imaging techniques, histochemistry and electron microscopy also work at the same level. By contrast, biochemistry measures changes in whole tissue and it is presently not possible to measure protein phosphorylation changes in single cells (except using antibodies to phosphorylated epitopes on specific endogenous substrates). This is partly because of a lack of sensitivity of current biochemical methods, and partly because of the difficulty of isolating from a whole tissue the particular cell or group of cells from which functional recordings had just been made. This problem can be solved in at least one of two ways. Either new sensitive biochemical techniques have to be developed or new physiological preparations possessing very large cells need to be developed. There is a need for both.

13. PROTOCOLS

The major factors that need to be taken into account when designing a protocol for *in vitro* phosphorylation have been outlined in section 6. We often use slightly different protocols for assaying the same enzyme activity depending on the tissue being assayed and the question being addressed. Some of these protocols have been published in an extended 'how to' format (8). What follows is a selection of protocols which we use routinely for assaying crude samples of brain tissue. The protocols can be adapted for other tissues or purified samples by checking the variables outlined in section 6.

13.1. Protein kinase assays using exogenous peptide substrates

The assays are based on the procedure developed by Roskoski (7) utilizing P81 phosphocellulose paper which, at acid pH, will separate the phosphorylated peptide substrate from the unhydrolysed ATP and free phosphate.

Phosphorylation assays are carried out in a total volume of 40 μl, usually in independent triplicate assay tubes. Because of the many enzyme activities present in tissues, the phosphorylation reactions are started by the addition of 16 μl of ice cold tissue sample to the rest of the incubation mixture (24 μl) which has been preincubated for 5 minutes at the assay temperature (30°C). The reaction is stopped by the addition of 8 μl of ice-cold 450 mM phosphoric acid. Before the start of the assay, sheets of P81 paper are marked into 2 cm numbered squares with a pencil (gloves should be worn so that the paper does not bind proteins from fingers and the paper should be kept dry). Aliquots (30 μl) of the stopped reaction mixture are pipetted into the centre of each square. The entire sheet (usually 9 × 6 squares) is washed for 2 × 10 minutes in 75 mM phosphoric acid (approximately 10 ml per square) followed by 1 × 5 minutes in ethanol and then dried under a stream of air. Individual squares are cut out and placed into scintillation vials and counted for radioactivity. For routine assays we count the squares dry; if greater sensitivity is needed water or scintillant may be added. For each experiment 10 μl aliquots from 6 randomly chosen tubes are also pipetted onto P81 squares but not washed in acid. This gives a measure of the total radioactivity in the incubation which is necessary to calculate the activity in absolute units.

For each of the enzymes below, the assay mixture contains the following at the stated final concentration:

30 mM Tris–HCl, pH 7.4
10 mM $MgCl_2$
200 μM ATP containing 2 μCi of radioactive ATP
0.05% Tween 80 (to minimize peptide absorption to the tube).

To this basic incubation mixture, the appropriate activators, inhibitors and peptide substrates are added (see below). If the tissue sample contains ions or inhibitors, care must be taken to

Phosphorylation

account for these when calculating what to add to the reaction mixture in order to achieve the desired final concentration during the assay.

cAMP stimulated protein kinase (PKA):

1 mM EGTA (to minimize the effects of Ca^{2+} stimulated kinases)
10 μM PL_{8-21} or 100 μM Kemptide (see *Table 4*)
± 10 μM cAMP
Reaction time: 2 minutes (30°C)

cGMP stimulated protein kinase (PKG):

1 mM EGTA (to minimize the effects of Ca^{2+} stimulated kinases)
10 μM PL_{8-21} (see *Table 4*)
± 10 μM cGMP
± 100 μg/ml PKI (see *Table 1*)
Reaction time: 2 minutes (30°C)

Ca^{2+}-phospholipid stimulated protein kinase (PKC):

0.2 mM $CaCl_2$
10 μg/ml phosphatidylserine
1 μg/ml diacylglycerol or 500 ng/ml phorboldibutyrate
± 10 μM NG_{28-43} (see *Table 4*)
Reaction time: 2 minutes (30°C)

Ca^{2+}-calmodulin stimulated protein kinase II (CaMPK-II):

0.2 mM $CaCl_2$
20 μg/ml calmodulin
± 10 μM autocamtide-2 (see *Table 4*)
Reaction time: 45 seconds (30°C)

Using crude extracts from forebrain and hippocampus we usually aim for a final protein concentration in the assay of 0.08–0.12 mg/ml Lowry protein.

13.2. Protein kinase assays using endogenous protein substrates

These assays utilize SDS-PAGE to resolve endogenous proteins that have been phosphorylated during the *in vitro* assay. Equal aliquots of the tissue extract are phosphorylated in the presence and absence of activators and/or inhibitors of individual protein kinases and the activity is measured by the change in the degree of radioactive phosphate incorporation into one or more endogenous protein which is known to be a substrate for the protein kinase in question. When measuring a variety of protein kinase activities in a tissue, the conditions chosen usually represent a compromise between the optimal conditions for the various kinases: not optimal for any single kinase but sufficiently good to enable all the kinases of interest to be measured relative to a single 'basal' or 'unstimulated' condition. The total reaction volume is 100 μl. The reaction is started by the addition of 40 μl of ice cold tissue extract to give a final protein concentration of 1 mg/ml. The reaction is stopped by the addition of 50 μl of (3X) SDS sample buffer (200 mM Tris–HCl, pH 6.8, 6.9% SDS, 30% glycerol, 15% 2-mercaptoethanol) and boiling of the sample for 2 min; 80 μl is analyzed by SDS-PAGE.

Basal condition:

30 mM Tris–HCl, pH 7.4
10 mM $MgCl_2$
1 mM EGTA
40 μM ATP containing 5 μCi radioactive ATP

Cyclic nucleotide stimulated protein kinases:

10 μM cAMP or cGMP ± 100 μg/ml PKI

Ca^{2+}-phospholipid stimulated protein kinase (PKC):

0.2 mM $CaCl_2$
10 μg/ml phosphatidylserine
1 μg/ml diacylglycerol or 500 ng/ml phorbol dibutyrate

Ca^{2+}-calmodulin stimulated protein kinase II (CaMPK-II):

0.2 mM $CaCl_2$
20 μg/ml calmodulin

The specificity of the phosphorylation may be determined by the addition of specific kinase inhibitors (see *Table 1*) or by the analysis of individual sites on endogenous proteins whose specificity of phosphorylation is known. In the above recipe the presence of EGTA in the basal conditions reduces the background labeling by Ca^{2+}-stimulated kinases which normally label many endogenous proteins and produce a high background making it difficult to identify bands whose phosphorylation is stimulated by other kinases such as PKA or PKG. If EGTA cannot be used, the concentration of $MgCl_2$ can be lowered to 1 mM to reduce the basal contribution of PKC and other Ca^{2+}-activated kinases.

Compared with those in the peptide substrate assays, the concentration of ATP is lower (to maximize the specific activity of the ATP since the concentration of endogenous substrates is low compared with exogenous peptide substrates) and the concentration of tissue protein is higher (in order to maximize the ability to detect individual protein bands) in these assays. Therefore, reaction times are shorter in order to remain within the linear phosphorylation range. If tissue extracts contain no inhibitor, cocktail phosphorylation can be carried out for 15–20 s at 37°C or for 20–30 s at 30°C. If inhibitors of ATPases or protein phosphatases are present in the tissue extract, the time of labeling can be extended to 45 s at 30°C. The linearity of the assay must be checked for each tissue extract.

Phosphorylation

13.3. Phosphorylation of intact synaptosomes

Synaptosomes are prepared using the Percoll gradient method (9), washed and resuspended in ice cold oxygenated (using 95% O_2/5% CO_2) isotonic Krebs-like buffer solution containing the following ingredients (final concentrations):

25 mM $NaHCO_3$
1.2 mM $MgCl_2$
118 mM NaCl
5 mM KCl
0.1 mM $CaCl_2$
10 mM glucose.

Using the same buffer the synaptosomes are diluted to 5 mg/ml protein; 750 μCi/ml of $^{32}P_i$ is added and the mixture is incubated for 45 min at 37°C with constant shaking. We routinely use 150 µl for the total volume of this incubation. For synaptosomes constant oxygenation during this incubation is not absolutely necessary but oxygenation is required if this procedure is used for tissue slices or whole cells. For the stimulus (e.g. depolarization) 70 µl of the preincubated synaptosome suspension is transferred to another tube containing an equal volume of a prewarmed oxygenated isotonic buffer containing the stimulating agent or control solution. After the required stimulus period the reactions are stopped by one of several methods (lysis in with SDS sample buffer, acid precipitation, boiling, detergent and immunoprecipitation) depending on the method of analysis to be used.

13.4. Protein phosphatase assays using exogenous protein substrates

The assay most commonly used is based on that developed by Cohen *et al.* (10) using ^{32}P-labeled glycogen phosphorylase as substrate. Dephosphorylation reactions are carried out at 30°C for 20 min in a total volume of 30 µl, comprising 10 µl assay buffer, 10 µl dilute tissue extract and 10 µl ^{32}P-labeled glycogen phosphorylase, in separate microfuge tubes in triplicate. The reaction is started by the addition of glycogen phosphorylase and stopped by the addition of trichloroacetic acid which precipitates the ^{32}P-labeled glycogen phosphorylase. The precipitate is separated from the supernatant by centrifugation and phosphatase activity quantitated by measuring the radioactivity in the supernatant which contains free phosphate removed by the endogenous phosphatases. In order to limit the dephosphorylation to PP1 and PP2A, the reaction is carried out in the absence of divalent cations. Dephosphorylation due to PP1 and PP2A is quantitated by the inclusion of either 2 nM okadaic acid (which inhibits only PP2A) or 0.2 µM Inhibitor-1 (which inhibits only PP1). Reactions are limited to <30% dephosphorylation to ensure linearity. Note that okadaic acid is a stoichiometric inhibitor of PP2A such that the concentration required to inhibit all the PP2A in the sample depends on the concentration of PP2A in that sample. Thus the source of endogenous phosphatase must be diluted at least 10-fold for okadaic acid to accurately distinguish between PP1 and PP2A. We find this is routinely achieved by using a protein concentration of ≤0.03 mg/ml (final), but new assays must always be calibrated by comparison with the Inhibitor-1 approach. In order to calculate the phosphatase activity in terms of nmoles phosphate per minute per mg, the total radioactivity in 10 µl ^{32}P-labeled glycogen phosphorylase is also counted. It should also be noted that proteoelytic activity may also lead to the presence of radioactivity in the supernatant by cleaving small peptides containing the phosphorylation site which are not precipitated by TCA. Wherever this is suspected, and when new assays are first carried out, the free phosphate in the supernatant must first be converted to a phosphomolybdate complex before counting.

Preparation of ^{32}P-labeled phosphorylase

Reaction buffer:

- 250 mM Tris, pH 8.2
- 16.7 mM $MgCl_2$
- 0.83 mM $CaCl_2$
- 133 mM sodium glycerophosphate
- 1.67 mM ATP

Solubilization buffer:

- 50 mM Tris, pH 7
- 0.1 mM EDTA
- 15 mM caffeine
- 1 mM DTT

Reagents:

- 100 mg/ml glycogen phosphorylase b
- 20 mg/ml phosphorylase kinase (500 units/mg)
- 0.5 mCi ^{32}P-ATP in 100 µl
- 90% saturated ammonium sulphate

Procedure:

- To 300 µl reaction buffer add 0.5 mCi ^{32}P-ATP
- Add 110 µl glycogen phosphorylase b
- Add 10 µl phosphorylase kinase
- Vortex well and incubate at 30°C for 60 min
- Remove 10 µl for determination of specific radioactivity
- Stop reaction with 500 µl ammonium sulphate
- Vortex, stand on ice for 30 min and spin in microfuge for 10 min
- Discard supernatant and resuspend pellet in 1 ml 45% ammonium sulphate
- Spin for 10 min in microfuge
- Repeat previous 2 steps twice
- Resuspend pellet in 2 ml solubilization buffer
- Transfer to an Amicon 30 concentrator and spin at 5000 g for 30 min at 20°C
- Collect labeled phosphorylase from upper reservoir and transfer to a clean vial
- Wash reservoir with 5 × 0.2 ml aliquots of solubilization buffer, adding washes to vial
- Measure protein concentration (A_{280} of 1 mg/ml is 1.31), adjust to 3 mg/ml with solubilization buffer and store at 4°C
- Calculate stoichiometry of phosphorylation which should approach 1 mole/mole. This is important for subsequent calculation of phosphatase activity.

Phosphatase assay procedure

Assay buffer:

- 50 mM Tris, pH 7.4
- 0.1 mM EDTA
- 2 mM EGTA
- 1 mM DTT
- ± 6 nM okadaic acid or 0.6 µM I-1 or I-2

Dilution buffer:

- 50 mM Tris, pH 7.4
- 0.1 mM EDTA
- 1 mg/ml BSA
- 1 mM DTT

Procedure ON ICE:

- Dilute tissue extracts to 0.03 mg/ml with dilution buffer
- Add 10 µl assay buffer ± okadaic acid to 3 microfuge tubes
- Add 10 µl dilute tissue extract to each tube
- Add 10 µl dilution buffer to 3 blank tubes containing 10 µl assay buffer
- Start reaction by adding 10 µl ^{32}P-phosphorylase to each tube at 10 second intervals
- Incubate for 20 min at 30°C
- Stop reaction (at 10 s intervals) by adding 90 µl 30% TCA
- Vortex tubes and stand on ice for 10 min
- Spin in microfuge for 4 min

Carefully remove 100 µl supernatant and add to a fresh tube containing 1 ml scintillant
Count tubes in a liquid scintillation counter set for ^{32}P
Also count 10 µl ^{32}P-phosphorylase from which the degree of dephosphorylation obtained during the assay can be calculated

Conversion of free phosphate to phosphomolybdate
To 100 µl supernatant add 0.2 ml acid phosphate (1.25 mM KH_2PO_4 in 0.5 M H_2SO_4)
Add 0.5 ml isobutantol/heptane (1:1 v/v) and 0.1 ml 5% ammonium molybdate
Vortex vigorously and stand at room temp for 10 min until the aqueous and organic layers separate
Transfer 0.3 ml of the top (organic) phase to fresh scintillation vials, add scintillant and count.

14. ACKNOWLEDGMENTS

We thank Peter Dunkley, Phillip Robinson, Phillip Marley, Imre Lengyel, Stephen Bunn and Troy Margrie for helpful suggestions on drafts of this manuscript. This work was supported by grants from the National Health and Medical Research Council of Australia.

15. REFERENCES

1. Dunkley, P.R. *et al.* (1986) *J. Neurochem.*, **46**, 1692.

2. Van der Geer, P. *et al.* (1993) in *Protein Phosphorylation: A Practical Approach* (D.G. Hardie, ed.), IRL Press, Oxford, p. 31.

3. Forn, J. and Greengard, P. (1978) *Proc. Natl Acad. Sci. USA*, **75**, 5195.

4. Dunkley, P. R. *et al.* (1988) *J. Neurochem.*, **51**, 57.

5. Vitale, N. *et al.* (1993) *J. Biol. Chem.*, **268**, 14715.

6. Knight, D.E. (1981) *Techn. Cell Physiol.*, **113**, 1.

7. Roskoski, R. (1983) *Meth. Enzymol.*, **99**, 3.

8. Jarvie, P.E. and Dunkley, P.R. (1995) *Sig. Transduct. Proto.*, **41**, 239.

9. Dunkley, P.R. *et al.* (1988) *Brain Res.*, **441**, 59.

10. Cohen, P. (1991) *Meth. Enzymol.*, **201**, 389.

11. Tischler, A.S. *et al.* (1990) *J. Neurochem.*, **55**, 1159.

12. Tokomitsu, H. *et al.* (1990) *J. Biol. Chem.*, **265**, 4315.

13. Sumi, M. *et al.* (1991) *Biophys. Biochem. Res. Comm.*, **181**, 968.

14. Smith, M.K. *et al.* (1992) *J. Biol. Chem.*, **267**, 1761.

15. Glass. *et al.* (1989) *J. Biol. Chem.*, **264**, 8802.

16. Dostmann, W.R.G. *et al.* (1990) *J. Biol. Chem.*, **265**, 10484.

17. Bouli, N.M. and Davis, M. (1990) *Brain. Res.*, **525**, 198.

18. Hidaka, H. *et al.* (1984) *Biochemistry*, **23**, 5036.

19. Chijiwa, T. *et al.* (1990) *J. Biol. Chem.*, **265**, 5267.

20. Kase, H. *et al.* (1987) *Biochem. Biophys. Res. Comm.*, **142**, 436.

21. Martiny-Baron, G. *et al.* (1993) *J. Biol. Chem.*, **268**, 9194.

22. Tamaoki, T. *et al.* (1990) *Adv. Second. Mess. Phosphoprotein. Res.*, **24**, 497.

23. Toullec, D. *et al.* (1991) *J. Biol. Chem.*, **266**, 15771.

24. House, C. and Kemp, B.E. (1987) *Science*, **238**, 1726.

25. Tamaoki, T. *et al.* (1986) *Biochem. Biophys. Res. Comm.*, **135**, 397.

26. Haystead, T.A.J., Sim, A.T.R., Carling, D., Honner, R. C., Tsukitani, Y., Cohen, P. and Hardie, D.G. (1989) *Nature*, **337**, 78.

27. Ishihara, H. *et al.* (1989) *Biochem. Biophys. Res. Comm.*, **159**, 871.

28. Mackintosh, C. and Klumpp, S. (1990) *FEBS Lett.*, **277**, 137.

29. Mackintosh, C. *et al.* (1990) *FEBS Lett.*, **264**, 187.

30. Matsushima, R. *et al.* (1990) *Biochem. Biophys. Res. Comm.*, **171**, 867.

31. Li, Y-M. *et al.* (1993) *Biochem. Pharmacol.*, **46**, 1435.

32. Cohen, P. *et al.* (1988) *Meth. Enzymol.*, **159**, 427.

33. Mackintosh, C. (1993) in *Protein Phosphorylation: A Practical Approach* (D.G. Hardie, ed.), IRL Press, Oxford, p. 197.

34. Liu, J. *et al.* (1992) *Biochemistry*, **31**, 3896.

35. Hashimoto, Y. *et al.* (1990) *J. Biol. Chem.*, **265**, 1924.

36. Hardie, D.G. *et al.* (1993) in *Protein Phosphorylation: A Practical Approach* (D.G. Hardie, ed.), IRL Press, Oxford, p. 61.

37. Kemp, B. *et al.* (1973) *J. Biol. Chem.*, **252**, 4888.

38. Pearson, R.B. and Kemp, B.E. (1991) *Meth. Enzymol.*, **200**, 62.

39. Yasuda, I. (1990) *Biochem. Biophys. Res. Comm.*, **166**, 1220.

40. Egan, J. *et al.* (1988) *Anal. Biochem.*, **175**, 552.

41. Chen, S.-J. *et al.* (1993) *Biochemistry*, **32**, 1032.

42. Spencker, T. *et al.* (1993) *Liebigs. Ann. Chem.*, **237**, 240.

43. Hashimoto, Y. and Soderling, T.R. (1987) *Arch. Biochem. Biophys.*, **252**, 418.

44. Hanson, J. *et al.* (1989) *Neuron*, **3**, 59.

45. Deanna, A.D. *et al.* (1990) *Biochim. Biophys. Acta*, **1051**, 199.

46. Agostinos, P. *et al.* (1990) *Eur. J. Biochem.*, **189**, 235.

47. Blumenthal, D.K. *et al.* (1986) *J. Biol. Chem.*, **261**, 8140.

48. Hvaldy, O. *et al.* (1994) *Proc. Natl Acad. Sci. USA*, **91**, 4761.

CHAPTER 11
RECEPTOR mRNA LOCALIZATION

G.D. Pratt

1. INTRODUCTION

A gene is a sequence of DNA carrying information that represents an RNA transcript which may thereafter be transcribed into a specific protein. The sequences encoding the protein (exons) are not necessarily continuous on the DNA since they may be interrupted by sequences that are not translated into protein (introns). Splicing out of the intron sequences therefore produces continuous exons or a protein coding sequence. This mature form of the RNA is the messenger RNA (mRNA) which accounts for up to 5% of the total RNA. A number of techniques are now available for studying RNA expression in tissue homogenates; however, expression at the cellular level requires the technique of *in situ* hybridization. This method, first described by Pardue and Gall (1), involves the hybridization of a complementary nucleic acid probe specific to a target nucleic acid *in situ*, permitting the determination of cells transcribing a particular gene.

In situ hybridization has subsequently proved invaluable in many areas of neurobiology and medicine with applications ranging from differential gene expression in animal models of ischemia (2) and in striatal transplants (3), to altered expression of heat shock protein-encoding genes in brains from patients with Alzheimer's disease (4). In addition, the identification of the chromosomal locus for a specific human glutamate receptor-encoding gene in the same region as that responsible for a form of amyotrophic lateral sclerosis (motor neurone disease) (5) may offer new insights into understanding the pathology of this disorder.

The continued evolution of molecular cloning techniques has revealed an overwhelming heterogeneity of receptors, receptor subunits, ion channels and neuropeptides. Receptor subtypes exist as functionally distinct entities comprising defined subunit combinations (6, 7) and possess specific anatomical and cellular distributions of their subunit-encoding mRNAs as revealed by *in situ* hybridization (8).

This chapter seeks to define the general principles of *in situ* hybridization whilst giving detailed protocols for the use of oligonucleotide probes for the radioactive detection of mRNA species in brain tissue, which is probably the simplest approach for the investigator embarking on *in situ* hybridization for the first time.

2. TISSUE PREPARATION FOR HYBRIDIZATION

In experimental animals, tissue preparation procedures can be optimized in order to limit the influences of pre- and post-mortem factors on the stability of brain mRNA. However, such standardization for human post-mortem samples is not so readily controlled, in particular those events comprising the pre-mortem phase (or agonal state) of the patient. These issues have been the subject of a recent review (9) which surprisingly concluded that abundant and biologically intact RNA is present in tissue that has experienced a post-mortem freezing delay of 36 hours or more. Furthermore, although pre-mortem factors may affect certain mRNA species and have implications for the quantitative analysis of transcript changes under pathological conditions, no correlation has yet been found between agonal state and the overall RNA level or activity.

2.1. Tissue preparation

(i) Sacrifice the animal by decapitation and rapidly remove fresh tissue.
(ii) Attach the required brain region to a thin cork slice using 'Tissue-tek' embedding medium.
(iii) Freeze the mounted brain in powdered dry-ice.
(iv) Allow to freeze for about 5 minutes and store at –80°C until use.

2.2. Pretreatment (subbing) of slides

The best slides to use for *in situ* hybridization are BDH Super Premium microscope slides with twin-frost ends rather than those with a colored, non-translucent end. Frosted, translucent glass greatly facilitates individual slide identification using pencil markings under red light darkroom conditions. Ensure that gloves are worn when handling subbed slides. Although the following protocol describes subbing with gelatin, slides may also be coated with either 3-(triethoxysilyl-)propylamine (TESPA) or poly-L-lysine for *in situ* hybridization.

(i) Clean slides by mounting in glass staining racks and soaking overnight in 5% Extran.
(ii) Rinse in running cold tap water for 2 hours.
(iii) Dry slides in an oven (90°C) and allow to cool at room temperature.
(iv) Freshly prepare a solution of 0.5% gelatin (2.5 g/500 ml) by warming to 40–50°C. When dissolved, add 0.05% chromium potassium sulfate (250 mg/500 ml).
(v) Immerse slides in the gelatin for around 5 seconds.
(vi) Drain excess gelatin and dry overnight at 90°C.
(vii) Store dust-free in a sealed box.

2.3. Cryosectioning

(i) Ensure that a good supply of glass slides has been allowed to equilibrate to the temperature of the cryostat; i.e. –12 to 20°C for brain).
(ii) Attach the cork to a cryostat chuck with 'Tissue-tek' embedding medium and allow to equilibrate to the temperature of the cryostat for about 30 minutes prior to sectioning.
(iii) Slice the tissue into 14 μm sections.
(iv) Thaw-mount the sections onto subbed glass slides.
(v) Allow the sections to adhere to the slide by bringing to room temperature and completely drying before being returned to the cryostat (or –20°C) to await fixation.

2.4. Postfixation of the tissue

The primary requirement for tissue fixation is to retain the viability of the target nucleic acid for hybridization while concomitantly maintaining morphological integrity and allowing good accessibility of probes. For this a cross-linking fixative such as paraformaldehyde is preferred (see *Table 1*).

Table 1. Formulation of phosphate-buffered saline and paraformaldehyde

Phosphate-buffered saline (PBS)	Solution I	$Na_2HPO_4.2H_2O$ (0.5 M)
	Solution II	$NaH_2PO_4.H_2O$ (0.5 M)
	Add:	15.75 ml Solution I
		4.25 ml Solution II
		8 g NaCl
	Make up to 1 liter with water and adjust to pH 7.4.	
Paraformaldehyde (PFA; 4% in PBS)	To 40 g PFA add 250 ml distilled water and heat to 40–50°C	
	Add 6 drops 10 M NaOH to dissolve the PFA	
	Add to Solution I, Solution II, NaCl and make up to 1 liter with water	
	Adjust to pH 7.4 and filter	

(i) Incubate sections for 20 min in 4% paraformaldehyde at room temperature.
(ii) Rinse sections in phosphate-buffered saline (PBS) at room temperature, three times.
(iii) Rinse sections rapidly (2 second dip) in distilled water, twice.
(iv) Dehydrate twice through 70%, 95% and 100% ethanol.
(v) Dry sections in air for a few minutes and store in a sealed slide box at –80°C until use.

3. PROBE SELECTION AND LABELING

3.1. Choice of probe and labeling

The choice of which of four types of probe to be hybridized will depend upon the experimenter's experience with molecular biology and evaluation of the advantages and disadvantages associated with each type (10). See *Tables 2 and 3*.

Table 2. Approaches to probe labeling

Type of probe	Probe labeling
Double-stranded DNA	Nick translation of a cloned DNA in a plasmid vector. DNase I initially makes cuts at random sites along the DNA molecule followed by replacement of the excised nucleotides with radioactive ones using DNA polymerase I. Higher specific activity DNA probes can be synthesized by random primer extension using the Klenow enzyme
Single-stranded DNA	DNA fragments to be cloned are ligated into the multiple cloning site (MCS) of an M13 bacteriophage
Single-stranded RNA	Labeling relies on the use of the bacteriophage RNA polymerase promotor regions, SP6 and T7, incorporated into a plasmid vector reverse transcription system containing a cDNA insert
Oligonucleotide	The 5′ terminus of the probe is end-labeled or 'tailed' through the action of T4 polynucleotide kinase which catalyses the transfer of a single radioactive gamma phosphate from $[^{32}P]\gamma ATP$. Probes with higher specific activities are produced by substitution of dNTPs with multiple α-$[^{35}S]$thiophosphate groups at the 3′ end using terminal deoxy-nucleotidyl transferase

mRNA Localization

Table 3. Advantages and disadvantages of probe types

Type of probe	Advantages/disadvantages
Double-stranded DNA	Resulting hybrids are much less stable than with RNA probes and reannealing to the sense strand can occur
Single-stranded DNA	Overcomes the problem of reannealing that occurs with double-stranded DNA
Single-stranded RNA	This enables the synthesis of sense and antisense RNA from a single construct; the antisense strand is complementary to the cellular RNA and serves as the hybridization probe while the sense strand acts as an appropriate negative control to define hybridization specificity. RNase enzymes can additionally be used in order to digest unhybridized probe and achieve low background signals
Oligonucleotide	No cloning is required. Oligonucleotide probes are stable, offer good target penetration and do not self-hybridize. Access to a DNA synthesizer is required

3.2. Oligonucleotide probes for *in situ* hybridization

If expertise in techniques of molecular biology is a limiting factor, the simplest type of probes to use for *in situ* hybridization are synthetic complementary oligodeoxyribonucleotide probes (oligoprobes). Provided that the sequence of a particular nucleic acid is known, short (30–60) nucleotide sequences, complementary to the mRNA to be analyzed, are selected with a specific GC content (usually around 55%). A published nucleotide sequence is written from the 5′ end to the 3′ end (i.e. 5′-AGTCCGAT-3′) The probe sequence is therefore complementary from 3′ to 5′ (i.e. 3′-TCAGGCTA-5′), although it is conventionally written in reverse (i.e. 5′-ATCGGACT-3′). The oligoprobe should be designed so as to avoid homologies with related sequences thus allowing hybridization to a region specific to this mRNA. Comparison of the sequence with reference to gene data banks facilitates this process (e.g. by using FastA analysis on the GCG program, University of Wisconsin). Non-conserved cytoplasmic loops of receptors may offer suitable sequences (membrane-spanning domains being most conserved regions between related receptors/subunits). Also large homonucleotidic clusters should be avoided. Oligoprobes are synthesized using a DNA sequencer (0.2 μM is a suitable amount to order) using specialized commercial outlets. Purifying an oligoprobe by polyacrylamide gel electrophoresis or HPLC does not necessarily improve the quality of the technique.

The following series of protocols are based essentially on those described by Persohn *et al.* (8) and Pratt *et al.* (11) and provide a detailed step-by-step account for the successful use of oligoprobes for *in situ* hybridization in brain tissue.

Nucleotide concentration determination

Oligoprobes are usually supplied as a lyophilized pellet which then requires reconstituting to the required concentration. This is achieved via spectrophotometric analysis using the following relationship:

1 optical density unit @ 260 nm ≈ 20 μg nucleotide/ml
Example: Dissolve oligonucleotide into 100 μl water
Assay 5 μl diluted into 1000 μl water (1:200 dilution)

OD_{260nm} of a 1/200 dilution = 1.025
$1.023 \times 20 \times 200$ = 4100 μg/ml
= 4.1 μg/μl.

Protocol for labeling oligonucleotides with [^{35}S]dATP

(i) To a sterile 1.5 ml Eppendorf tube add:
- 100 ng oligoprobe in 27 μl autoclaved water
- 10 μl 5x tailing buffer
- 5.5 μl [^{35}S]dATP
- 1 μl bovine serum albumin (2 mg/ml; Boehringer)

(ii) Vortex the above mixture. This step is vital since it will affect the efficiency of the labeling reaction which in turn determines the specific activity of the oligoprobes. After vortexing add:
- 6.5 μl terminal deoxynucleotidyl transferase (TdT; 15 units/μl).

(iii) Incubate the reaction mixture for 8 min at 37°C.
(iv) Stop the reaction by adding 5 μl EDTA (500 mM).
(v) Denature the enzyme by incubating the solution for 10 min at 70°C.
(vi) Prepare a Biogel P30 spin column (Bio-Rad) by ensuring that the gel and buffer are well-mixed.
(vii) Prespin the column by centrifuging at 1100 *g* for 1 min to remove excess gel buffer.
(viii) Purify the labeled oligonucleotide from unincorporated oligonucleotides by loading the sample onto the column and centrifuging at 1100 *g* for 2 min. Approximately 65 μl eluate is expected.
(ix) Assay a 2 μl aliquot of the eluate by scintillation counting. Between $0.5 - 2 \times 10^6$ cpm/2 μl are expected.
(x) Add dithiothreitol (DTT) from a 1 M stock solution to a final concentration of 40 mM.
(xi) Store the probe at –80°C until use (stable for around 2 weeks but ideally should be used as 'fresh' as possible).

4. HYBRIDIZATION

In situ hybridization procedures require a critical understanding of nucleic acid hybridization theory encompassing those factors that contribute to the stability of hybrid interactions and the kinetics of their formation (12).

Both hydrogen bonding and hydrophobic interactions stabilize the two strands of a DNA molecule which can be disrupted by increases in temperature. Hybridization depends on the ability of denatured DNA to reanneal with complementary strands just below their T_m. The term T_m describes the melting temperature of DNA in solution at which 50% of duplex molecules dissociate (melt) into single strands and provides a useful index of hybridization temperature *in situ*. Although the rate of DNA reannealing is maximal at T_m minus around 25°C (13) (10–15°C for RNA-DNA hybrids), a number of factors that influence the rate and specificity of hybridization need to be considered when empirically determining the optimal hybridization and washing conditions (stringency). See *Table 4*.

Table 4. Factors influencing hybridization rate and specificity.

Factor	Rationale
Sodium ions	Hybrid stability depends on electrostatic repulsion between negatively charged phosphate groups on opposing strands of the duplex which can be reduced in the presence of increasing concentrations of Na^+ ions. Therefore, post-hybridization washes are generally performed at low salt concentrations (high stringency) to dissociate duplexes of low homology (non-specific) while retaining those with high homology (specific)
Formamide	An organic solvent which reduces the T_m and hence thermal stability by disrupting hydrogen bonding
GC pairing	Hybrids are more stable with a higher frequency of G–C pairs since they are coupled by three hydrogen bonds, unlike A–T pairs which comprise only two
Probe length	The longer the probe, the greater the stability of the duplex owing to an increased incidence of hydrogen bonding between complementary base pairs. Shorter oligoprobes should therefore have a minimal percentage base pair mismatch to maximize stability
Probe concentration	Affects the rate at which the first few base pairs anneal which is increased in the presence of dextran sulphate. In aqueous solution, dextran becomes highly hydrated, reducing the availability of hydrating water to the nucleic acid and thus 'increasing' the local probe concentration.

4.1. Hybridization solution

The stock reagents shown in *Table 5* should be made up in advance of the *in situ* hybridization experiment. Sterile distilled water should be used throughout.

Since dextran sulphate is highly viscous in solution, it is recommended that around 30% excess fluid is made up to account for the loss that occurs due to retention during filtration and through adhering to Falcon tubes.

(i) Freshly prepare the following hybridization solution exactly as described in a 15 ml Falcon tube:
- 1 g dextran sulphate
- 200 µl Denhardt's solution (× 50)
- 2 ml saline sodium citrate (SSC × 20)
- 1 ml DTT (1 M)
- complete the volume to 4 ml with sterile distilled water.

(ii) Dissolve the dextran sulphate by heating to 45°C in a water-bath, vortexing approximately every 15 min. This will require approximately 1 h for the initially granular dextran suspension to become a more uniform opaque suspension.

(iii) Filter through 5 µm, 1.2 µm and 0.45 µm disposable filter holders (see *Table 6* for details).

(iv) Take 800 µl of the filtrate and add:
- 100 µl tRNA (5 mg/ml)
- 50 µl poly-adenylic acid (poly-A; 10 mg/ml)

- 50 µl denatured herring or salmon sperm DNA (10 mg/ml)
- 1 ml deionized formamide.

Table 5. Formulation of reagents for *in situ* hybridization

Reagent	Formulation
Denhardt's reagent × 50	Add: 5 g Ficoll 5 g polyvinylpyrollidone 5 g BSA (fraction V) Dissolve in 500 ml H_2O and store in 500 µl aliquots at –20°C
Formamide	Add: 500 ml formamide 25 g amberlite resin Stir at room temperature for 30 min and filter. Store in 10 ml aliquots at –20°C
Polyadenylic acid	Dissolve 100 mg/10 ml H_2O by heating at 60°C for 15 min. Store in 250 µl aliquots at –20°C
Saline sodium citrate SSC × 20	Add: 175.3 g NaCl (3 M) 88.2 g $C_6H_5Na_3O_7.2H_2O$ (300 mM) tri-sodium citrate dehydrate Dissolve in 800 ml water and adjust to pH 7.0 with a few drops of 10 M NaOH. Make up to 1 liter
tRNA	Dissolve in H_2O at a concentration of 5 mg/ml and store in 500 µl aliquots at –20°C.

Table 6. Reagents for *in situ* hybridization

Reagent		Reference number	Supplier
Amberlite resin		MB-1A	Sigma
Autoradiographic	31–883 nCi/g	RPA 504	Amersham
^{14}C microscales	0.1–100 nCi/g	RPA 511	
Biogel P30 spin column		732-6006	Bio-Rad
Bovine serum albumin		711 454	Boehringer
Bovine serum albumin (fraction V)		A-3059	Sigma
Chromium potassium sulphate		C-5926	Sigma
Cresyl violet acetate		C-1791	Sigma
[^{35}S]dATP		SJ1334	Amersham
		NEG 034H	NEN
Dextran sulphate (sodium salt)		17-0340-01	Pharmacia
Denatured herring sperm		D-7656	Sigma
Denatured salmon testis DNA		D-7290	Sigma
Disposable filter holders	5 µm	462 520	Schleicher & Schuell,
	1.2 µm	462 261	P.O. Box 4, D-37582
	0.45 µm	462 100	Dassel, Germany
Dithiothreitol		708 984	Boehringer
EDTA (0.5 M)		E-7889	Sigma
Eukitt		36189	BDH
Extran 300 concentrate		560363P	BDH
Ficoll (Type 400)		F-9378	Sigma

Table 6. Continued

Reagent	Reference number	Supplier
Formamide	551UB	Gibco BRL
Gelatin (type B from bovine skin)	G-9382	Sigma
Polyadenylic acid	P-9403	Sigma
Polyvinylpyrollidone	P-5288	Sigma
Sodium acetate	S-9513	Sigma
Terminal deoxynucleotidyl transferase	18008-011	Gibco BRL
tRNA	109 541	Boehringer
Super premium microscope	406/0181/02	BDH

4.2. Hybridization of oligoprobes to tissue samples

Example calculation

Use 50 µl hybridization fluid containing 250 000 dpm probe per section of 1 cm^2 surface area.

4 cm^2 tissue area/slide requires 200 µl hybridization fluid/slide.
An experiment using 25 slides therefore requires a total of 5 ml hybridization fluid (make up a 10% excess to allow for any wastage).

If labeled probe [radioactive] = 1.5×10^6 dpm/2 µl, then
5.5 ml @ 250 000 dpm/50 µl = 27.5×10^6 dpm, so
amount of probe to add = 36.7 µl/ 5.5 ml.

Note: cDNA or cRNA probes usually offer higher specific activities than oligoprobes with the implication of shorter exposure times. This can be overcome by using cocktails of oligoprobes (using nucleotide sequences from differing loci of the cDNA) whose effects are additive.

(i) Place hybridization solution onto a thawed section and cover with a piece of 'Seal-on' film (a type of 'Parafilm' obtainable from CamLab Ltd, Nuffield Road, Cambridge, CD4 1TH). Ensure that the dimensions of the 'Seal-on' film are less than the size of the slide. This will prevent seepage of hybridization fluid through capillary action.
(ii) Incubate for 20 h at 43°C in a sealed box lined with filter paper soaked in SSC × 4.

4.3. Post-hybridization washing

(i) Float off the 'Seal-on' film in a glass staining trough containing SSC × 1 at room temperature.
(ii) Wash the slides successively under the following conditions:
- 2 × 15 minutes at 55°C in SSC × 1 + 10 mM DTT
- 2 × 15 minutes at 55°C in SSC × 0.5 + 10 mM DTT
- 1 × 15 minutes at room temperature in SSC × 0.5 + 10 mM DTT.

(iii) Dip the slides briefly in distilled water.
(iv) Dehydrate the sections twice through 70%, 95% and 100% ethanol.
(v) Appose dried sections to β-max 'Hyperfilm' and expose for between 7 and 21 days depending upon the abundance of analyzed transcript under investigation.

4.4. Processing β-max 'Hyperfilm'

(i) Develop in Kodak D-19 developer (160 g/1 liter water) for 4 min.
(ii) Fix in Kodak 'Unifix' (1 part plus 4 parts water) for 7 min (or twice the clearing time).

(iii) Wash for at least 5 min in cold running water to remove traces of processing chemicals.
(iv) Allow films to dry vertically in air or in a drying cabinet set at 50°C for 30 min.

4.5. Controlling hybridization specificity

Essential to *in situ* hybridization methodology is the need to define the specificity of the hybridization signal and although there is no single definitive test, a number of control measures add to its integrity.

(i) Competition of the labeled probe with excess (100-fold) unlabeled probe or hybridizing with labeled sense probes (i.e. with nucleotide sequences identical to the mRNA under test) should produce little or no hybridization signal.
(ii) Conversely, cross-hybridization in the presence of a related but non-complementary sequence should not interfere with the signal.
(iii) Incubating sections in RNase A (100 μg/ml in 2 × SSC) for 60 min at 37°C removes endogenous RNA and should therefore prevent hybridization from occurring.
(iv) Multiple probe analyses using sequences taken from different loci of the genome should yield similar hybridization patterns.
(v) Correlation of the mRNA distribution with that of its corresponding protein using either receptor autoradiography or immunohistochemistry (14).
(vi) In instances where a probe derived from the genome of a species differing from the one under investigation produces a large percentage nucleotide mismatch, reduced specificity achieved under conditions of lower stringency would be preferred.

5. VISUALIZATION OF SIGNALS

Depending on the level of resolution, sensitivity and exposure period required, the most common isotopes used for probe labeling are ^{35}S, ^{32}P and ^{33}P, ^{35}S being the most common isotope employed for *in situ* hybridization (15).

More rapid detection (hours rather than days) is achieved using non-isotopic labeling for which there are two methodological approaches. Firstly, direct labeling of a detectable reporter molecule to the nucleic acid probe (e.g. fluorescein) permits the immediate microscopic examination of its target nucleic acid hybridization. Secondly, indirect labeling offers a more sensitive approach since the reporter molecule is rendered detectable by immunocytochemistry using specific antibodies. The most widely employed is dUTP-biotin which functions by virtue of its high affinity for the naturally-occurring proteins, avidin and streptavidin. Digoxigenin-linked uridine nucleotides provide an alternative indirect labeling approach and are detected by enzyme-conjugated antibodies such as alkaline phosphatase and peroxidase.

Different conjugates allow mixing of the two detection systems without the risk of cross-reaction, invaluable for visualizing two mRNA species within one cell. Double labeling of mRNAs for brain-derived neurotrophic factor and its receptor, TrkB, has enabled the visualization of cells co-expressing these two transcripts (16), while triple labeling to incorporate immunocytochemistry within the same brain section is also feasible (17).

Non-isotopically labeled probes require a chromogen for their microscopic visualization. Biotinylated enzymes such as alkaline phosphatase are detected by incubating with β-chloroindolyl phosphate (BCIP) and nitroblue tetrazolium, while peroxidase utilizes hydrogen peroxide with diaminobenzidine as chromogen.

6. MICROAUTORADIOGRAPHIC VISUALIZATION OF SIGNALS

The microscopic cellular localization of a hybridized radiolabeled probe is resolved by dipping sections/slides into a liquid photographic emulsion. The β-particle energy emitted from the isotope reduces Ag^+ ions in the emulsion to metallic silver, resulting in a latent image of silver grains which are visualized using photographic development and fixation. Counter-staining enables those cell populations associated with the grains to be revealed.

Emulsion dipping protocol
Either LM-1 (Amersham) or K5 (Ilford) nuclear emulsions are used for microautoradiography. It is important to be aware that these two preparations differ in terms of their gelatin and silver contents which will affect the thickness of the emulsion film and its sensitivity to the radioactive energy. K5 emulsion has gelatin and silver contents which are approximately double those of the Amersham preparation, which means that it should be diluted 1:1 with distilled water to achieve the appropriate concentration.

Optimal darkroom conditions are essential for the appropriate handling of liquid emulsions for which the level of red lighting should be kept to an absolute minimum. Open exposure of the emulsion should only occur in the presence of *one* beehive darkroom lamp containing a 15 watt bulb and a Kodak number 2 red safety filter. The lamp should be aimed at the darkroom ceiling and dipped slides should not be positioned in the direct path of the light. Greater illumination than this risks increasing the level of background grain signals.

(i) Pre-heat aliquots of the emulsion in a water-bath at 43°C for 30 min. LM-1 emulsion can be melted directly from stock. For K5 emulsion, transfer the solid emulsion into a plastic 50 ml Falcon tube using a spatula. Compress the emulsion until the volume in the tube approximates to 20 ml. Add the same volume of 2 × SSC and heat. The emulsion will require periodic gentle stirring with a glass rod to ensure homogeneity of the suspension.
(ii) Pour some of the warmed emulsion into a glass dipping vessel (supplied by Amersham) which should be 3/4 submerged in the water-bath.
(iii) Hold a slide vertically with the section facing towards you and dip in the vessel, holding it in place for 5 s. Slowly withdraw the slide from the vessel lightly pressing the 'back' of the slide against the wall of the dipping vessel. This removes excess emulsion from the back of the slide which is superfluous to requirements.
(iv) Place the slide vertically into a draining rack to dry.
(v) Repeat this procedure for all the slides ensuring that the rate of slide withdrawal from the emulsion remains as uniform as possible. This will help to keep variations in emulsion thickness to a minimum.
(vi) Allow the slides to dry completely (24 to 48 hours) before transferring them to a slide rack placed in a black plastic slide box containing a small sachet of silica gel.
(vii) Seal in a plastic bag and store at 4°C.

Developing and fixing of exposed emulsion
The exposure time for dipped hybridized slides is largely dependent upon the amount of nucleic acid present within the tissue sample and delay between labeling of the probe and its exposure to the emulsion. Since the half-life of ^{35}S isotopes is 87 days, the longer this delay, correspondingly longer exposure periods will be required. Generally, exposure periods of between 4 and 8 weeks are sufficient to allow visualization of silver grains over specifically hybridized cells.

Figure 1. Macroscopic autoradiographic hybridization signals of [^{35}S]-labeled oligonucleotide probes, complementary to the GABA$_A$ receptor α1 (A), α2 (B), α4 (C) and α6 (D) subunit-encoding mRNAs in horizontal sections of rat brain. The images are reverse negatives taken from β-max 'Hyperfilm'. Note the clear heterogeneity in the distribution patterns of these subunits which share around a 70% overall amino acid identity. The α1 subunit mRNA is the most ubiquitously expressed (A) with strong hybridization signals being observed in the neocortex (Ctx), cerebellar granule (gran) and Purkinje (PC) cells with medium levels throughout various thalamic nuclei (thal) and the inferior colliculus (IC). The highest α2 expression was apparent in the dentate gyrus (DG) and pyramidal cell layer (pyr) of the hippocampal formation (B) whilst that of the α4 subunit predominated in the thalamus (C). α6 subunit mRNA was restricted solely to granule cells of the cerebellum (D).

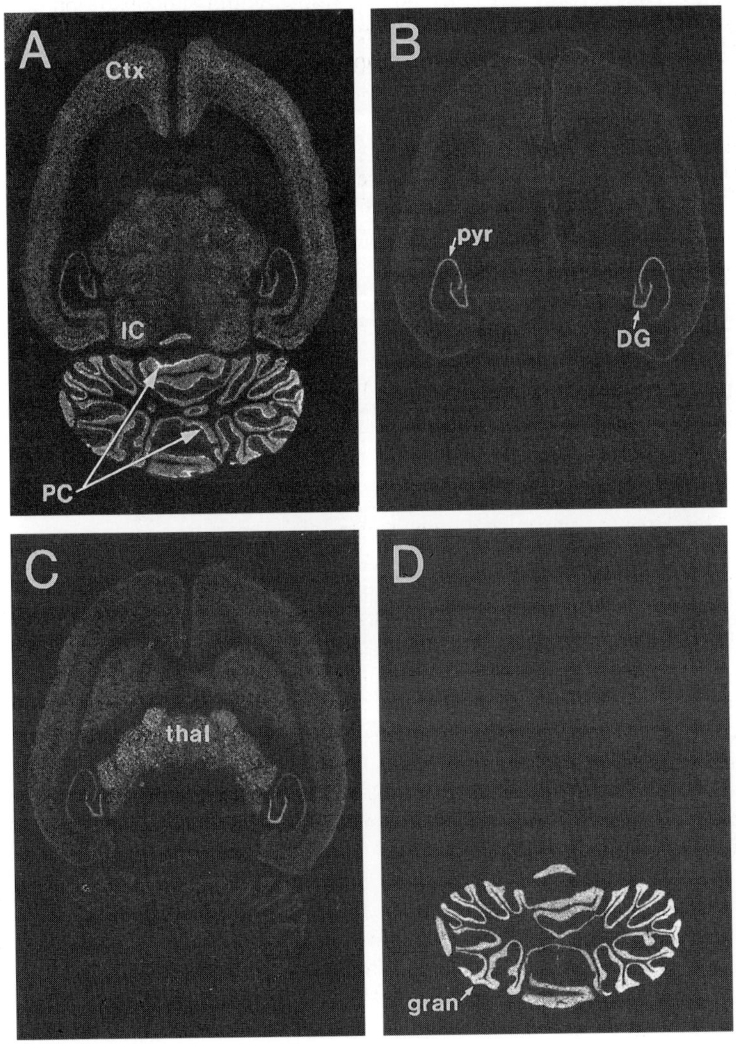

Figure 2. Hybridization signals of [^{35}S]-labeled oligonucleotide probes, complementary to the α1 (A), α2 (B), β2 (C) and γ2 (D) subunit mRNAs of the $GABA_A$ receptor in rat cerebellum. The cellular localization of the mRNA *in situ* was visualized by dipping sections in photographic emulsion, exposing for 1 month followed by developing, fixation and cresyl violet counter-staining. Differential expression patterns (as denoted by clusters of black grains) are evident especially in Purkinje cells with intense hybridization signals for the α1 subunit mRNA (A), moderate expression of the β2 and γ2 subunits (C and D), while expression of the α2 transcript is absent in these cells (B; arrowheads). The α1 and γ2 subunit mRNAs are also present in putative stellate cells (arrows) in the molecular layer (MOL). Much lower levels of hybridization intensity are associated with granule cells (GRAN).

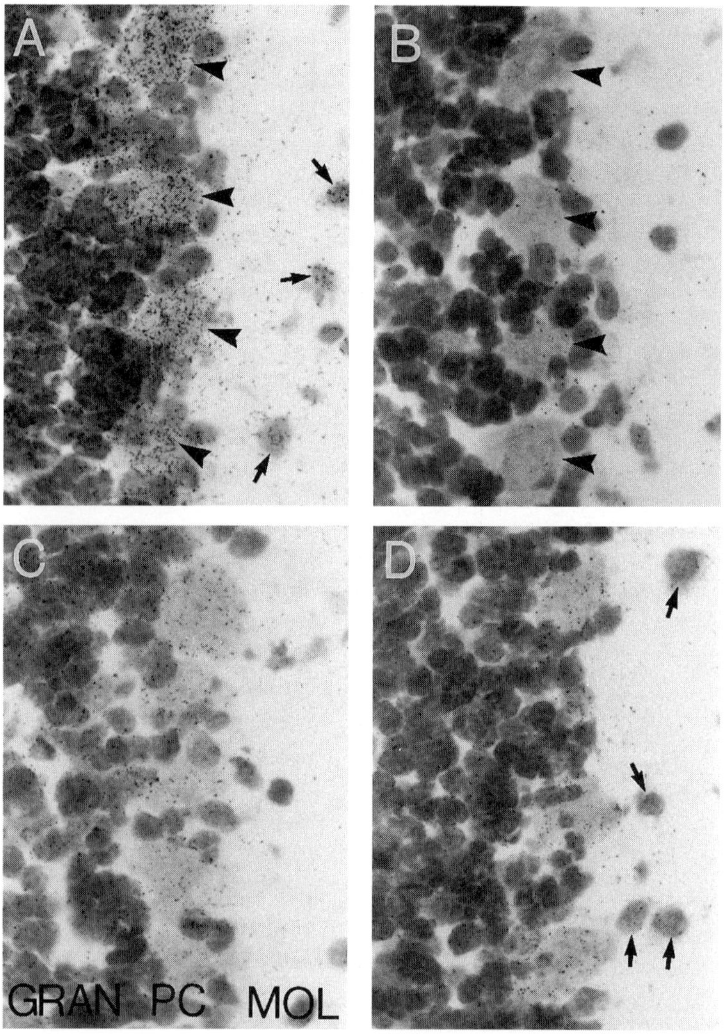

Similarly, the development time following exposure must be empirically determined. The optimal development time is that which produces the greatest signal to noise ratio. The development process distinguishes between those crystals in the emulsion either with or without a latent image center only by the differential rates of reduction of Ag^+ ions to metallic silver. Ultimately if sufficient time is allowed the developer will also reduce crystals without a latent image centre thereby increasing the level of background signal.

It is important to ensure that the temperatures of the developer, stop solution and fixative are the same and between 15°C and 20°C. Make up the following solutions.

Developer
Dissolve 160 g Kodak D-19 developer in 1 l water and dilute 1:1.
Stop solution
Dilute 1 part Ilford IN-1 stop solution plus 39 parts water.
Fixative
Dilute 1 part Kodak 'Unifix' plus 4 parts water.

The thickness of emulsion on a slide will determine the rate at which the developer and fixative become exhausted. For this reason, it is recommended that 250 ml aliquots of developer and fixative are used for 12–15 slides and then discarded.

(i) Remove slides from their 4°C environment and allow to equilibrate to room temperature before proceeding further. This will prevent any sudden changes in temperature from cracking the emulsion.
(ii) Using appropriate darkroom lighting conditions, remove dipped slides from their storage boxes and arrange in slide racks.
(iii) Using either a test or blank slide, develop for 1 min.
(iv) Place in the stop solution for 30 s.
(v) Transfer to the fixative. The fixation time is generally expressed as twice the time required for clearing the emulsion; i.e. for the emulsion to pass from opaque to transparent. Generally, at least 10 min should be allowed for this to occur and to take into account any subtle variations in emulsion thickness on a slide which will clear at a different rate to thinner regions.
(vi) Gently rinse in running tap water for around 15 min to remove traces of the processing chemicals.
(vii) At this stage, in the absence of any counter-staining, viewing the slide microscopically under dark-field illumination will reveal the intensity of the background signal (blank slide) in relation to specific 'hot-spots' on test slides with mounted sections.
(viii) The development time can be increased to between 2 and 4 min to increase signal intensity although periods in excess of 4 min should not be necessary since higher noise levels will result. Once satisfied with the intensity of the signal, the remaining slides may be processed.

Counter-staining with the Nissl stain – cresyl violet
Prepare the following reagents:

Cresyl violet
0.2% stock = 2 g/500 ml distilled water

Acetate buffer (0.1 M; pH 3.7)
Acidic component
3 ml glacial acetic acid in 500 ml distilled water

Basic component
1.36 g sodium acetate crystals ($CH_3OONa.3H_2O$) in 100 ml distilled water.

Titrate acetate buffer to pH 3.7 by adding approximately 180 ml acidic component to 20 ml basic component. Combine equal proportions of stock 0.2% cresyl violet and acetate buffer, making a final concentration of 0.1% and filter.

(i) Arrange slides in glass staining racks having carefully removed any excess drained emulsion.
(ii) Stain sections in 0.1% cresyl violet for 10 min.
(iii) Differentiate the intensity of the violet coloration by processing the samples twice through 70%, 95% and 100% ethanol. Approximately 2 min in each solution should be sufficient. The final stained color of cells should be light violet rather than intense purple. This is important especially when simultaneously photographing cells and silver grains using black and white film since cell populations stained intensely purple will appear very dark, which will impair the visualization of 'black' silver grains associated with those cells. This is especially true for densely packed cells such as those of the cerebellar and dentate gyrus granule cell layers.
(iv) Process slides through xylene (3×5 min).
(v) Cover-slip slides using 'Eukitt' mounting medium.

7. SIGNAL QUANTIFICATION

For both macroscopic and microscopic quantification, access to appropriate image analysis software is essential. One of the most versatile packages available for the neuroscientist is the MCID system supplied by Imaging Research (St Catherines, Ontario, Canada).

Quantification of hybridization signals, however, is at best semi-quantitative since an accurate extrapolation of autoradiographic grey levels to absolute cellular RNA concentrations is difficult. The most reliable approach is to make brain paste standards containing a range of [^{35}S]dATP concentrations from the same stock radioactivity as used for hybridizing sections. This will ensure that any variations attributable to the half-life of the isotope will be consistent and minimized. A standard curve to known amounts of radioactivity is plotted, optical density values converted into radioactivity units enabling molar quantities of label hybridized per unit area to be expressed. An alternative approach is to use commercially-available plastic autoradiographic microscales such as those supplied by Amersham (see *Table 6*). Given that the β-emission energies of ^{35}S and ^{14}C isotopes are similar (0.167 MeV and 0.156 MeV, respectively), ^{14}C-plastic standards can be used to calibrate ^{35}S-labeled hybridization signals (18). In view of the different half-lives of these two isotopes, only relative (rather than absolute) levels of signal can be expressed. For macroscopic analysis, since optical density is linear between 0 and 0.8 optical density units, the duration of film exposure should be such that the radioactive signal is not allowed to saturate the film since this will lead to over-estimated values.

Software packages for the microscopic quantification of silver grains have greatly advanced over recent years. Thresholding the image so that background signals can be eliminated enables only those grains specifically associated with cell populations to be recognized, thus providing a grain density per unit area determination.

8. REFERENCES

1. Pardue, M.L. and Gall, J.G. (1969) *Proc. Natl Acad. Sci.*, **64**, 600.

2. Comelli, M.C., Guidolin, D., Seren, M.S., Zanoni, R., Canella, R. *et al.* (1993) *Neuroscience*, **55**, 473.

3. Campbell, K., Wictorin, K. and Bjorklund, A. (1992) *Proc. Natl Acad. Sci.*, **89**, 10489.

4. Harrison, P.J., Proctor, A.W., Exworthy, T., Roberts, G.W., Najlerahim, A. *et al.* (1993) *Neuropathol. Appl. Neurobiol.*, **19**, 10.

5. Gregor, P., Reeves, R.H., Jabs, E.H., Yang, X., Dackowski, W. *et al.* (1993) *Proc. Natl Acad. Sci.*, **90**, 3053.

6. Mohler, H., Sigel, E., Malherbe, P., Richards, J.G., Persohn, E. *et al.* (1991) in *Transmitter Amino Acid Receptors: Transduction and Models for Drug Development*, vol. 6 (E. Costa and E.A. Barnard, eds), Raven Press, New York, p. 23.

7. Nakanishi, S. (1992) *Science*, **258**, 597.

8. Persohn, E., Malherbe, P. and Richards, J.G. (1992) *J. Comp. Neurol.*, **326:** 193.

9. Barton, A.J.L., Perason, R.C.A., Najlerahim, A., Harrison, P.J. (1993) Pre- and postmortem influences on brain mRNA. *J. Neurochem.*, **61**, 1.

10. Hofler, H. (1990) in *In Situ Hybridization: Principles and Practice* (J.M. Polak and J.O'D. McGee, eds), Oxford University Press, Oxford, p. 15.

11. Pratt, G.D., Kokaia, M. Bengzon, J., Kokaia, Z., Fritschy, J.-M., Mohler, H. and Lindvall, O. (1993) *Neuroscience*, **57**, 307.

12. Tecott, L.H., Eberwine, J.H., Barchas, J.D. and Valentino, K.L. (1987) in *In Situ Hybridization: Applications to Neurobiology* (K.L. Valentino, J.H. Eberwine and J.D. Barchas, eds), Oxford University Press, Oxford, p. 3.

13. Bonner, T.L., Brenner, D.J., Neufeld, B.R. and Britten, R.J. (1973) *J. Mol. Biol.*, **81**, 123.

14. McCabe, J.T. and Pfaff, D.W. (1989) in *Methods in Neuroscience*, vol. 1 (P.M. Conn, ed.), Academic Press, London, p. 98.

15. Emson, P.M. (1993) *Trends Neurosci.*, **16**, 9.

16. Kokaia, Z., Bengzon, J., Metsis, M., Kokaia M., Perssohn, H. and Lindvall, O. (1993) *Proc. Natl Acad. Sci.*, **90**, 6711.

17. Trembleau, A., Roche, D. and Calas, A. (1993) *J. Histochem. Cytochem.*, **41**, 489.

18. Miller, J.A. (1991) *Neuroscience Lett.*, **121**, 211.

CHAPTER 12

GEL ELECTROPHORESIS, WESTERN, SOUTHERN AND NORTHERN BLOTTING

S. Fazeli

Electrophoretic separation of macromolecules is one of the most widely used procedures in biology today. The aim of this chapter is to provide an introduction to some of the methods often used by molecular biologists, in a simple format. In most instances detailed protocols are given. A variety of high-quality reliable kits are available which simplify some procedures. Where, applicable these are noted.

1. POLYACRYLAMIDE GEL ELECTROPHORESIS (PAGE)

The technique of polyacrylamide gel electrophoresis was introduced by Laemmli in 1970. This technique has become one of the most widely used methods in biology. The basic mechanism of separation is size fractionation of proteins which have been rendered negatively charged by the presence of excess amounts of the detergent sodium dodecyl sulfate (SDS). An electric field is applied to the samples and the negatively charged proteins migrate to the anode. Larger proteins migrate at a slower rate than smaller ones, which therefore leads to the separation of a complex mixture of proteins into discrete bands under the influence of a discontinuous buffer system.

The gel system is divided into two different units, the separating and the stacking gel. The separating gel, as the name suggests, is where the protein separation occurs. The pore size of this part of the gel depends on the concentration of acrylamide used. The smaller the molecular size of the protein components of interest, the higher the percentage of the gel should be. The stacking gel, often made with a lower percentage of acrylamide, provides the wells into which the samples are loaded and also, by virtue of its buffer system and pH, provides a means of 'focusing' the proteins into a sharp band. See *Table 1*.

Preparation of the gel assembly will vary depending on the source of the apparatus.

Tables 2–5 illustrate the preparation of gel solutions. Note that the volumes apply to the Biorad protean II gel electrophoresis set up; for other systems the volumes may have to be adjusted.

Table 1. 5X sample buffer

Constituents	Volume/weight
1 M Tris-base (pH 6.8 using HCl)	4 ml
Sodium dodecyl sulfate (SDS)	1 g
Glycerol	4 ml (5 g)
Dithiothreitol (DTT)	0.771 g
Bromophenol blue (BPB)	0.01 g
Ultrapure water (UPW)	To 10 ml

Table 2. Buffers

Solution	Preparation notes	Precautions	Storage
A-Sep	18.4 g Tris (base) 0.4% SDS in 100 ml UPW, pH 8.7 with conc. HCl	Wear mask during preparation	4°C
A-Stack	6 g Tris (base) 0.4% SDS in 100 ml UPW, pH 6.8 with conc. HCl	Wear mask during preparation	4°C
B	38.4 g acrylamide 1.2 g bis-acrylamide in 100 ml UPW	Acrylamide is a neurotoxin: wear mask during preparation	4°C in a dark bottle
C	UPW		
D	TEMED[a]		4°C
E	10% ammonium persulfate		Must be made fresh prior to use

[a]TEMED = *N,N,N′,N′*-tetramethylethylenediamine.

Table 3. Gel solution

Running gel (%)	Volume of buffers					Application[a]
	A-Sep (ml)	B (ml)	C (ml)	D (μl)	E (ml)	
7.5	11.25	8.6	25.2	30	0.3	High-molecular-weight proteins > 80 kDa
10	11.25	11.3	22.5	30	0.3	Suitable for 50–150 kDa
12.5	11.25	14.3	19.5	30	0.3	Suitable for 30–130 kDa
15	11.25	17.1	16.1	30	0.3	Suitable for 10–90 kDa
Stacking gel[b]	2.5	0.75	6.75	10	0.2	

[a]If separation of proteins with very similar molecular weights is desired, the user should refer to ref. 2, which details the use of Tricine buffers. Also, halving the amount of bis-acrylamide in solution B could help.
[b]Astac.

Table 4. 10X concentrate electrode buffer

Constituents	Volume/weight
Tris-base	30.28 g
Glycine	144.14 g
UPW	2 liters

Table 5. Sample preparation and gel running

Step	Action	Special procedures	Notes
1	Add 1 part sample buffer to 4 parts sample; heat to 95°C for 3–4 min; freeze at –70°C if required	To lyse cells directly into sample buffer, use 1X diluted buffer	If the sample is viscous due to DNA, sonicate for about 10 min; sample should be blue; if yellow or purple, add NaOH or HCl respectively
2	Prepare desired separating gel solution adding ammonium persulfate when ready to pour the gel; pour into gel assembly	Fill to within 2 cm of top of glass plates to allow room for the stacking gel; cover the gel solution with a layer of water-saturated butanol	Allow 30 min to set
3	Wash the top of the polymerized gel with UPW, removing excess using a piece of 3MM filter paper		Wear gloves to avoid contact with unpolymerized acrylamide solution
4	Pour stacking gel; place combs in position and leave to polymerize for 30 min		Wear gloves to avoid contact with unpolymerized acrylamide solution
5	Remove combs and rinse the wells with UPW; flick gel assembly to empty wells		
6	Place gel assembly into running tank; fill the inner and outer chambers with 1X electrode buffer *with added SDS*	Add 5 ml 10% SDS for each 500 ml of electrode buffer	
7	Load samples into wells using a blunted Hamilton syringe		Load one well with prestained or normal protein standards

2. TWO-DIMENSIONAL GEL ELECTROPHORESIS (2D-PAGE)

O'Farrell (3) introduced this improved technique for high-resolution two-dimensional polyacrylamide gel electrophoresis (2D-PAGE) in 1975. Since then it has become one of the most powerful tools for the separation of complex protein mixtures. Recent advances in the production of high-quality and consistent ampholines (the chemicals that make up the pH gradient), and of sophisticated equipment, have made the technique applicable to quantitative analysis. High-powered computer software and laser scanning equipment have made the analysis of 2D gels relatively easy. 2D-gels can be used to detect amino acid substitutions, changes in post-translational modifications such as phosphorylation, and up-and-down regulation of the level of a protein in a complex mixture. This section provides the basic procedures required for the production of 'home-made' 2D-gels. However, a number of pre-cast first-dimension gels with immobilized pH gradients are available and produce good results. The final quality of the gels depends on consistent technique and the use of high-quality reagents.

The principle of operation of this gel system is the separation of proteins on the basis of two different characteristics. Proteins are first separated on a pH gradient according to their isoelectric points (the pH at which the protein has no net charge), usually on a long rod-shaped gel, and then by conventional PAGE according to their molecular size.

See *Table 6 and 7* for an illustration of the technique.

Table 6. First-dimension isoelectric focusing

Solution	Procedure		Notes
Sample overlay	8 M urea containing 1% ampholines; this should be a mixture of 0.8% pH 5–7 and 0.2% pH 3–10 ampholines		Urea should be dissolved by warming in hot water; then the ampholines can be added
30% acrylamide	28.38% acrylamide 1.62% bis-acrylamide		Wear mask and gloves during preparation of this solution
Equilibration buffer	8 ml	10% SDS	
	2 ml	Tris-base (pH 6.8 with HCl)	
	14.4 ml	UPW	
	600 μl	BPB (0.05%)	
	0.13 g	DTT	
Lysis buffer	5.4 g	urea	This should be aliquoted as desired and frozen at –20°C; use within 1 month, avoiding freeze-thawing
	2.5 ml	UPW	
	1 ml	DTT (10%)	
	0.3 ml	ampholines (5–7)	
	0.2 ml	ampholines (3–10)	NP40[a] can be replaced by CHAPS[b]; add 0.4 g per 10 ml of buffer
	1.6 ml	NP40 (25%)	
Gel solution	1.57 ml	UPW (degassed)	
	1.33 ml	acrylamide stock (30%)	
	2 ml	NP40 (10%)	
	5.4 g	urea	
	0.5 ml	ampholines 40% (pH 5–7)	
	0.2 ml	ampholines 40% (pH 3–10)	

[a]NP40 = Nonidet P40.

[b]CHAPS = (3-[(3-cholamidopropyl)-dimethyl(ammonio]-1-propane-sulfonate.

Table 7. Sample preparation and gel running

Step	Action	Special procedures	Notes
1	Homogenize soft tissues direct into urea lysis buffer using a Dounce homogenizer; then centrifuge at 12 000 *g* for 5 min		As a rule of thumb, 20 mg (wet weight) of brain tissue into 300 µl of buffer Do not allow the buffer to heat up since this will result in protein amidation by the urea; it is for this reason that samples should be used as fresh as possible, but samples can be kept at –70°C for up to a month
2	If samples are already in solution, add 54 mg solid urea per 56 µl of sample; add 100 µl of lysis buffer	Warm in palm of hand to dissolve the urea	
3	Wash glass rods with acid, rinse and silane treat		
4	Seal the bottom of rods with 2 or 3 layers of parafilm; mark the position to which the gels will be poured		Making 10–20 gels at once is recommended to allow consistency
5	Make gel solution, add 7 µl TEMED[a] and 10 µl 10% ammonium persulfate		
6	Fill each tube with gel solution using a drawn-out glass Pasteur pipette; fill from the bottom up		Tap all tubes to remove any trapped air bubbles
7	Overlay the gels with 10 µl lysis buffer and allow to polymerize for 1 h		Keep tubes away from cold draughts to prevent crystallization of the urea
8	Remove parafilm and place tubes in the electrophoresis chamber		
9	Fill lower chamber with 10 mM phosphoric acid	The buffer should be made fresh	Remove any air bubbles from the bottom of the tubes using a bent needle and syringe to flush them out
10	Remove lysis buffer from the top of gels, rinse with water, apply fresh lysis buffer and then fill with 20 mM NaOH (cathode) which should also fill the top chamber	Prepare the NaOH freshly and degas before use	

Table 7. Continued

Step	Action	Special procedures	Notes
11	Pre-run the gels at 200 V for 15 min, 300 V for 30 min, and 400 V for 1 h	At the end of the pre-run the current should *not* be more than 2 mA per 12 gels	Pre-running ensures the formation of stable pH gradient, and any excess urea is removed
12	Remove all spent cathode buffer and replace with fresh buffer; add 10 µl sample overlay to each tube; inject samples under the overlay		Run one dummy gel for the measurement of pH gradient
13	Run gels at 400 V for 18 h; turn up the power to 1000 V for 1 h before the end; do not exceed 10 000 V		
14	Remove tubes; squeeze gels out into a trough containing equilibrium buffer; leave for 2–3 min	Remove gels by attaching a silicone tube to the end and gently force out by applying pressure with a syringe filled with UPW; be very gentle near the end, since the top of the gel often shrinks	Gels can be removed into a 15 ml tube containing equilibrium buffer and frozen at –20°C for about a week; thaw and use immediately
15	Apply gel to the top of a slab (second-dimension) gel; this is prepared as described above (Polyacrylamide gel electrophoresis); stacking gel is made with only one well in which molecular weight may be applied.		
16	Fix the gel rod into position using 1% molten agarose in equilibrium buffer		
17.	Cut the dummy gel rod into 1 cm pieces and soak in 1 ml UPW overnight; read the pH the following day		This gives a rough idea of the pH gradient; pH gradient markers can also be purchased

[a]TEMED = *N,N,N′,N′*-tetramethylenediamine.

3. GEL STAINING

One- and two-dimensional gels can be stained by a variety of methods. Most commonly used are Coomassie blue and silver stains. Coomassie blue stains are very simple and can be used to visualize proteins in gels that are then used for autoradiographic exposures. The stain, though relatively quantitative, is not very sensitive.

Silver staining may be more desirable if sensitivity is more important than quantitative ability. Gels to be used for autoradiography cannot be stained with silver. There are a number of silver staining kits commercially available which may be more suitable for applications where reproducibility is important. The kits are also relatively quantitative. These are, however, quite expensive and therefore a protocol is provided here which gives good-quality stains with low background and should be adequate for most purposes. The stained gels can be stored in sealed plastic bags containing a small amount of water for extended periods.

Tables 8 and 9 describe the solutions used.

Table 8. Silver staining solutions

Reagent	Amount	Time (min; with agitation)
Methanol Trichloroacetic acid Copper chloride Formaldehyde	50% (v/v) 20% (w/v) 2% (w/v) 0.1% (v/v)	25
Ethanol Acetic acid	10% (v/v) 5%	15
Potassium permanganate	0.01% (w/v)	15
Ethanol Acetic acid	10% (v/v) 5% (v/v)	15
Ethanol	10% (v/v)	15
UPW		15
Silver nitrate	0.1% (w/v)	15
UPW		Rinse
Potassium carbonate	10% (w/v)	1
Formaldehyde Potassium carbonate	0.02% (v/v) 2% (w/v)	Observe while bands appear
Ethanol Acetic acid	10% 5%	2
UPW		15 (3 changes)

Gel Electrophoresis

Table 9. Coommassie blue

Constituents	Volume/weight	Notes
Staining solution		
Coommassie blue R	0.25 g	Dissolve the dye in methanol first
Methanol	125 ml	
Acetic acid	25 ml	
UPW	100 ml	
Destaining solution		
Methanol	400 ml	
Acetic acid	50 ml	
UPW	550 ml	

Gel staining procedure

This protocol for silver staining and staining with Coommassie blue, if followed exactly, should give high-quality low-background stains. The purity of the reagents (both those used for the preparation of the gels and the silver stain reagents) are of particular importance in achieving low background levels.

(i) Place gels into the staining solution in a sandwich box. Place on an orbital shaker for at least 1 h. The staining solution can be used until the staining becomes too weak. Increase the incubation time gradually with use.
(ii) Remove staining solution and rinse with water.
(iii) Add destaining solution and agitate until the desired signal is seen. Change the destaining solution at least once.
(iv) Gel can be dried onto Whatmann 3MM paper or stored in sealed plastic bags with a small amount of water or 70% acetic acid.

4. WESTERN BLOTTING

By virtue of the fact that proteins in the gel have a net negative charge, they can be eluted out of the gel onto a solid support matrix such as nitocellulose. This process is termed western blotting (4). Following electrophoresis gels are sandwiched with a piece of nitrocellulose membrane between sheets of Whatmann filter paper and a current is applied across the gel, which results in the gradual migration of proteins onto the nitrocellulose membrane. The actual preparation of the sandwich is usually described in the booklet that accompanies the apparatus. *Tables 10 and 11* illustrate the technique.

Table 10. Preparation of the sandwich

Step	Action	Special procedures	Notes
1	Soak sponge, six pieces of 3MM paper and a piece of nitrocellulose in transfer buffer		Transfer buffer is the same as electrode buffer without SDS
2	Prepare the sandwich as follows: sponge/ 3×3MM paper/gel/nitrocellulose/3×3MM paper/sponge	It is crucial to note the order of the components; use of a color-coded cassette helps; the gel should be nearer the black/ negative terminal	Remove air bubbles after each layer is added using a glass rod or pipette; do not press too hard, to avoid distortion of the gel
3	Place assembly in the tank, noting the orientation		Make sure the buffer covers the entire gel assembly
4	Apply current of 100 mA to the assembly, keeping the voltage constant; this is left overnight		Current should reach about 150 mA
5	Remove the nitrocellulose into the appropriate buffer (see below)	The membrane can be stained using Ponceau S; follow the manufacturer's recommendation; membranes can be stained before or after immunoblotting	Prestained molecular weight markers are of tremendous help in localizing immunoreactive bands

Table 11. Protein detection in the western blotting method

Step	Action	Special procedures	Notes
1	Dissolve 4 g casein in TBS or PBS[a] by heating to about 50°C; add phenol red and adjust pH with NaOH until pink, i.e. about pH 7.0	Heat slowly, otherwise the casein will denature	Aliquot into 30 ml fractions and freeze at –20°C; can be frozen and thawed a number of times 4% BSA[a] or powdered milk is also regularly used, but some antibodies show high background signal in milk
2	Following transfer, place the membrane in 4% casein for about 1 h at room temperature		If mini gels are used, the membrane will fit into a 50 ml Falcon; use 5 ml of 4% casein and place on a roller mixer
3	Change to primary antibody diluted in 4% casein at 1:100–1:1000 dilutions, depending on the antibody; shake for 1 h		The dilution will depend on the titre of the antibody
4	Wash 3 times in PBS or TBS for 10 min		
5	Apply HRP-labeled[a] secondary antibody in 4% casein; shake for 1 h		Again the dilution will depend on the quality and titre of the antibody; it should be fine at 1:1000; good batches should be used at 1:3000
6	Repeat step 4; follow by a 10 min wash in UPW		
7	Develop immunoreactive band(s)		Use enhanced chemiluminescence, following the manufacturer's protocol

[a]TBS = Tris-buffered saline; PBS = phosphate-buffered saline; HRP = horseradish peroxidase; BSA = bovine serum albumin.

5. SOUTHERN BLOTTING

Agarose gel electrophoresis

The technique of southern blotting involves separation of DNA fragments by standard agarose gel electrophoresis, transfer to a solid support such as nitrocellulose or nylon membranes, followed by the detection of specific sequences of interest. In this section a simple method is provided which we have routinely used for the analysis of genomic DNA fragments. See *Tables 12 and 13*.

Table 12. Solutions for southern blotting

Solution	Constituents	Amounts	Notes
5X TBE[a]	Tris-base	54 g	
	Boric acid	27.5 g	
	0.5 M EDTA (pH 8.0)	20 ml	
	UPW	1 liter	
6X DNA sample buffer	BPB[a]	0.25%	
	Glycerol	30%	
	EDTA (pH 8.0)	1 mM	
Denaturation buffer	NaCl	1.5 M	
	NaOH	0.5 M	
Neutralization buffer	Tris-base, pH 7.4	1 M	
	NaCl	1.5 M	
20X SSC	NaCl	175 g	pH 7.0 with a few drops of 10 M NaOH
	Na citrate	88.2 g	
	UPW	To 1 liter	Autoclave
Hybridization buffer	SSC[a]	6×	
	Denhardts	5×	
	SDS[a]	0.5%	
	Salmon/herring sperm DNA	100 μg/ml	
Denhardts solution (50X)	Ficoll (Type 400, Pharmacia)	5 g	Filter and store at –20°C
	Polyvinyl pyrrolidone	5 g	
	Bovine serum albumin	5 g	
	UPW	to 500 mls	

[a] TBE = Tris–borate–EDTA; BPB = bromophenol blue; EDTA = ethylenediaminetetracetic acid; SSC = sodium chloride/sodium citrate; SDS = sodium dodecyl sulfate.

Table 13. Gel running

Step	Action	Special procedure	Notes
1	Dissolve agarose at 1% in 0.5X TBE	Use a microwave; 600 ml will require approximately 8 min to dissolve	After dissolving place on a magnetic stirrer and mix slowly (no air bubbles) until the container is not hot to touch
2	Add 1 vol ethidium bromide (10 mg/ml) to 10 000 vol of gel solution		Ethidium bromide is a toxin; do not allow contact with skin
3	Pour gel into gel cast		
4	Add 6X sample buffer to DNA digests		
5	Place set gel into a tank filled with 0.5X TBE; load samples into wells		Leave one lane free for molecular weight markers
6	Run gel at 100 V		Small gels should not be run at more than 50 V; for good separation the gel should be run at 20–30 V overnight
7	Photograph gel with a ruler		

Transfer of DNA fragments

A number of methods have been used for the transfer of DNA fragments from agarose gels onto membranes for subsequent analysis. We favor the capillary transfer method for its simplicity and reliability. See *Table 14*.

Table 14. Transfer of DNA fragments

Step	Action	Notes
1	Trim gel; cut the left hand corner to aid orientation	
2	Place gel in denaturation buffer for 45 min on a rotary shaker	In this step the DNA strands are denatured by the alkali, which aids transfer
3	Rinse gel in UPW	The gel is quite fragile at this stage
4	Shake in neutralization buffer for 15 min	
5	Change to fresh neutralization buffer and shake for further 30 min	
6	Transfer the gel onto a platform which has been already covered with two layers of Whatmann filter paper soaked in 10X SSC	Each end of the filter paper, hanging from the platform, should be submerged in 10X SSC; orient the gel such that the top surface is now facing down
7	Gently remove any air bubble by squashing the gel by hand	
8	Cover the free surfaces of the filter paper with clingfilm or lengths of parafilm	These surfaces must not come into contact with the filter paper which will cover the membrane (step 9–10), which would short-circuit the transfer
9	Layer nylon membrane previously soaked in UPW onto the gel; remove any air bubbles with a glass rod	Pall membranes (Biodyne A transfer membrane (0.2 μm)) gives very satisfactory results
10	Layer two pieces of filter paper soaked in UPW on top of the membrane and remove any air bubbles	
11	Cover with many layers of absorbent paper towels (hand towels) and compress down with a 1 pound weight; leave overnight	
12	Mark the position of a few wells with a pen; cut off the appropriate corner of the membrane; air dry membrane for 10 min	
13	Crosslink the transferred DNA according to manufacturer's recommendations; the membrane can be stored indefinitely	For Pall membranes, wrap in clingfilm and expose on a UV box for 5 or 6 min

Pre-hybridization

(i) Soak the membrane in 6X SSC for 1 min.

(ii) Place in hybridization tube and add hybridization mix for 3–4 h at 65°C. Normally a dedicated rotary hybridization oven is used. Hybritubes (Gibco) can also be used very effectively and cost considerably less.

Preparation of probe

This should begin about 1.5 h prior to the end of the pre-hybridization step. There are a number of methods which can be used for the preparation of radiolabeled or non-radioactive probes. We have routinely used the random primed labeling procedure which results in very highly specific activity probes. This procedure is not suitable for use with probes that are smaller than 200 bp in length. We find that the Random Primed kit from Germany gives highly reliable labeling of DNA fragments and this system is routinely used in our laboratory. See *Table 15*.

Table 15. Preparation of probe

Step	Action	Notes
1	Heat DNA (100–200 ng in 10 µl) to denature strands, then snap chill on ice	
2	Add 6 µl labeling mix, then 5 µl radiolabeled nucleotide followed by 1 µl Klenow enzyme (1 unit)	Observe all usual precautions while using radioactivity
3	Incubate in dedicated water-bath for 1–1.5 h	
4	Add 350 µl Tris–EDTA (pH 8.0) and 50 µl salmon/herring sperm DNA (from a solution of 10 mg/ml)	
5	Immerse in boiling water for 5 min and then snap chill on ice	This separates the synthesized and template strands
6	Spin for 10 s and add to a tube containing 10–15 ml of hybridization buffer	
7	Remove the pre-hybridization buffer and the probe; leave overnight at 65°C	
8	Carefully remove membrane into a large sandwich box containing 300 ml of 2X SSC/0.5% SDS; place on a rotary shaker at 65°C for 15 min	
9	Repeat step 8	
10	Reduce the SSC strength to 1X and wash for another 15 min	It is often not necessary to wash with higher stringency buffer; if in doubt expose the membrane for a day and if necessary wash further
11	Wrap the wet membrane in clingfilm and expose at –70°C	Wrap well so as to avoid leakage of buffer

6. NORTHERN BLOTTING

Northern blotting is similar to southern blotting, except that in this case it is RNA molecules that are of interest rather than DNA.

One of the major difficulties encountered in northern analysis is RNA degradation. All solutions that come into contact with RNA samples should therefore be free from contaminating RNAses. Most standard chemicals are not free from RNAse, and for northern blotting applications they should be of special purity. All glassware should be autoclaved and all solutions should be made up with RNAse-free water. Thus, water should be treated with diethylpyrocarbonate (DEPC; 0.1%), and then autoclaved. All equipment should be washed with either mild acid or detergent, then soaked in DEPC water (which is only active for about 2–3 h after preparation) for 2 h. It is preferable to have designated items for northern blotting, which are not used for other purposes.

Gloves should be worn at all times, since human skin is a rich source of RNAse. DEPC should not be used for treatment of Tris buffers and other 'amines'. These agents should be kept aside and used only for RNA gel preparations.

There are a number of methods for the preparation of total RNA (see ref. 5). This section describes a simple and efficient method for total RNA preparations which should give high-quality RNA for most purposes.

One-step AGPC RNA purification
This method (6) can be used to isolate total RNA from small or large quantities of tissue or cells. Guanidinium thiocyanate is a strong inhibitor of RNAses, as well as an effective protein denaturant. In this procedure, the method of Chirgwin *et al.* (7) has been modified to give a rapid protocol involving a single-step extraction followed by isopropanol precipitation. *Table 16* describes the necessary solutions, and *Table 17* illustrates tissue preparation.

If the RNA under examination is of low abundance, it may be advisable to use mRNA preparations rather than total RNA. For this purpose, the poly-A tail of mRNA is utilized as a tool for its specific purification. A detailed protocol is provided in ref. 5. There are also a number of kits available for the purification of mRNA from mammalian cells which give high-quality mRNA preparations.

See *Table 18* for a sample preparation, and *Table 19* for a description of gel electrophoresis of RNA.

Table 16. Solutions for northern blotting

Solutions	Constituents	Amounts	Notes
Denaturing solution	Guanidinium thiocyanate	250 g	Dissolved in 293 ml DEPC-treated UPW; add β-mercaptoethanol or DTT to 0.1 M prior to use
	0.75 M soldium citrate (pH 7)	17.6 ml	
	10% Sarcosyl (dissolved at 65°C)	26.4 ml	
20X formaldehyde running buffer (FRB)	Morpholino-propanesulfonic acid (MOPS)	41.8 g	
	Sodium acetate	6.8 g	pH 7.0 with 10 M NaOH
	Di-sodium EDTA (0.5 M stock)	40 ml	
	DEPC-treated UPW	To 500 ml	
Denaturing buffer	20X Formaldehyde running buffer (FRB)	0.1 ml	
	Formaldehyde (37%)	0.7 ml	
	Formamide	2 ml	
	DEPC-treated UPW	0.3 ml	

Table 17. Tissue preparation

Step	Action	Special procedures	Notes
1	Chop tissue coarsely in a petri dish using a scalpel	Maintain at 4°C; place petri dish on ice	Minimize the time for which tissue is left exposed as it will degrade rapidly
2	Transfer to a 50 ml Falcon tube containing a small volume of denaturing solution		The volume required will vary depending on how much tissue is to be used; as a rule use 10 ml/g of tissue
3	Homogenize tissue in denaturing solution until mixture is homogenous		
4	Sequentially add: (i) 0.1 vol of 2 M soldium acetate (pH 4) (ii) 1 vol of phenol (water-saturated)		Mix thoroughly by inversion after each addition
	(iii) 0.2 vol of chloroform/isoamyl alcohol (49:1)	After the final addition shake vigorously for 10 s	Incubate on ice for 15 min
5	Centrifuge at 4000 *g* for 20 min		Transfer aqueous phase (top) to fresh tube, without taking any of the interface
6	Add 1 vol of isopropanol; incubate at –20°C for at least 1 h		
7	Centrifuge at 12 000 *g* for 20 min		Redissolve the pellet in 0.3 vol of denaturing solution; add 1 vol of isopropanol and incubate at –20°C for at least 1 h
8	Centrifuge at 12 000 *g* for 20 min; wash pellet with 70–80% ethanol, recentrifuge and allow to air dry		Do not over-dry; 10 min should be ample
9	Dissolve the pellet in a suitable volume of 0.5% SDS at 65°C		Alternatively, dissolve in 10 mM Tris (pH 7.5)/1 mM EDTA

Table 18. Sample preparation

Step	Action	Notes
1	Add 31 μl of denaturing buffer to every 9 μl of sample	If total RNA is being used, 15–30 μg needs to be loaded; mRNA preparations can be used at 1–10 μg
2	Mix and heat at 55°C for 30 min; immediately place on ice for at least 5 min	
3	Add gel loading buffer (1:10 v/v denatured sample); store on ice until use	

Table 19. Gel electrophoresis of RNA

Step	Action	Notes
1	Prepare gel solution: agarose and DEPC-treated UPW	
2	Boil in microwave for 2–3 min to melt; cool to 50–60°C, then add: 5 ml 20X FRB and 18 ml formaldehyde (37%)	
3	Pour into mold	Once gel has set, cover with 1X FRB and carefully remove combs
4	Pre-run the gel for 10 min at 100 V (about 85 mA)	
5	Fill first well with 5–10 μg of RNA ladder	This is treated exactly the same as samples, except add 1/10 vol of a solution 5 mg/ml ethidium bromide after the addition of gel loading buffer
6	Load samples	
7	Run gel for 2–3 h at 100 V	
8	Visualize stained ladder under UV and photograph adjacent to a ruler	

Transfer

Follow the procedure for southern blots, except that denaturation/neutralization of gel is not required prior to transfer. Probe preparation and hybridization are also identical.

7. REFERENCES

1. Laemmli, U.K. (1970) *Nature*, **227**, 680.

2. Schagger, H. and von Jagow, G. (1987) *Anal. Biochem.*, **166**, 368.

3. O'Farrell, P.H. (1975) *J. Biol. Chem.*, **250**, 4007.

4. Towbin, H., Staehelin, T. and Gordon, J. (1979) *Proc. Natl Acad. Sci. USA*, **76**, 4350.

5. Sambrook, J., Fritsch, E.F. and Maniatis, T. (1989) *Molecular Cloning: A Laboratory Manual*, vol. 1, 2nd edn, Cold Spring Harbor Laboratory Press, Cold Spring Harbor.

6. Chomczynski, P. and Sacchi, N. (1987) *Anal. Biochem.*, **162**, 156.

7. Chirgwin, J.M., Przybyla, A.E., MacDonald, R.J. and Rutter, W.J. (1979) *Biochemistry*, **18**, 5294.

CHAPTER 13
SAMPLING OF EXTRACELLULAR FLUIDS *IN SITU*: A COMPARISON OF HPLC TECHNIQUES AND MICRODIALYSIS FOR THE ANALYSIS OF NEUROTRANSMITTERS IN BIOLOGICAL FLUIDS

I.C. Kilpatrick

1. INTRODUCTION

Most biologists will need to sample biological fluids at some point during their research. As far as neurobiology is concerned, the sampling of fluids both within and surrounding the nervous tissue of mammals has been severely hampered by the anatomical protection commonly but necessarily afforded to these delicate structures. Indeed, the inaccessibility of neural tissue and its environment has often meant that clinical diagnoses of nervous system disorders are little more than 'best guesses', typically founded on analysis of biopsies and/or blood constituents. Now, it is a common axiom that the physiological functioning of the nervous system in most mammals relies on the release and sequestration of chemical messenger molecules from and by both neurons and glial cells. It is therefore clear that if we are even to begin to understand the workings of the nervous system at the cellular level, then we must employ techniques in the laboratory – and eventually in the clinic – that allow us to monitor this chemical interplay between cells. Electrophysiological recording from individual neurons partly offers this possibility already. However, rarely do such experiments provide definitive evidence for the existence of one or more substances in the extracellular milieu.

This chapter is intended to show that the methods available for sampling the neuronal environment have come a long way since the 'gross biopsy' method yielded to progressively less invasive techniques. Specifically, it concentrates on the use of microdialysis (1), a method which exploits chemical concentration gradients and relies on the fluxes of substances across a semi-permeable membrane in an approximate proportion to their concentration gradients. The term 'approximate' is used here because it is well-known that individual substances display different but not necessarily unique diffusional characteristics across a given type of dialysis membrane. However, it is not the purpose of this chapter to provide a major theoretical treatment of the method, and readers are referred to several excellent chapters in the current microdialyser's 'bible' (2–4).

Since small-diameter, hollow dialysis fibers became routinely available, the technique has continuously grown in popularity since its application was realistically offered just over a decade ago. In combination with increasingly more sensitive analytical techniques that both stemmed from and include high-performance liquid chromatography (HPLC), the microdialysis method is now allowing us to glimpse some of the chemical events that occur in the nervous system and even to correlate these events with changes not only in groups of neurons but in the behavior of the whole animal (5).

The concept and principles of using dialysis to sample extracellular fluids in a variety of tissues are some 30 years old (6) and the method's development is charted in *Table 1*. Use of the latest combined microdialysis–electrode arrays (13–15) will be mentioned later.

2. APPLICATIONS OF MICRODIALYSIS

Microdialysis is used increasingly in the basic science research environment and has been exploited for the measurement of a host of neurotransmitter compounds, both accepted and suspected. These have traditionally included catecholamines, indoleamines and their metabolites, amino acids and acetylcholine, but more recently the technique has been refined to permit measurement of several neuropeptides (see ref. 16 for example). This chapter concentrates on the application of microdialysis to monoamine and amino acid analysis, although there are several useful articles that detail the methodology required to analyse microdialysis samples for acetylcholine by chromatography and electrochemical detection (17, 18) and neuropeptides by chromatography and radioimmunoassay (19–22). It should be noted that microdialysis is now being applied to the clinic for studies of brain chemistry linked to head injury, trauma and neurological disease including epilepsy (23, 24). It is not difficult to foresee that such techniques will eventually be of both diagnostic value and functionally restorative.

3. MONOAMINES AND METABOLITES: THEIR SEPARATION AND DETECTION

3.1. Electrochemical detection

Extensive treatment of these detection methods and their practical operation has already been given elsewhere (25–28) and so only a brief description is warranted here. *Tables 2 and 3* outline some key features of modern analysis of catecholamines, indoleamines and their metabolites, stemming from the fact that such compounds are electroactive and therefore amenable to electrochemical detection at suitable electrodes. This fortuitous molecular nature of the monoamines and their metabolites is probably not well-appreciated outside the physical chemistry laboratory but is fundamental to the success and continued development of the analysis techniques. It comes as no surprise that as the use of the microdialysis technique has burgeoned, so has the demand for increasingly low detection limits. As a result, the last five years or so have seen considerable improvements in sensitivity of electrochemical detectors.

3.2. The separation process

Since publication of earlier articles on this topic (27, 28), detection enhancements in commercially-available systems have been achieved in part through improvements in electronic circuit components, and in part through reductions in the detector cell volume and/or changes in flow profile within the detector. However, the most consistent (and guaranteed) improvement in detection sensitivity results from the use of narrower columns (*Table 4*).

Traditionally (in the 1970s and 1980s), the stationary phase was packed into columns whose internal diameter was 4.6 mm. Together with the elution of a sample's components using a mobile phase that was pumped through them, such 'normal-bore' columns were (and probably still are!) the mainstay of chromatographers in neuroscience. Although 1.0 mm-diameter columns had been shown to be of great promise for enhancing detection limits of monoamines and metabolites as early as 1984 (29), these so-called 'microbore' columns have only really been exploited in the current decade (30–32). Apart from a lack of readily-available hardware (with consequently poor marketing) and to a certain extent a genuine ignorance on behalf of the end-users, it is puzzling why biologists were not introduced to such columns much earlier.

Table 1. Development of microdialysis methodology

Procedure	Collection	Advantage(s)	Disadvantage(s)	References
Implantation of dextran/ saline-filled sacs (Visking®tubing) into dog cerebral cortices and neck skin	Leave for 10 weeks; remove; analyse contents for amino acids	Introduced idea of two compartments whose contents would equilibrate across the dialysis membrane	Equilibrium data; single time interval measurement only	6
Gaddum's push–pull cannulae	On-line collection; off-line analysis	Continuous sampling; time-course feasible	Hard to optimize; damage to brain tissue due to high fluid flow rates	7
Cortical cup perfusions	As above	As above	As above, but also limited to superficial brain structures (see refs 10, 11)	8, 9
Dialytrode (Delgado and colleagues)	A push–pull cannula with a small, permeable bag enveloping both tips; forerunner of today's on-line probes.	Small-scale; electrode alongside cannula permitted quasi-seizure recording from monkey amygdala after infusing a 3.0 μl bolus of 1.0 M L-glutamate.	Width of probe too large for some smaller nuclei	12
Microdialysis–electrode array	Continuous, on-line; the latest combination	AgCl electrode permits sampling of DC potential at point of exposed dialysis membrane		13–15

Table 2. Electrochemical detection of catecholamines, indoleamines and their metabolites

Basis of detection	Advantage(s)	References
Compounds are electroactive; readily undergo voltage-dependent oxidation and subsequent reduction	Required electrical potential is relatively low; this permits low detector noise and good compound selectivity with little interference from compounds requiring greater potentials for oxidation or reduction	25–28
Liberation of electrons during oxidation is proportional to amount of electroactive species	Changes in concentration of analyte yield linear alterations in detector response.	

Table 3. Types of EC detection

(i)	**Amperometric** – only a portion of the analyte is subjected to the oxidative or reductive field	Analyte typically flows *over* the electrode surface; can produce low noise whilst maintaining high detection sensitivity
(ii)	**Coulometric** – the intention is to electroactively convert the total analyte sample	Analyte typically flows *through* the electrode; can produce larger analyte signal as long as detection electronics are suitably noise-free

Table 5 summarizes the characteristics and inherent benefits (some only gained with hindsight) of smaller-bore columns whose use can be a real boon to microdialysers, particularly those who have regularly complained about a lack of detection sensitivity. Such columns also offer a way of counteracting the oft-voiced criticism of poor temporal resolution in microdialysis experiments.

The picture is not completely rosy, however. One of the more frustrating aspects is the general unsuitability of electrochemical detection for use with solvent gradient chromatography. The perturbations in solvent composition created by such gradients are often too great to permit detector baseline stability, especially at high detector gains, and so single-solvent (isocratic) conditions are rarely deviated from. Further, while the benefits of smaller-bore chromatography are an enhanced detection sensitivity and a reduced solvent consumption, the trade-off is that the volume of sample that can be loaded onto the column is restricted (*Table 5*).

3.3. The way forward: capillary electrophoresis?

More striking where small volumes are concerned, however, is the adoption and development (36–38) of capillary electrophoresis (CE) and capillary zone electrophoresis (CZE) in which a sample is introduced into a glass or fused-silica capillary (typically 30–100 µm i.d.) that has been equilibrated with a buffer solution and which may also contain a detergent, this latter being essential for the separation of neutral compounds. The capillary is exposed to a high-voltage field (10–30 kV) and the sample's components are separated both electrically (due to

Table 4. Examples of conventional and newer liquid chromatography (LC) methods of analysis for monoamines and metabolites

Analyte(s)	Separation method	Detection method	Sample volume	Detection limits	References
DA, NA, 5-HT and metabolites	Conventional HPLC	EC (coulometric array)	20 µl	25 fmol/20 µl	26, 28, 34
DA, NA, 5-HT and metabolites	Microbore HPLC	EC (amperometric)	1 µl	1 fmol/µl	29, 32
DA and metabolites	Capillary LC	EC (amperometric)	0.1–2.0 µl	0.25 fmol/µl	33
DA, NA, A	Conventional HPLC	Fluorometric PCD with DPE / KFeCN	10 µl	2–5 fmol/10 µl	35

5-HT = 5-hydroxytryptamine; A = adrenaline; DA = dopamine; DPE = 1,2-diphenylethylenediamine; EC = electrochemical; KFeCN = potassium ferricyanide; NA = noradrenaline; PCD = pre-column derivatization.

Table 5. The separation process with smallbore and microbore columns (small volume of the stationary phase)

Consequence	Advantage(s)	Disadvantage(s)	Comments	References
(i) Small volume of mobile phase	Reduced solvent consumption		Both solvent usage reduction and detector signal increase are proportional to the ratio of the squares of the column radii; ~ 20-fold for 1.0 mm *vs* 4.6 mm-diameter columns	29–32
Reduced dilution of analyte by column, so increased concentration of analyte at detector	Enhanced detection sensitivity			
(ii) Sample volume must be small (0.05–5.0 μl) and ideally ≤ 1.0 μl		Submicrolitre volumes are difficult to handle	Fused-silica capillary columns may be beneficial	33
		Typical microdialysis sampling times of 10–20 min are not appropriate	Reduced loading capacity of microbore column can be used to provide two advantages: (i) A reduced probe perfusion rate will enhance relative recovery and concentration of collected material. (ii) Sampling more frequently will provide a vastly improved temporal resolution.	

differences in isopotential) and if a detergent is present, chromatographically (due to differential distribution of the analytes between a mobile phase that is electro-osmotically 'pumped' and a micellar phase, created by the detergent). Electrochemical detection is accomplished with miniature electrodes placed either inside the capillary or at a 'joint' created by fracturing the capillary towards its migratory end.

Whilst the main difference between CE and CZE is the flow profile generated during the separation, their common characteristics are small sample volumes, high separation efficiencies and rapid separations. Since separation speed is a function of the field strength, extremely fast separations can be achieved. These can be as low as tens of milliseconds, but 5–10 minutes is a more common time-frame. Owing to the nature of the capillary loading procedure (pressure, vacuum or electrokinesis), sample volumes are sometimes difficult to determine with accuracy. Typical injection volumes are between 0.1 and 10 nl, but in practice much larger sample volumes are used owing to the difficulty of manipulating such small volumes. A 0.5 μl sample thus provides sufficient scope for several hundred repeat injections if required! Coupling a CE apparatus to an electrochemical detector offers all the attendant advantages of such detection, i.e. selectivity and the ability to use voltammetric characterization of unknown components in the microdialysate or other sample. Most impressive, however, are the absolute detection limits of this technique. Whilst the *concentration* detection limits are similar to if only marginally better than conventional HPLC and electrochemical detection (typically 20–50 nM), the *mass* detection limits are staggeringly low. Attomoles (10^{-18} mol) are the common units here, although as described in section 4.2, even these barriers have been broken using CE.

The CE technique is currently in its infancy and is not without its teething troubles, but we are likely to see more and more microdialysis laboratories adopting this as their quest for improved temporal resolution grows.

3.4. Other forms of detection of monoamines and their metabolites

Whilst electrochemical detection remains the major weapon for monoamine analysis in the analytical chemist's armamentarium, other methods are continually being developed. In view of the fact that catecholamines were first detected in brain tissue in the late 1950s and early 1960s by fluorescent methods, it is somewhat ironic that one of the most recent developments for catecholamine analysis by HPLC involves a fluorescent method (35). A slightly complicated pre-column derivatization procedure (though its automation was described) with 1,2-diphenylethylenediamine and potassium ferricyanide yielded highly fluorescent catecholamine derivatives whose detection limits were between 2 and 5 fmol (35). This represents a 2- to 4-fold improvement on electrochemical detection methods using similar chromatographic conditions. Perhaps more importantly, the relative ease with which fluorescence detectors can be maintained in comparison with electrochemical detectors (see refs 27, 28 and 39) may encourage some laboratories to adopt this technique.

4. AMINO ACIDS: THEIR SEPARATION AND DETECTION

4.1. Fluorescence detection

Owing to a lack of sensitivity, the traditional cation exchange-based chromatographic separations of free amino acids and their spectrophotometric detection by post-column derivatization with ninhydrin are inappropriate for analysis of amino acids collected by microdialysis. However, seemingly countless articles exist on the application of derivatization chemistries to convert amino acids to compounds that (i) are easily resolved using reversed-phase HPLC and (ii) can be detected with high sensitivity and selectivity by fluorimetry. The majority of these methods deploy pre-column derivatization of amino acids and, as already discussed (39, 40), the author's experience resides almost exclusively with their pre-column derivatization using *ortho*-phthaldialdehyde (OPA) and a reducing agent, usually 2-mercaptoethanol.

From the author's point of view, a major benefit of the OPA technique rests in its inability to create derivatives of anything other than primary amines. This may not be sufficient for clinically-based analyses where, for example, proline determinations are required, but it certainly minimizes the number of peaks that appear on the chromatogram and is a genuine aid to unknown peak identification (39–41).

The technique is sufficiently sensitive for most neurotransmitter-related amino acids present in dialysates, but a number of modifications/alternatives have appeared which have resulted in major increases in detection sensitivity. These are summarized in *Table 6*. Of course, not everyone has access to laser-based systems as employed by Roach and Harmony (42), but they would certainly be a boon to the neuroscience community. For comparison, *Table 7* illustrates an array of pre-column derivatization agents that have often been employed together with some of the features of each method.

4.2. Separation of amino acid derivatives

For the most part, efficient separation on reversed-phase columns of amino acid derivatives created by pre-column treatment of the sample is only possible with gradient chromatography; and unlike electrochemical detectors, fluorimeters are thankfully extremely tolerant to changes in solvent composition. Fluorimetric analysis of such derivatives of the more common amino acids, including those regarded as neurotransmitter candidates (i.e. aspartate, glutamate, glycine and 4-aminobutanoate (GABA)), can normally be accomplished in a single chromatographic run of 20–30 min using a binary gradient (39–41, 43–46). The imposition of a solvent elution gradient which increases the proportion of organic modifier (usually methanol or acetonitrile) with time is required to cope with the varied hydrophobicity of some of the amino acid derivatives present within the sample. Indeed, the most hydrophobic of these would not be eluted within an acceptable run time by use of isocratic solvent conditions. More complex analyses may benefit from ternary gradients, although in the author's experience, the simpler the gradient, the more transferable the method between laboratories. Typical capacity factors for some OPA-based amino acid derivatives using a phosphate buffer *versus* methanol gradient are given in *Table 8*.

The use of narrow-bore columns has been exploited for fluorescence-based analysis of the tripeptide, glutathione, in cortical microdialysates (47). Not surprisingly, CE- and CZE-based separations of amino acid derivatives have been in continuous development since the 1980s resulting in commercially-available CE/laser-induced fluorescence equipment. Both OPA derivatives (48, 49) and NDA–CN derivatives (50–52) have been analyzed in this manner and the explosive increase in detection sensitivity afforded with the latter technique is unlikely to be surpassed for practical purposes. In brain microdialysates, for example, detection of NDA–CN derivatives of glutamate by laser-induced fluorescence has been achieved at a 10 nM concentration (52). However, the limit of mass detection of the glutamate derivative has been measured by Hernandez and colleagues (51) in the *zeptomole* (10^{-21} mol) range (equating to low nanomolar glutamate concentrations) and if a 15-μm i.d. capillary was used, the migration time of glutamate from a brain dialysate could be all of 60 seconds! If this is not sufficiently stunning, a microdialysis study of extracellular cortical glutamate changes in ischemia by Hernandez' group has used CE and laser-induced fluorescence of the fluorescein isothiocyanate derivative to explore such alterations in the *yoctomole* (10^{-24} mol) range (53). Detection at this order of magnitude could begin to raise questions about Avogadro's number!

Table 6. Modifications to analysis of fluorescent derivatives of amino acids

Traditional	Modification	Sensitivity increase	Comments	References
Excitation of isoindole fluorophore with halogen/xenon lamp	Excitation of isoindole fluorophore with an argon-ion laser	20- to 100-fold depending on the amino acid	Useful for less abundant amino acids, e.g. excitatory sulfur-containing amino acids	40–42
OPA reagent with halogen/xenon lamp	NDA–CN with halogen/xenon lamp	10- to 30-fold for primary amines	Produces highly fluorescent cyanobenz[f]isoindoles; di-, tri- and pentapeptides successfully derivatized	42
OPA reagent with halogen/xenon lamp	NDA–CN with argon-ion laser	Up to 500-fold for primary amines; absolute sensitivity of 100–400 aM per amino acid	All this despite the use of conventional-bore columns!	42

NDA–CN = 2,3-naphthalenedialdehyde/sodium cyanide; OPA = *o*-phthaldialdehyde.

Table 7. Examples of pre-column derivatization methods for conventional LC or CE analysis of amino acids

	DANSYL	DABSYL	FMOC	OPA	NDA–CN	AQC
Reaction time	30 min	25 min	< 30 s	0.5–2.0 min	15 min	1 min
Reaction temperature	37°C	70°C	20°C	20°C	20°C	20°C
Detection mode	Fluorescence	Visible	Fluorescence	Fluorescence	Fluorescence	Fluorescence
Detection limits						
LC	1–10 pmol	2–10 pmol	100 fmol	50 fmol	10–20 fmol	40–300 fmol
CE/laser				4–10 fmol	0.1–0.4 fmol	
Secondary amines	Yes	Yes	Yes	No	No	Yes
Multiple product formation	Yes	Yes	Yes	No	No	No
Interference by derivatization reagent	Yes	Yes	Yes	No	No	No
Stability of derivative	Stable	2 days	6 days	min–h	Stable	Stable
References	61	62	63, 64	39–46	42, 50–53	65

AQC = 6-aminoquinolyl-*N*-hydroxysuccinimidyl carbamate; DABSYL = 4-(dimethylamino)azobenzene-4′-sulphonyl chloride; DANSYL = 5-dimethylaminonaphthalene-1-sulphonyl chloride; FMOC = 9-fluorenylmethylchloroformate; NDA-CN = 2,3-naphthalenedialdehyde/sodium cyanide; OPA = *o*-phthaldialdehyde.

Table 8. Typical capacity factors of some OPA-amino acid derivatives[a]

Derivative	Capacity factor (k′)
Aspartate	0.94
Glutamate	2.28
Asparagine	3.22
Serine	4.17
Glutamine	5.34
Glycine	6.45
Threonine	7.17
Arginine	9.56
Alanine	10.67
Taurine	12.23
Tyrosine	15.28
GABA	16.16

[a]All data were obtained using the chromatographic conditions detailed in refs 39 and 46.
$k' = (t_c - t_o)/t_o$, where t_c is the analyte retention time (min) and t_o is the void volume retention time (min). In this case, $t_o = 2.45$ min.

4.3. Other forms of detection of amino acids

Some of the fluorescent derivatives of amino acids generated by pre-column derivatization are also electrochemically active, and the ability to immobilize enzymes to various substrates has allowed development of methods for the continuous electrochemical detection of some amino acids, particularly glutamate. These features have been exploited by several groups and are outlined in *Table 9*. As far as enzyme-based glutamate assays are concerned, detection limits of 500 nM (59) and 200 nM (60) are obviously well-suited to measurements of glutamate in brain microdialysates.

Table 9. Other forms of detection of amino acids

Detection	Comments	References
Electrochemical detection of OPA derivatives	Concurrent electrochemical and fluorimetric detection	34, 54–57
Electrochemical detection of OPA-based derivatives using sulfite ions instead of thiol	Combats unpleasant odour of commonly used thiols; can be used (57) to measure GABA in microdialysates (10 fmol/20 μl)	55–57
Electrochemical detection of NDA–CN derivatives after separation by CE	Reported detection limits of 80 nM for aspartate and 100 nM for glutamate	58
Electrochemical detection of oxidative products after enzyme immobilization	Immobilization of glutamate oxidase to silica beads (59) or inside a tubular platinum electrode (60) permits the continuous monitoring of glutamate following the production of ferrocinium ion (59) or H_2O_2 (60)	59, 60

NDA–CN = 2,3-naphthalenedialdehyde/sodium cyanide; OPA = *o*-phthaldialdehyde.

5. SOME NEWER PRACTICAL ASPECTS OF MICRODIALYSIS

It is not the remit of this chapter to detail the minutiae of the microdialysis method since discussions on the choice of dialysis membrane (16, 66), and the membrane perfusion medium (67, 68) can already be consulted. However, there are some recent methodological advances that are noteworthy and these include the design of a loop-probe, spinal dialysis catheter for chronic placement within the intrathecal space of the unanesthetized rat (69). The use of the spinal loop probe has been verified by measuring extracellular glutamate and glucose concentrations in response to ischemia, compression and noxious afferent stimulation amongst others. This is yet another example of a system that allows contemporaneous measurement of neurochemistry and behavior.

Another recent application was alluded to earlier. This involves the insertion of a microelectrode within a microdialysis probe so that the DC potential of the surrounding tissue can be recorded concurrently in response to the transfer of depolarizing agents from within the probe (13–15). Such microdialysis–electrode arrays have been used to good effect in determining, for example, that high extracellular potassium ions and not glutamate are required for propagation of spreading depression within the rat cerebral cortex (15). Similar studies have revealed that the very presence of the microdialysis probe can *prevent* such propagation (14), an effect that was both independent of perfusion flow rate and probe-induced tissue damage. Such data infer that the microdialysis probe is a very efficient buffer of transient changes in the composition of extracellular fluid. In certain situations (e.g. clinical explorations) this may have a detrimental action.

The recent findings of Westergren and colleagues should also be borne in mind (70). Their evidence indicates that the use of intracerebral dialysis allows access from plasma to brain of drugs and other compounds that do not normally cross the blood–brain barrier.

Finally, one elegant study has recently combined microdialysis, amino acid analysis and histology to reveal that increases in extracellular glutamate by inhibition of its uptake are not sufficient to cause neuronal destruction (presumably by an excitotoxic action) in the rat caudate–putamen (71).

6. QUANTIFICATION IN MICRODIALYSIS

As universally recognized, the estimation of substance recovery via a probe measured *in vitro* is almost certainly going to be different when that probe is used *in vivo* to recover those same substances (2, 3, 16, 72). This makes calculation of actual extracellular analyte concentrations unreliable at best, and it is common to see dialysate data given simply as 'amount per sample', about which there is no dispute.

Among the various methods available for determining *in vivo* recovery, and hence resting extracellular concentration, of a substance (see ref. 11), one of the simplest is the 'zero net flux' method first developed by Lönnroth and colleagues (73). Here, a point is obtained at which a known concentration of substance, when added to the medium perfusing the implanted probe, does not diffuse out into the tissue. Rarely is such an exact concentration of substance fed through the perfusate, however. More frequently, a series of concentrations above and below this equilibrium point are used to plot the differences between the supplied and recovered concentrations *versus* the supplied concentration. Since the procedure uses a constant perfusion rate, Justice (74) points out the value of this technique over variable-flow procedures. These are unable to use either linear regression (to produce a straight line fit whose gradient represents the *in vivo* recovery) or interpolation (to obtain the no net flux, equilibrium concentration). There are slight disadvantages to this method, including the fact that constant intra-tissue perfusion may remove other low-molecular-weight molecules that may affect the data. Nevertheless, application of this method to both the caudate–putamen (74) and nucleus accumbens (75) has estimated extracellular dopamine concentrations to be in the same, low nM range in each area.

7. CONCLUSION

It is clear that neurotransmitter analysis has come a long and valuable way since the 'pluck it and see' approach that dominated the first half of this century. However, it is possible that technology has now grown too successfully for its benefits to be reaped in the short term by the neuroscientist. The vanishingly small quantities of material that can now be analyzed in a single sample are posing difficult questions to the experimentalist. For example, should he or she change the design of their experiment just because equipment can measure ever-smaller changes? As always, the answer will depend on the desired end-point of the data, the ingenuity of the laboratory scientist and the extent to which thoroughness can be sacrificed for a desire to test the latest gizmo!

8. ACKNOWLEDGMENTS

The author is extremely grateful to the Medical Research Council, the Royal Society and the Wellcome Trust for their financial support. Sincere thanks to Dr Nicky French for her patience and helpful tabular suggestions.

9. REFERENCES

1. Robinson, T.E. and Justice, J.B. (1991) *Microdialysis in the Neurosciences: Techniques in the Behavioral and Neural Sciences*, vol. 7, Elsevier, Amsterdam.

2. Morrison, P.F., Bungay, P.M., Hsiao, J.K., Mefford, I.N., Dykstra, K.H. and Dedrick, R.L. (1991) in *Microdialysis in the Neurosciences: Techniques in the Behavioral and Neural Sciences*, vol. 7 (T.E. Robinson and J.B. Justice, eds), Elsevier, Amsterdam, p. 47.

3. Benveniste, H. and Hansen, A.J. (1991) in *Microdialysis in the Neurosciences: Techniques in the Behavioral and Neural Sciences*, vol. 7 (T.E. Robinson and J.B. Justice, eds), Elsevier, Amsterdam, p. 81.

4. Kissinger, P. (1991) in *Microdialysis in the Neurosciences: Techniques in the Behavioral and Neural Sciences*, vol. 7 (T.E. Robinson and J.B. Justice, eds), Elsevier, Amsterdam, p. 103.

5. Westerink, B.H.C. (1995) *Behav. Brain Res.*, **70**, 103.

6. Bito, L., Davson, H., Levin, E., Murray, M. and Snider, N. (1966) *J. Neurochem.*, **13**, 1057.

7. Gaddum, J.H. (1961) *J. Physiol. London*, **155**, 1–2P.

8. Mitchell, J.F. (1963) *J. Physiol. London*, **165**, 98.

9. Pepeu, G. and Bartolini, A. (1968) *Eur. J. Pharmacol.*, **4**, 254.

10. Ungerstedt, U. (1984) in *Measurement of Neurotransmitter Release In Vivo. IBRO Handbook Series: Methods in the Neurosciences*, vol. 6 (C.A. Marsden, ed.), Wiley, Chichester, p. 81.

11. Ungerstedt, U. (1991) in *Microdialysis in the Neurosciences: Techniques in the Behavioral and Neural Sciences*, vol. 7 (T.E. Robinson and J.B. Justice, eds). Elsevier, Amsterdam, p. 3.

12. Delgado, J.M.R., De Feudis, F.V., Roth, R.H., Ryugo, D.K. and Mitruka, B.M. (1972) *Arch. Int. Pharmacodyn. Ther.*, **198**, 9.

13. Obrenovitch, T.P., Urenjak, J. and Zilkha, E. (1994) *Br. J. Pharmacol.*, **113**, 1295.

14. Obrenovitch, T.P., Zilkha, E. and Urenjak, J. (1995) *J. Neurochem.*, **64**, 1884.

15. Obrenovitch, T.P. and Zilkha, E. (1995) *J. Neurophysiol.*, **73**, 2107.

16. Kendrick, K.M. (1991) in *Neuroendocrine Research Methods* (B.D. Greenstein, ed.), Harwood Academic, Chur, p. 249.

17. Damsma, G., Westerink, B.H.C., de Vries, J.B., van den Berg, C.J. and Horn, A.S. (1987) *J. Neurochem.*, **48**, 1523.

18. Damsma, G., Westerink, B.H.C., de Boer, P., de Vries, J.B. and Horn, A.S. (1988) *Life Sci.*, **43**, 1161.

19. Maidment, N.T., Siddall, B.J., Rudolph, V.R., Erdelyi, E. and Evans, C.J. (1991) *Neuroscience*, **45**, 81.

20. Evans, C.J., Maidment, N.T. and Newcomb, R. (1993) in *High Performance Liquid Chromatography in Neuroscience Research. IBRO Handbook Series: Methods in the Neurosciences*, vol. 15 (R.B. Holman, A.J. Cross and M.H. Joseph, eds), Wiley, Chichester, p. 265.

21. You, Z.-B., Herrera-Marschitz, M., Brodin, E., Meana, J.J., Morino, P., Hökfelt, T., Silveira, R., Goiny, M. and Ungerstedt, U. (1994) *J. Neurochem.*, **62**, 76.

22. You, Z.-B., Nylander, I., Herrera-Marschitz, M., O'Connor, W.T., Goiny, M. and Terenius, L. (1994) *Neuroscience*, **63**, 415.

23. Meyerson, B.A., Linderoth, B., Karlsson, H. and Ungerstedt, U. (1990) *Life Sci.*, **46**, 301.

24. Kanthan, R., Shuaib, A., Goplen, G. and Miyashita, H. (1995) *J. Neurosci. Meth.*, **60**, 151.

25. Refshauge, C., Kissinger, P.T., Dreiling, R., Blank, L., Freeman, R. and Adams, R.N. (1974) *Life Sci.*, **14**, 311.

26. Matson, W.R., Langlais, P., Volicer, L., Gamache, P.H., Bird, E. and Mark, K.A. (1984) *Clin. Chem.*, **30**, 1477.

27. Kilpatrick, I.C., Jones, M.W. and Phillipson, O.T. (1986) *J. Neurochem.*, **46**, 1865.

28. Kilpatrick, I.C. (1991) in *Neuroendocrine Research Methods* (B.D. Greenstein, ed.), Harwood Academic, Chur, p. 527.

29. Caliguri, E.J. and Mefford, I.N. (1984) *Brain Res.*, **296**, 156.

30. Pettit, H.O. and Justice, J.B. (1991) in *Microdialysis in the Neurosciences: Techniques in the Behavioral and Neural Sciences*, vol. 7 (T.E. Robinson and J.B. Justice, eds), Elsevier, Amsterdam, p. 117.

31. Carter, A.J. (1994) *J. Chromatogr. B: Biomed. Appl.*, **660**, 158.

32. Cheng, F.C. and Kuo, J.S. (1995) *J. Chromatogr. B: Biomed. Appl.*, **665**, 1.

33. Ruban, V.F. (1993) *J. Chromatogr. B: Biomed. Appl.*, **619**, 111.

34. Gamache, P., Ryan, E., Svendsen, C., Murayama, K. and Acworth, I.N. (1993) *J. Chromatogr. B: Biomed. Appl.*, **614**, 213.

35. Kehr, J. (1994) *J. Chromatogr. A*, **661**, 137.

36. Ewing, A.G., Wallingford, R.A. and Olefirowicz, T.M. (1989) *Anal. Chem.*, **61**, 292A.

37. Novotny, M.V. (1989) *Science*, **246**, 51.

38. Ewing, A.G., Mesaros, J.M. and Gavin, P.F. (1994) *Anal. Chem.*, **66**, 527A.

39. Kilpatrick, I.C. (1991) in *Neuroendocrine Research Methods* (B.D. Greenstein, ed.), Harwood Academic, Chur, p. 555.

40. Kilpatrick, I.C., Waller, S.J., Harrison, P.S. and Evans, R.H. (1993) in *High Performance Liquid Chromatography in Neuroscience Research. IBRO Handbook Series: Methods in the Neurosciences*, vol. 15 (R.B. Holman, A.J. Cross and M.H. Joseph, eds), Wiley, Chichester, p. 139.

41. Waller, S.J., Kilpatrick, I.C., Chan, M.W.J. and Evans, R.H. (1991) *J. Neurosci. Meth.*, **36**, 167.

42. Roach, M.C. and Harmony, M.D. (1987) *Anal. Chem.*, **59**, 411.

43. Hodgin, J.C. (1979) *J. Liq. Chromatogr.*, **2**, 1047.

44. Jones, B.N. and Gilligan, J.P. (1983) *J. Chromatogr.*, **266**, 471.

45. Lindroth, P. and Mopper, K. (1979) *Anal. Chem.*, **51**, 1667.

46. Kilpatrick, I.C. and Mozley, L.S. (1986) *Neurosci. Lett.*, **72**, 189.

47. Yang, C.S., Chou, S.T., Lin, N.N., Liu, L., Tsai, P.J., Kuo, J.S. and Lai, J.S. (1994) *J. Chromatogr. B: Biomed. Appl.*, **661**, 231.

48. Cheng, Y.-F. and Dovichi, N.J. (1988) *Science*, **242**, 562.

49. Liu, J., Cobb, K.A. and Novotny, M. (1989) *J. Chromatogr.*, **468**, 55.

50. Oates, M.D. and Jorgenson, J.W. (1989) *Anal. Chem.*, **61**, 432.

51. Hernandez, L., Escalona, J., Verdeguer, P. and Guzman, N.A. (1993) *J. Liq. Chromatogr.*, **16**, 2149.

52. Dawson, L.A., Stow, J.M., Dourish, C.T. and Routledge, C. (1995) *J. Chromatogr. A*, **700**, 81.

53. Rada, P., Tucci, S. and Hernandez, L. (1995) *Soc. Neurosci. Abstr.*, **21**, 610.

54. Joseph, M.H. and Davies, P. (1983) *J. Chromatogr.*, **277**, 125.

55. Pearson, S.J., Czudek, C., Mercer, K. and Reynolds, G.P. (1991) *J. Neural Transm.*, **86**, 151.

56. Smith, S. and Sharp, T. (1994) *J. Chromatogr. B: Biomed. Appl.*, **652**, 228.

57. Rowley, H.L., Martin, K.F. and Marsden, C.A. (1995) *J. Neurosci. Meth.*, **57**, 93.

58. O'Shea, T.J., Weber, P.L., Bammel, B.P., Lunte, C.E., Lunte, S.M. and Smyth, M.R. (1992) *J. Chromatogr.*, **608**, 189.

59. Boutelle, M.G., Fellows, L.K. and Cook, C. (1992) *Anal. Chem.*, **64**, 1790.

60. Berners, M.O., Boutelle, M.G. and Fillenz, M. (1994) *Anal. Chem.*, **66**, 2017.

61. Bayer, E., Grom, E., Kaltenegger, B. and Uhmann, R. (1976) *Anal. Chem.*, **48**, 1106.

62. Chang, J., Knecht, R. and Braun, D.G. (1981) *Biochem. J.*, **199**, 547.

63. Einarsson, S., Josefsson, B. and Lagerkvist, S. (1983) *J. Chromatogr.*, **282**, 609.

64. Einarsson, S. (1985) *J. Chromatogr.*, **348**, 213.

65. Cohen, S.A. and Michaud, D.P. (1993) *Anal. Biochem.*, **211**, 279.

66. Hamberger, A., Jacobson, I., Larsson, S., Lönnroth, P., Nyström, B. and Sandberg, M. (1991) in *Microdialysis in the Neurosciences: Techniques in the Behavioral and Neural Sciences*, vol. 7 (T.E. Robinson and J.B. Justice, eds), Elsevier, Amsterdam, p. 407.

67. Westerink, B.H.C., Hofsteede, H.M., Damsma, G. and de Vries, J.B. (1988) *Naunyn-Schmiedeberg's Arch. Pharmacol.*, **337**, 373.

68. Moghaddam, B. and Bunney, B.S. (1989) *J. Neurochem.*, **53**, 652.

69. Marsala, M., Malmberg, A.B. and Yaksh, T.L. (1995) *J. Neurosci. Meth.*, **62**, 43.

70. Westergren, I., Nyström, B., Hamberger, A. and Johansson, B.B. (1995) *J. Neurochem.*, **64**, 229.

71. Massieu, L., Morales-Villagran, A. and Tapia, R. (1995) *J. Neurochem.*, **64**, 2262.

72. Kehr, J. (1993) *J. Neurosci. Meth.*, **48**, 251.

73. Lönnroth, P., Jansson, P.-A. and Smith, U. (1987) *Am. J. Physiol.*, **253**, E228.

74. Justice, J.B., Jr. (1993) *J. Neurosci. Meth.*, **48**, 263.

75. Parsons, L.H. and Justice, J.B. (1992) *J. Neurochem.*, **58**, 212.

CHAPTER 14
RECEPTORS IN THE NERVOUS SYSTEM

G.J. McBean

1. INTRODUCTION

Cell-surface receptors in the nervous system are a set of integral membrane proteins that form the essential link between release of a neurotransmitter from the pre-synaptic terminal and initiation of the desired response in the target cell. Activation of a receptor is achieved by the binding of a neurotransmitter to its specific receptor (1), which induces a conformational change in the receptor protein, and thus evokes a response in the post-synaptic cell.

Long before the structural identity of individual receptors was known, the function of a receptor was viewed largely in terms of its ability to provide a mechanism of propagation of nervous impulses. This definition is acceptable as an explanation of the mechanism of signal propagation associated with fast synaptic transmission, where the receptor is coupled to a ligand-gated ion channel, but it falls short of explaining the mechanism of signal propagation associated with slow responses in the nervous system. There are now a far greater number of receptors identified in the CNS that are linked to second messenger systems rather than to ion-channels, as the data in *Table 1* show. A small number of receptors (e.g. the NMDA subtype of glutamate receptor) can, by allowing the entrance of calcium as well as sodium ions into the post-synaptic cell, contribute to both membrane depolarization and the production of second messengers via stimulation of calcium-sensitive processes (2).

The full range of receptors located in the brain extends far beyond receptors associated exclusively with the 'classical' neurotransmitters. Several other compounds are physiological regulators of cellular activity whose action is mediated via stimulation of their own receptors. A good example is the nerve growth factor (NGF) receptor which, upon activation by NGF, undergoes autophosphorylation due to its intrinsic tyrosine kinase activity. This process culminates in the internalization of the NGF which then travels by retrograde transport from the site of the receptor on the axon to the cell body. *Table 1* shows a classification of receptors according to structure and the second messenger pathway with which the receptor is linked.

The continuous development and introduction of novel pharmaceutical agents with selectivity for a subset of receptors for a specific activator has led to the recognition that there may be several subtypes of receptor for a particular neurotransmitter in the CNS, as the data in *Table 2* show. In most cases, the application of molecular biological techniques to the study of receptors has confirmed that the differences in receptor activity, obtained from data involving pharmacological techniques, does have a basis in different structures of the receptor proteins themselves. The isolation of cDNAs coding for the glutamate receptor, for instance, has revealed that there exist at least three families of glutamate receptor-activated ion channels, as well as a family of G-protein linked receptors, which regulate both adenyly cyclase activity as well as several isoforms of phospholipase (see *Table 3*, and ref. 87). It is highly probable that future work will lead to the identification of even more receptor subtypes, not only for

➤p. 227

Table 1. Overview of classification of receptors by structure and second messenger pathways (see notes at end of table)

Receptors with a single transmembrane segment per subunit				
Receptors for	**Subtype**	**Structure**	**Linked to**	**Refs**
NGF	-	Hetero-oligomer (33 kDa)	Tyrosine kinase	3, 4
Receptors with four or more transmembrane segments per subunit functioning as a receptor ion-channel complex[a]				
Receptors for	**Subtype**	**Structure**	**Effective pathway**	**Refs**
Acetylcholine (nicotinic)	Muscle	Pentamer $((\alpha)_2\beta\gamma\delta)$	$Na^+/K^+/Ca^{2+}$	5–7
	Ganglionic	Multisubunit	$Na^+/K^+/Ca^{2+}$	8
	Neuronal CNS	Pentamer $((\alpha)_2(\beta)_3)$	$Na^+/K^+/Ca^{2+}$	
	α7 neuronal	Multisubunit	$Na^+/K^+/Ca^{2+}$	
GABA	$GABA_A$	α, β and γ subunits	Cl^- (competitive site)	9–12
Glycine	–	Pentamer	Cl^-	13
5-HT	$5\text{-}HT_3$	487 aa (mouse)	Cation	14
Receptors with seven transmembrane segments linked to second messengers via G-proteins				
Receptors for	**Subtype**	**Structure**	**Second messenger**	**Refs**
Acetylcholine (muscarinic)	M_1	460 aa (human)	$\uparrow IP_3/DAG$	15
	M_3	590 aa (human)	$\uparrow IP_3/DAG$	
	M_5	532 aa (human)	$\uparrow IP_3/DAG$	
	M_2	466 aa (human)	↓cAMP[c]	
	M_4	479 aa (human)	↓cAMP[c]	
Adenosine[b]	A_1	326 aa (canine)	↓cAMP[c]	16–20
	A_3	320 aa (rat)	↓cAMP[c]	
	A_{2A}	409 aa (human)	↑cAMP	
	A_{2B}	332 aa (human)	↑cAMP	
	AT_2	363 aa (human)	↓cGMP[d]	

Angiotensin	AT_1	359 aa (human)	↑cAMP	21–23
Cholecystokinin	CCK_A	444 aa (rat)	↑cAMP	7, 24
	CCK_B	447 aa (human)	↑cAMP	
Dopamine	D_1	446 aa (human)	↑IP_3/DAG	25–29
	D_1, D_5	477 aa (human) (D_5)	↑cAMP	
	D_2	443 aa (human)	↓cAMP[c]	
	D_3	446 aa (rat)	↓cAMP[c]	
	D_4	387 aa (human)	↓cAMP[c]	
Epinephrine (adrenaline) (adrenergic receptors)	α_{1A}	466 aa (human)	↑cAMP	30–34
	α_{1B}	515 aa (human)	↑cAMP	
	α_{1D}	560 aa (human)	↑cAMP	
	α_{2A}	450 aa (human)	↑cAMP	
	α_{2B}	450 aa (human)	↑cAMP	
	α_{2C}	461 aa (human)	↑cAMP	
GABA	$GABA_B$	Not determined	↓cAMP[c]	35
Glutamate	$mGlu_1$	1199 aa (rat)	↑IP_3/DAG	36–38
	$mGlu_2$	872 aa (rat)	↑cAMP	
	$mGlu_3$	879 aa (rat)	↑cAMP	
	$mGlu_4$	912 aa (rat)	↑cAMP	
	$mGlu_5$	1171 aa (rat)	↑IP_3/DAG	
	$mGlu_6$	871 aa (rat)	↑cAMP	
	$mGlu_7$	915 aa (rat)	↑cAMP	
Histamine	H_1	487 aa (human)	↑IP_3/DAG	39, 40
	H_2	359 aa (human)	↑cAMP	
5-HT	5-HT_{1A}	421 aa (human)	↓cAMP[c]	14, 41, 42
	5-HT_{1B}	386 aa (rat)	↓cAMP[c]	
	5-HT_{1D}	377 aa (human)	↓cAMP[c]	
	5-HT_{1E}	365 aa (human)	↓cAMP[c]	
	5-HT_{1F}	366 aa (human)	↓cAMP[c]	

Table 1. Continued

Receptors for	Subtype	Structure	Second messenger	Refs
5-HT – continued	5-HT_{2A}	471 aa (human)	$\uparrow IP_3/DAG$	
	5-HT_{2B}	479 aa (human)	$\uparrow IP_3/DAG$	
	5-HT_{2C}	458 aa (human)	$\uparrow IP_3/DAG$	
	5-HT_4	387 aa (rat)	↑cAMP	
	5-HT_6	436 aa (rat)	↑cAMP	
	5-HT_7	445 aa (human)	↑cAMP	
Norepinephrine (noradrenaline) (adrenergic receptors)	β_1	477 aa (human)	↑cAMP	43, 44
	β_2	413 aa (human)	↑cAMP	31, 45
	β_3	402 aa (human)	↑cAMP	
Neuropeptide Y	Y_1	384 aa (human)	↓cAMP[c]	46–48
	Y_2	Not determined	↓cAMP[c]	
	Y_3	Not determined	↓cAMP[c]	
Neurotensin	–	418 aa (human)	$\uparrow IP_3/DAG$	49, 50
Opioid peptides	μ	400 aa (human)	↓cAMP[c]	51
	δ	372 aa (human)	↓cAMP[c]	
	κ	380 aa (human)	↓cAMP[c]	
Oxytocin and vasopressin	V_{1A}	418 aa (human)	$\uparrow IP_3/DAG$	52, 53
	V_{1B}	424 aa (human)	$\uparrow IP_3/DAG$	
	V_2	371 aa (human)	↑cAMP	
Purines	P_{2U}	375 aa (human)	$\uparrow IP_3/DAG$	54
	P_{2Y}	Not determined	$\uparrow IP_3/DAG$	
Somatostatin	sst_1	391 aa (human)	↓cAMP[c]	55
	sst_2	369 aa (human)	↓cAMP[c]	
	sst_3	418 aa (human)	↓cAMP[c]	
	sst_4	388 aa (human)	↓cAMP[c]	
	sst_5	364 aa (human)	↓cAMP[c]	

Tachykinin	NK_1	407 aa (human)	$\uparrow IP_3/DAG$	56, 57
	NK_2	398 aa (human)	$\uparrow IP_3/DAG$	
	NK_3	368 aa (human)	$\uparrow IP_3/DAG$	
VIP	VIP_1	457 aa (human)	$\uparrow$cAMP	58, 59
	VIP_2	437 aa (rat)	$\uparrow$cAMP	
	PACAP	525 aa (human)	$\uparrow$cAMP	

[a] The ionotropic glutamate receptors were originally believed to have 4TM, but current evidence is controversial. The NMDA subtype is linked to $Na^+/K^+/Ca^{2+}$, whereas both AMPA and kainate receptors are linked to Na^+/K^+ (37, 75, 76). The ionotropic P_2 purinergic receptors (P_{2T}, P_{2X} and P_{2Z}) are linked to integral cation channels. Recent cloning experiments on two P_{2X}-like receptors have suggested that they contain two transmembrane domains (95, 96).

[b] Also known as P_1 purinoceptors.

[c] These receptors are coupled to the Gi/o family, and may also $\uparrow K^+$ channels; $\downarrow Ca^{2+}$ channels.

[d] It is not definitely known whether a G-protein is involved in this mechanism.

Table 2. Pharmacology of receptors in the nervous system (see notes at end of table)

Receptor for	Selective agonists	Selective antagonists[a]	Radioligands[b]	Endogenous activator	Refs
Acetylcholine					
Muscarinic					
M_1	–	Pirenzepine (8.0) Telenzepine (9.1)	[^{3}H]pirenzepine	Acetylcholine	5, 7, 8, 60–62
M_2	–	Methocetramine (7.9) AFDX 116 (7.3)	–	Acetylcholine	
M_3	–	hexahydrosiladifendiol (8.0) *p*-fluorohexahydro-Siladifendiol (7.8)	– –	Acetylcholine	
M_4	–	Tropicamide (7.8)	–	Acetylcholine	
M_5	–	–	–	Acetylcholine	
Nicotinic					
Muscle	–	α-bungarotoxin	[^{3}H] or [^{125}I]-α-bungarotoxin	Acetylcholine	
Ganglionic	–	Neosurugatoxin	[^{3}H]epibatine	Acetylcholine	
CNS (neuronal)	Nicotine	–	[^{3}H]cytosine [^{3}H]nicotine	Acetylcholine	
α7 neuronal	–	α-bungarotoxin	[^{3}H] or [^{125}I]-α-bungarotoxin	Acetylcholine	
Adenosine					
A_1	N6-cyclopentyladenosine	DPCPX (8.3–9.3)	[^{3}H]DPCPX (0.4 nM)	Adenosine	18, 63
A_{2A}	CGS21680	ZM241385 (9.0)	[^{3}H]CGS21680	Adenosine	
A_{2B}	–	–	–	Adenosine	
A_3	APNEA	I-ABOPX (8.1)	[^{125}I]APNEA	Adenosine	
Angiotensin					
AT_1	–	Losartan (8.1) SKF108566 (9.7)	[^{3}H]losartan (3–6 nM) [^{3}H]SKF108566 (0.8 nM)	Angiotensin	22, 64
AT_2	CGP42112A	PD123177 (Ki 30 nM)	[^{125}I]CGP42112A (0.1 nM)	Angiotensin	

Cholecystokinin					
CCK_A	A71623	Devazepide (9.8) Lorglumide (7.2)	[^{3}H]devazepide (0.2 nM)	CCK_4, CCK_8, CCK_{33} and gastrin	24, 65
CCK_B	Desulfated CCK_8	L365260 (7.5–8.7)	[^{3}H]L365260 (2.0 nM)	CCK_4, CCK_8, CCK_{33} and gastrin	
Dopamine					
D_1	SKF38393	SCH23390	[^{3}H]SCH23390 (0.2 nM)	Dopamine	29, 66–69
D_2	Fenoldopam	(−)-sulpiride	[^{125}I]iodosulpiride (0.5 nM)	Dopamine	
D_3	Bromocriptine	–	–	Dopamine	
D_4	–	–	–	Dopamine	
D_5	–	–	–	Dopamine	
Epinephrine/norepinephrine (adrenaline/noradrenaline)					
α_1 *adrenoceptors*					
α_{1A}	–	5-methyluradipil (9.2)	–	Norepinephrine and epinephrine	43–45
α_{1B}	–	Spiperone (8.9)	–	Norepinephrine and epinephrine	
α_{1D}	–	BMY7378 (8.4)	–	Norepinephrine and epinephrine	
α_2 *adrenoceptors*					
α_{2A}	Oxymetazoline (partial agonist)	–	–	Norepinephrine and epinephrine	
α_{2B}	–	Prazosin (7.5)	–	Norepinephrine and epinephrine	
α_{2C}	–	Prazosin (7.5)	–	Norepinephrine and epinephrine	
β-*adrenoceptors*					
β_1	Noradrenaline[c]	Bexatolol (8.5) Atenolol (7.0)	[^{3}H]bisoprolol (10 nM)	Norepinephrine and epinephrine	
β_2	Procaterol	Butaxamine (6.2) ICI118551 (8.3–9.2), α-methylpropranolol (8.5)	[^{3}H]ICI118551 (10 nM)	Norepinephrine and epinephrine	

Table 2. Continued

Receptor for	Selective agonists	Selective antagonists[a]	Radioligands[b]	Endogenous activator	Refs
β_3	BRL37344	–	[^{125}I]iodocyanopindolol (0.5 nM)	Norepinephrine and epinephrine	
GABA					
$GABA_A$					
Competitive site	Isoguvacine, muscimol	Bicuculline (6.0) SR95531	[^{3}H]muscimol [^{3}H]SR95531	GABA, taurine	12, 13, 35, 70, 71
Modulatory site	Flunitrazepam, zolpidem	Flumazenil	[^{3}H]flunitrazepam (2 nM) [^{3}H]flumazenil (0.9 nM) [^{3}H]zolpidem (7 nM)	GABA, taurine	
$GABA_B$	L-baclofen 3-aminopropyl-phosphinic acid	Saclofen (5.3) CGP35348 (5.0)	[^{3}H]L-baclofen [^{3}H]3-aminopropyl-phosphinic acid	GABA, taurine	
Glutamate					
Ionotropic					
1. NMDA[d]					
Competitive site	NMDA	D-AP5 (5.2–5.9) CGS19755 (6.0) CGP37849	[^{3}H]CPP [^{3}H]CGS19755 [^{3}H]dizocilpine	Glu, Asp[e]	2, 37, 72–77
Channel blockers		Dizocilpine, CPP			
Modulatory site	Glycine, serine	5,7-dichlorokynurenate MNQX	[^{3}H]glycine	Gly, Ser	
2. AMPA	AMPA	NBQX (7.1)	[^{3}H]AMPA [^{3}H]CNQX	Glu, Asp	
3. Kainate	Kainate, domoate	–	[^{3}H]kainate	Glu, Asp	
Metabotropic[g]					
*m*Glu$_1$	DHPG	4CPG, MCPG	–	Glu, Asp	
*m*Glu$_2$	DCG-IV	MCCG, MCPG	–	Glu, Asp	

mGlu$_3$	DCG-IV	MCCG	–	Glu, Asp	
mGlu$_4$	L-AP4	MAP4	–		
mGlu$_5$	DHPG	4CPG	–	Glu, Asp	
mGlu$_6$	L-AP4	MAP4	–	Glu, Asp	
mGlu$_7$	L-AP4	MAP4	–	Glu, Asp	
mGlu$_8$	L-AP4	–	–	Glu, Asp	
Histamine					
H_1	2-(*m*-fluorophenyl)-histamine	Mepyramine (9.1)	[^{3}H]mepyramine (0.8 nM)	Histamine	39, 40
H_2	Dimaprit, impromidine	Ranitidine (7.2) Tiotidine (7.8)	[^{3}H]tiotidine (15 nM)	Histamine	
5-Hydroxytryptamine					
5-HT$_{1A}$	8-OH-DPAT	WAY100635 (8.7)	[^{3}H]WAY100635 (0.1 nM)	5-HT	14, 21, 78–81
5-HT$_{1B}$	CP93129	SDZ21009	[^{3}H]CP93129	5-HT	
5-HT$_{1D}$	L694247	GR127935 (8.5)	[^{3}H]L694247	5-HT	
5-HT$_{1E}$	–	–	[^{3}H]5-HT	5-HT	
5-HT$_{1F}$	–	–	[^{125}I]LSD	5-HT	
5-HT$_{2A}$	α-methyl 5-HT	Ketanserin (8.5–9.5)	[^{3}H]ketanserin	5-HT	
5-HT$_{2B}$	α-methyl 5-HT	SB200646 (7.5)	[^{3}H]5-HT	5-HT	
5-HT$_{2C}$	α-methyl 5-HT	Mesulergine (9.1)	[^{3}H]mesulergine	5-HT	
5-HT$_3$	2-methyl 5-HT	Granisteron (10)	[^{3}H]zacopride	5-HT	
5-HT$_4$	Renzapride	GR113803 (9–9.5)	[^{3}H]GR113803	5-HT	
5-HT$_{5A}$	–	–	[^{125}I]LSD	5-HT	
5-HT$_{5B}$	–	–	[^{125}I]LSD	5-HT	
5-HT$_6$	–	–	[^{125}I]LSD	5-HT	
5-HT$_7$	–	–	[^{125}I]LSD	5-HT	
Neuropeptide Y					
Y_1	[Pro^{34}] NPY	BIPB3226 (Ki 7 nM)	[^{125}I] or [^{3}H]NPY	NPY PYY PP	46–48
Y_2	NPY	–	[^{125}I] or [^{3}H]NPY	PP	

Table 2. Continued

Receptor for	Selective agonists	Selective antagonists[a]	Radioligands[b]	Endogenous activator	Refs
Neurotensin					
	Neurotensin	SR48692 (Ki 10 nM)	[^{3}H]SR48692	Neurotensin Neuromedin	49, 50
Opiate peptides					
μ	Sufentanil, DAMGO	CTOP (Ki 0.18 nM)	[^{3}H]DAMGO	[met]enkephalin [leu]enkephalin β-endorphin Dynorphin A and B	82, 83
δ	DPDPE	Naltrindole (9.7)	[^{3}H]DPDPE	As above	
κ	U69593	Nor-binaltorphimine (10.3)	[^{3}H]U69593	As above	
Oxytocin and vasopressin					
V_{1A}	–	SR49059 (9.4)	[^{3}H]SR49059 (0.6–2.5 nM)	Oxytocin arginine vasopressin	52
V_{1B}	d[D3Pal^2]VP	–			
V_2	d[DArg^8]VP	OPC31260 (Ki 26 nM)	[^{3}H]d[DArg^8]VP	As above	
OT	[Thr^4,Gly^7]OT	d(CH$_2$)$_5$[Tyr(Me)^2,–Thr^4,Orn^8]OT$_{1-8}$ (7.7)	Note *f*	As above	
Purines					
P_{2T}	2-methylthioADP	2-chloro-ATP (5.2)	β[^{32}P]2-methylthioADP	ADP	54, 84–86
P_{2X}	α,β-methyleneATP	Suramin (5.0)	[^{3}H]α,β-methyleneATP	ATP>ADP>AMP	
P_{2Z}	3′-O-benzoylbenzoylATP	–	–	ATP	
P_{2U}	UTP-γ-S	–	–	UTP, ATP	
P_{2Y}	2-methylthioATP	Suramin (5.0)	[^{35}S]ADPβS	ATP, ADP	
Somatostatin					
sst_1	–	–	[^{125}I]somatostatin	Somatostatin	55
sst_2	Seglitide	–	[^{125}I]seglitide	Somatostatin	
sst_3	Seglitide	–	[^{125}I]somatostatin	Somatostatin	

sst_4	–	–	[^{125}I]somatostatin	Somatostatin	
sst_5	Seglitide	–	[^{125}I]somatostatin	Somatostatin	
Tachykinin					
NK_1	SP methylester	CP99994 (Ki 0.5–50 nM)	[^{125}I]BH-SP	SP, NKA, NKB NPK, NPγ	56, 57
NK_2	[b-Ala^8]$NKA_{4–10}$	SR48968 (9.0)	[^{125}I]NKA	As above	
NK_3	Senktide	SR142802 (9.2)	[^{3}H]senktide	As above	
Vasoactive intestinal peptide					
VIP_1	–	–	[^{125}I]VIP	VIP, PACAP, PHI, PHM	58, 59
VIP_2	–	–	[^{125}I]VIP	As above	
PACAP	–	$PACAP_{6–27}$	[^{125}I]PACAP	As above	

[a] The values in parentheses are *pA*2 estimates on human and rat receptors, if available. Where possible, selective agonists and antagonists have a selectivity of at least two orders of magnitude.
[b] The values in parentheses are estimates of *Kd* values from human and rat receptors (if available) for antagonists.
[c] Selective relative to β_2 adrenoceptors.
[d] The voltage-dependent magnesium blockade of the NMDA receptor prevents activation until the membrane is depolarized.
[e] Endogenous sulfur-containing amino acids are also agonists at glutamate receptors.
[f] [^{125}I]d$(CH_2)_5$[Tyr(Me)^2,Thr^4,Orn^8,Tyr^9-NH2]OT (0.03 nM).
[g] There are no agonists or antagonists that are selective for a single *m*GluR.

Table 3. Key to abbreviations and code names

Abbreviation	Chemical name
4CPG	S-4-carboxyphenylglycine
5-HT	5-hydroxytryptamine (serotonin)
8-OH-DPAT	8-hydroxy-2-(di-*n*-propylamino)tetralin
A71623	Boc-Trp-Lys(*o*-Me-Phe-NH)-Asp-(NMe)Phe-NH_2
AFDX116	11-([2-{(diethylamino)methyl}-1-piperidinyl]acetyl)-5,11-dihydro-6H-pyridol[2,3-*b*][1,4]benzodiazepine-6-one
AMPA	D,L-α-amino-3-hydroxy-5-methyl-4-isoxazolepropionic acid
APNEA	N6-2-(4-aminophenyl)ethyladenosine
BIPB3226	R-N^2-(diphenylacetyl)-*N*-(4-hydroxyphenyl)methyl-argininamide
BMY7378	8-{2-[-(2-methoxyphenyl)-1-piperazinyl]ethyl}-8-azaspiro[4,5]decae-7,9-dione dihydrochloride
BRL37334	Sodium-4-(2-[2-hydroxy-{3-chlorophenyl}ethylamino]propyl)-phenoxyacetate
CCK	Cholecystokinin
CGP35348	p-(3-aminopropyl)-P-diethoxymethylphosphinic acid
CGP37849	D,L-(*E*)-2-amino-4-methylphosphono-3-pentanoic acid
CGP42112A	Nicotinic acid-Tyr-(*N*-benzoylcarbonyl-Arg)Lys-His-Pro-Ile-OH
CGS19755	4-phosphonomethyl-2-piperidinecarboxylic acid
CGS21680	(2-*p*-carboxyethyl)phenylamino-5′-*N*-methyluronamide
CP93129	5-hydroxy-3-(4-1,2,5,6-tetrahydropyridyl)-4-azaindole
CP99994	(+)-(2S,3S)-3-(2-methyoxybenzylamino)-2-phenylpiperidine
CPP	phencyclidine
CTOP	DPhe-Cys-Tyr-DTrp-Lys-Thr-Pen-Thr-NH_2
D-AP5	D-amino-5-phosphonopentanoate
DAG	Diacylglycerol
DAMGO	Tyr-DAla-Gly-*N*-[MePhe]-$NH(CH_2)_2$-OH
DCG-IV	1s,1′R,2′R,3′R)-2-(2′,3′-dicarboxycyclopropyl)glycine
DCPDX	1,3-dipropyl-8-cyclopentylxanthine
DHPG	2,3-dihydrophenylglycine
DPDPE	[DPen^2,DPen^5]enkephalin
GABA	γ-amino butyric acid
GR113803	[1-[2-[(methylsulphonyl)amino]ethyl]-4-piperidinyl]methyl-1-methyl-1*H*-indole-3-carboxylate
I-ABOPX	3-(3-iodo-4-aminobenzyl)adenosine
ICI118551	Erythro-DL-1-(7-methylindan-4-yloxy)-3-isopropyl-aminobutane-2-ol
IL-2	Interleukin 2
IP_3	Inositol trisphosphate
L-AP4	L-amino-4-phosphonobutanoate
L365260	3R(+)-*N*-(2,3,-dihydro-1-methyl-2-oxo-5-phenyl-1*H*-1,4-benzodiazepin-3-yl)-*N*′-3-methylphenyl urea
L694247	2-(5-3-{4-methylsulphonylaminobenzyl}-1,2,4-oxadiazol-5-yl]1H-indole-3-yl)ethylamine
MAP4	α-methyl-L-AP4
MCCG	2s,1′s,2′s-2-methyl-2-(2′-carboxycyclopropyl)glycine
MCPG	α-methyl-4-carboxyphenylglycine
MNQX	5,7-dinitroquinoxoline-2,3-dione
NBQX	6-nitro-7-sulphamobenzo(*f*)quinoxaline-2,3-dione
NGF	Nerve growth factor

Table 3. Continued

Abbreviation	Chemical name
NKA	Neurokinin A
NKB	Neurokinin B
NMDA	*N*-methyl-D-aspartate
NPg	Neuropeptide g
NPK	Neuropeptide K
NPY	Neuropeptide Y
OPC31260	5-dimethylamino-1-(4-[2-methylbenzoylamino]benzoyl)-2,3,4,5-tetrahydro-1*H*-benzazepine
PACAP	Pituitary adenylate cyclase activating polypeptide
PD123177	1-(4-amino-3-methylphenyl)methyl-3-diphenyl-acetyl-4,5,6,7-tetrahydro-III-imidazo-(4,5*O*)pyridine-6-carboxylic acid
PHM	Peptide histidine methionineamide
PP	Pancreatic polypeptide
PYY	Peptide YY
SB200646	*N*-(1-methyl-5-indolyl)-*N′*-(3-pyridyl)urea HCl
SCH23390	7-chloro-2,3,4,5-tetrahydro-3-methyl-5-phenyl-1*H*-3-benzazepine-7-ol
SKF108566	(E)-α-([2-butyl-1-{(4-carboxyphenyl)methyl}-1*H*-imidazol-5-yl]-methylene)-2-thiophenepropanoic acid
SKF38393	2,3,4,5-tetrahydro-7,8-dihydroxy-1-phenyl-1*H*-3-benzazepine HCl
SP	Substance P
SR142802	(S)-(N)-(1-(3-(1-benzoyl-3-(3,4-dichlorophenyl)piperidin-3-yl)propyl)-4-phenylpiperidin-4-yl)-*N*-methylacetamide
SR48692	2-9[1-{7-chloro-4-quinolinyl}-5-{2,6-dimethoxyphenyl}pyrazole-3-yl] carbonylamino)trichloro(3.3.1.3.7)decan-2-carboxylic acid
SR48968	(S)-*N*-methyl-*N*-(4-acetylamino-4-phenylpiperidino)-2-(3,4-dichlorophenyl)betylbenzamide
SR49059	(2s)1-([2R,3s]-[5-chloro-3-{chlorophenyl}-1-{3,4,-dimethoxysulphonyl}-3-hydroxy-2,3-dihydro-1*H*-indole-2-carbonyl]-pyrrolidine-2-carboxamide
SR95531	2-(3′-carboxy-2′-propyl)-3-amino-6-*p*-methoxyphenylpyridazinium bromide
U69593	5α,7α,β-(–)-*N*-methyl-*N*-[7-(1-pyrrolidinyl)-1-oxaspiro(4,5)dec-8-yl]-benzene acetamide
VIP	Vasoactive intestinal peptide
WAY100535	*N*-tert-butyl-3,4-(2-methoxyphenyl)piperazin-1-yl-2-phenyl-propanamide dihydrochloride
ZM241385	4-(2-[7-amino-2-{2-furyl}{1,2,4}triazolo{2,3-a}{1,3,5}triazin-5-yl amino]ethyl)phenol

glutamate receptors, but also for other neurotransmitters and neuromodulators. The continual discovery of new receptor subtypes enormously increases the scope for the development of novel therapeutic agents targeted at a specific response in the nervous system.

2. STRUCTURES OF RECEPTORS IN THE NERVOUS SYSTEM

There are a number of structural similarities between the various receptors activated by first messengers in the CNS. They are all glycoproteins which contain a variable number of transmembrane (TM) segments embedded in the lipid bilayer, which are α-helices, comprising a hydrophilic interior and hydrophobic exterior. A receptor may also contain either one or more subunits, each subunit having its own set of TM units. While receptors with a single TM segment are rare in the nervous system, there being only the example of the tyrosine kinase-linked NGF receptor (*Table 1*), many receptors fall into categories containing multiple TM segments (88). Receptors linked to ion channels frequently contain four or more TM segments, and there is, for example, a high degree of both sequence and secondary structural similarity between $GABA_A$, nicotinic acetylcholine (nAchR), glycine and 5-hydroxy-tryptamine type 3 ($5\text{-}HT_3$) receptors (see *Table 1*, and ref. 89).

In the case of GTP-binding protein (G-protein)-coupled receptors (90), seven TM segments are common, and this is by far the largest group of receptors classified by structure. G-protein linked receptors can be further subdivided into those which activate adenyly cyclase, those which inhibit this enzyme, or those which are linked to phosphatidylinositol bisphosphate (PIP_2) hydrolysis. Several receptors also contain extensive intracellular domains, which often include a number of tyrosine and serine residues which are the targets for phosphorylation, and hence regulation, by kinases.

3. MULTIPLE BINDING SITES ON RECEPTORS

Both the $GABA_A$ and the NMDA glutamate receptor are characterized as having a number of locations to which additional modulators can bind (see *Table 2*), in addition to binding sites for the endogenous neurotransmitter. In the case of the $GABA_A$ receptor, binding of the clinically-important benzodiazepines (for example, flunitrazepam) to a high-affinity site results in an allosteric modulation of the binding of sub-maximal concentrations of GABA to the receptor. The benzodiazepine binding site is located on the α-subunit of the protein, but is physically distinct from the site to which GABA binds. The $GABA_A$ receptor also has binding sites for a number of other agonists such as barbiturates, some steroids and also channel blocking agents (12, 71). The NMDA glutamate receptor is equally complex, and contains a distinct site at which glycine exerts a modulatory influence on the activity of the receptor (74), as do polyamines. The kynurenates are competitive antagonists at the glycine site. Magnesium ions exert a voltage-sensitive blockade of the ion channel itself (see *Table 2*, and ref. 67). Other ligand-gated ion-channel receptors may also be subject to allosteric regulation, for example, calcium or steroids for the nAchR (60) and Zn^{2+} in the case of the glycine receptor (91).

4. ANALYSIS OF RECEPTOR ACTIVITY

The classical approach to characterizing receptors in the nervous system relies on the preparation of membranes from brain tissue homogenates and analysis of the characteristics of binding of either the radiolabeled transmitter itself or a close structural analog. Displacement studies have revealed information on the structural specificity of the receptor population being analyzed. The principles of receptor binding assays, and analysis of the data to provide information of receptor function and activity, are subjects to which entire books have been devoted, and space would not permit an adequate treatment of this subject here (92).

There are at least two reasons why the analysis of receptor proteins by conventional protein purification techniques is not always a reasonable option to anyone wishing to investigate the structure of these proteins. The first is that because receptors are by their very nature, integral membrane proteins with an extremely lipophilic exterior, solubilization from the membrane is difficult; and the second is that the percentage of the total neuronal protein that comprises the receptor protein is very small. Despite these difficulties information on the structure of a wide range of receptor proteins has been gained using these techniques (see *Table 2*), although there is still a need in many cases for more specific antagonists with which to differentiate between receptor subtypes mediating a particular response.

More recently, the application of molecular genetic techniques to the analysis of receptors has yielded a great deal of valuable information. Isolation of cDNA or genomic clones for receptors has largely replaced the need to work on purified receptors themselves, and many details of the structure of receptors have come from the expression of these proteins in systems that readily lend themselves to analysis, for example, *Xenopus* oocytes. Much information has been gained on previously-unidentified muscarinic acetylcholine receptors (93), dopamine receptors (28) and β_1 and β_2 adrenergic receptors (9, 94). In the case of the dopamine receptor, cloning experiments have revealed that there exists a structural and pharmacological similarity between D_1 and D_5 receptors, whereas the D_3 and D_4 subtypes are recognized as being similar to the D_2 receptor, which has led to the reappraisal of the classification system for dopamine receptors into the D_1-like and D_2-like subfamilies (28). To date, the cDNA of an ever-increasing number of receptor proteins in the nervous system has been isolated. Molecular cloning studies of the muscarinic acetylcholine receptor, for instance, have revealed the existence of five molecularly distinct receptor proteins in the mammalian nervous system (designated M_1–M_5, see *Table 2*). Furthermore, the availability of cloned receptor subtypes has enabled the genetic manipulation of the protein in order to identify functionally important domains (62).

5. CLONING STRATEGIES

The experimental procedures for cloning receptors are similar among different types of receptors. In the case of the adenosine A_1 and A_2 receptors, for example, the cloning of these receptors began with an homology-screening protocol whereby oligonucleotides were designed according to the similarities among genes encoding the superfamily of proteins with seven transmembrane helices (17). Amplification of cDNA fragments from human thyroid tissue and identification of the encoded proteins led to the identification of specific classes of the adenosine receptor subtypes. Localization of the receptors to specific sites in the brain was assisted by the identification of mRNA in particular regions of the brain; notably, in the case of the adenosine receptor, in the striatum. Ligand binding and signal transduction studies, using the existing pharmacological profile of the receptor, are used to confirm the identity of the encoded protein as the receptor in question.

6. LOCALIZATION OF RECEPTORS IN THE BRAIN AND NERVOUS SYSTEM

Despite the enormous strides that have been made regarding the recognition of different receptor subtypes and the structure of the proteins comprising these receptors, there is still much information to be gained regarding the localization of these proteins within the brain. The technique of *in situ* hybridization using cDNA of various cloned receptors, as well as the more traditional methods of autoradiography or immunohistochemistry, have provided information on the localization of a wide range of receptor subtypes in the brain; for example, $GABA_A$ (71), 5-HT receptor subtypes (42), adenosine receptors (16), angiotensin receptors (64), and mGluRs (38).

One of the key questions about receptors in the brain which remains unresolved is why should there be such a vast diversity of receptor subtypes for a particular neurotransmitter or modulator? The existence of cDNA or mRNA for a specific protein in the brain does not guarantee that that protein is essential for the normal physiological activity of the nervous system. 'Gene knockout' experiments are perhaps the only way to answer this question conclusively. This strategy enables the generation of mutant animals (normally mice) which lack a specific gene, and therefore, the corresponding protein. Mutant mice have been generated that lack the 5-HT_{1B} and 5-HT_{2C} receptor, for example. Behavioral studies with these animals have suggested that 5-HT_{1B} receptors are involved in the modulation of aggressive behavior, whereas the 5-HT_{2C} receptor exerts an influence over feeding behavior (42). One will have to wait until 'gene knockout' experiments with other receptor proteins provide the answers to similar questions regarding the physiological function of the vast array of cell surface receptors found in the CNS.

7. REFERENCES

1. Siegel, G.J., Agranoff, W., Albers, R.W. and Molinoff, P.B. (1994) *Basic Neurochemistry*, 5th edn, Raven Press, New York.

2. Jahr, C.E. (1994) *Semin. Neurosci.*, **6**, 127.

3. Johnson, D., Lanahan, A., Buck, C.R., Seghal, A., Morgan, C., Mercer, E., Bothwell, M. and Chao, M.V. (1986) *Cell*, **47**, 545.

4. Welcher, A.A., Bitter, C.M., Radeke, M.J. and Shooter, E.M. (1991) *Proc. Natl Acad. Sci.*, **88**, 159.

5. Albuquerque, E.X., Pereira, E.F.R., Castro, N.G. and Alkondon, M. (1995) *Semin. Neurosci.*, **7**, 91.

6. Betz, H. (1991) *Trends Neurosci.*, **14**, 458.

7. Lindstrom, J.M. (1994) in *Handbook of Receptors and Ion Channels: Ligand-Gated and Voltage-Gated Ion Channels* (R.A. North, ed.), CRC Press, Boca Raton, p. 153.

8. Wu, W.C.S. and Raftery, M.A. (1981) *Biochemistry*, **20**, 694.

9. Dixon, R.A.F. *et al.* (1986) *Nature*, **321**, 75.

10. Dunn, S.M., Bateson, A.N. and Martin, I.L. (1994) *Int. Rev. Neurobiol.*, **36**, 51.

11. Schofield, P.R. *et al.* (1987) *Nature*, **328**, 221.

12. Smith, G.B. and Olsen, R.W. (1995) *Trends Pharmacol. Sci.*, **16**, 162.

13. Bittiger, H., Froestl, W., Mickel, S.J. and Olpe, H.R. (1993) *Trends Pharmacol. Sci.*, **14**, 391.

14. Martin, G.R. and Humphrey, P.P.A. (1994) *Neuropharmacology*, **33**, 261.

15. Caulfield, M.P. (1993) *Pharmacol. Ther.*, **58**, 319.

16. Linden, J., Taylor, H.E., Robeva, A.S., Tucker, A.L., Stehle, J.H., Rivkees, S.A., Fink, J.S. and Reppert, S.M. (1993) *Mol. Pharmacol.*, **44**, 524.

17. Linden, J., Tucker, A.L. and Lynch, K.R. (1991) *Trends Pharmacol. Sci.*, **12**, 326.

18. Linden, J. (1994) *Trends Pharmacol. Sci.*, **15**, 298.

19. Salvatore, C.A., Jacobson, M.A., Taylor, H.E., Linden, J. and Johnson, R.G. (1993) *Proc. Natl Acad. Sci. USA*, **90**, 10365.

20. Tucker, A.L., Linden, J., Robeva, A.S., D'Angelo, D.D. and Lynch, K.R. (1992) *FEBS Lett.*, **297**, 107.

21. Nahmias, C. and Strosberg, A.D. (1995) *Trends Pharmacol. Sci.*, **16**, 223.

22. Timmermans, P.B. *et al.* (1993) *Pharmacol. Rev.*, **45**, 205.

23. Timmermans, P.B., Wang, P.C., Chiv, A.T. and Herblin, W.F. (1991) *Trends Pharmacol. Sci.*, **12**, 55.

24. Lindefors, N., Linden, A., Brene, S., Sedvall, G. and Persson, H. (1993) *Prog. Neurobiol.*, **40**, 671.

25. Bunzow, J.R., Van Tol, H.M.H., Gandy, D.K., Albert, P., Salmon, J., Christie, M., Machida, C., Neve, K.A. and Civelli, O. (1988) *Nature*, **336**, 783.

26. Gingrich, J.A. and Caron, M.G. (1993) *Ann. Rev. Neurosci.*, **16**, 299.

27. O'Dowd, B.F. (1993) *J. Neurochem.*, **60**, 804.

28. Sokoloff, P. and Schwartz, J.C. (1995) *Trends Pharmacol. Sci.*, **16**, 270.

29. Strange, P. (1993) *Neurochem. Int.*, **22**, 223.

30. Harrison, J.K., Pearson, W.R. and Lynch, K.R. (1991) *Trends Pharmacol. Sci.*, **12**, 62.

31. Kobilka, B.K. (1992) *Ann. Rev. Neurosci.*, **15**, 87.

32. Regan, J.W., Kobilka, T.S., Yang-Feng, T.L., Caron, M.G., Lefkowitz, R.J. and Kobilka, B.K. (1988) *Prog. Natl Acad. Sci, USA*, **85**, 6301.

33. Summers, R.J. and McMartin, L.R. (1993) *J. Neurochem.*, **60**, 10.

34. Yarden, Y. *et al.* (1986) *Proc. Natl Acad. Sci. USA*, **83**, 6795.

35. Bowery, N. (1993) *Ann. Rev. Pharmacol. Toxicol.*, **33**, 109.

36. Miller, R.J. (1994) *Semin. Neurosci.*, **6**, 105.

37. Nakanishi, S. (1994) *Neuron*, **13**, 1031.

38. Schoepp, D.D. and Conn, P.J. (1993) *Trends Pharmacol. Sci.*, **14**, 13.

39. Hill, S.J. (1990) *Pharmacol. Rev.*, **42**, 45.

40. Schwartz, J.C., Arrang, A.M., Garborg, M., Pollard, H. and Ruat, M. (1991) *Physiol. Rev.*, **71**, 1.

41. Hartig, P.R., Branchek, T.K. and Weinshank, R.L. (1992) *Trends Pharmacol. Sci.*, **13**, 152.

42. Lucas, J.J. and Hen, R. (1995) *Trends Pharmacol. Sci.*, **16**, 246.

43. Emorine, L., Blin, N. and Strosberg, A.D. (1994) *Trends Pharmacol. Sci.*, **15**, 3.

44. Ford, A.D.P.W., Williams, T.J., Blue, D.R. and Clarke, D.E. (1994) *Trends Pharmacol. Sci.*, **15**, 167.

45. Mackinnion, A.C., Spedding, M. and Brown, C.M. (1994) *Trends Pharmacol. Sci.*, **15**, 119.

46. Dumont, Y., Cadieux, A., Pheng, L.H., Fournier, A., St-Pierre, S. and Quiron, R. (1994) *Mol. Brain Res.*, **26**, 320.

47. Michel, M. (1991) *Trends Pharmacol. Sci.*, **12**, 389.

48. Rudolf, K., Eberlain, W., Engel, W., Wieland, H.A., Willim, K.D., Entzeroth, M. and Wienan, W. (1994) *Eur. J. Pharmacol.*, **271**, R11.

49. Gully, D. *et al.* (1993) *Proc. Natl. Acad. Sci. USA*, **90**, 65.

50. Vita, N. *et al.* (1993) *FEBS Lett.*, **317**, 139.

51. Reisine, T. and Bell, G.I. (1993) *Trends Neurosci.*, **16**, 506.

52. Hruby, V.J., Chow, M.S. and Smith, D.D. (1990) *Ann. Rev. Pharmacol. Toxicol.*, **30**, 501.

53. Lolait, S.J., O'Carroll, A.M., McBride, O.W., Konig, M., Morel, A. and Brownstein, M.J. (1992) *Nature*, **357**, 336.

54. Barnard, E.A., Burnstock, G. and Webb. T.E. (1994) *Trends Pharmacol. Sci.*, **15**, 67.

55. Hoyer, D., Bell, G.I., Berelowicz, M. *et al.* (1995) *Trends Pharmacol. Sci.*, **16**, 86.

56. Otsuka, M. and Yoshioka, K. (1993) *Physiol. Rev.* **73**, 229.

57. Regoli, D., Boudon, A. and Fauchere, J.L. (1994) *Pharmacol. Rev.*, **46**, 551.

58. Christophe, J. (1993) *Biochem. Biophys. Acta.*, **1154**, 183.

59. Harmer, T. and Lutz, E.M. (1994) *Trends Pharmacol. Sci.*, **15**, 97.

60. Bertrand, D. and Changeux, J.P. (1995) *Semin. Neurosci.*, **7**, 75.

61. Maelicke, A., Schrattenholz, A. and Schroder, H. (1995) *Semin. Neurosci.*, **7**, 103.

62. Wess, J. (1993) *Trends Pharmacol. Sci.*, **14**, 308.

63. Collis, M.G. and Hourani, S.M.O. (1993) *Trends Pharmacol. Sci.*, **14**, 360.

64. Steckelings, U.M., Bottari, S.P. and Unger, T. (1992) *Trends Pharmacol. Sci.*, **13**, 365.

65. Wank, S.A., Pisegna, J.R. and de Weerth, A. (1994) *Ann. NY Acad. Sci.*, **713**, 49.

66. Pulvirenti, L. and Koob, G.F. (1994) *Trends Pharmacol. Sci.*, **15**, 374.

67. Ruppersberg, J.P., Kitzing, E.V. and Schoepfer, R. (1994) *Semin. Neurosci.*, **6**, 87.

68. Seeman, P. and van Tol, H.H.M. (1994) *Trends Pharmacol. Sci.*, **15**, 264.

69. Strange, P. (1991) *Trends Neurosci.*, **14**, 43.

70. Bonnano, G. and Raiteri, M. (1993) *Trends Pharmacol. Sci.*, **14**, 259.

71. Darlison, M.G. and Albrecht, B.E. (1995) *Semin. Neurosci.*, **7**, 115.

72. Ben-Ari, Y. and Aniksztejn, L. (1995) *Semin. Neurosci.*, **7**, 127.

73. Choi, D. (1994) *Semin Neurosci.*, **6**, 127.

74. Kemp, J.A. and Leeson, P.D. (1993) *Trends Pharmacol. Sci.*, **14**, 20.

75. Nakanishi, S. (1992) *Science*, **258**, 579.

76. Trussell, L.O., Raman, I.M. and Zhang, S. (1994) *Semin. Neurosci.*, **6**, 71.

77. Watkins, J. and Collingridge, G. (1994) *Trends Pharmacol. Sci.*, **15**, 333.

78. Baxter, G., Kennett, G., Blaney, F. and Blackburn, T. (1995) *Trends Pharmacol. Sci.*, **16**, 105.

79. Bockaert, J., Fozard, J.R., Dumuis, A. and Clarke, D.E. (1992) *Trends Pharmacol. Sci.*, **13**, 141.

80. Fletcher, A., Cliffe, I.A. and Dourish, C.T. (1993) *Trends Pharmacol. Sci.*, **14**, 441.

81. Peters, J.A., Malone, H.M. and Lambert, J.J. (1992) *Trends Pharmacol. Sci.*, **13**, 391.

82. Knapp, R.J. and Yamamura, H.I. (1992) *Biochem. Pharmacol.*, **44**, 1687.

83. Takemori, Y. and Portoghese, P.S. (1992) *Ann. Rev. Pharmacol. Toxical.*, **32**, 329.

84. Cusack, N.J. (1993) *Drug Devel. Res.*, **28**, 244.

85. Dalziel, H.H. and Westfall, D.P. (1994) *Pharmacol. Rev.*, **46**, 449.

86. Humphries, R.G., Robertson, M.J. and Leff, P. (1995) *Trends Pharmacol. Sci.*, **16**, 179.

87. Mayer, M.L. (1994) *Semin. Neurosci.*, **6**, 67.

88. Watson, S. and Girdlestone, D. (1995) *Receptor and Ion Channel Nomenclature Supplement Trends Pharmacol. Sci.*, Elsevier, Amsterdam.

89. Grennington, G., Rienitz, A., Schmitt, B., Methfessel, C., Zenisen, M., Bayreuther, K., Gundelfinger, E. and Betz, H. (1984) *Nature*, **328**, 215.

90. Watson, S. and Arkinstall, S. (1994) *The G Protein-Linked Receptor Facts Book*, Academic Press, New York.

91. Bloomenthal, A.B., Goldwater, E., Pritchett, D.B. and Harrison, N.L. (1994) *Soc. Neurosci. Abstr.*, **20**, 463.3.

92. Hulme, E.C. (ed.) (1990) *Receptor Biochemistry: A Practical Approach*, IRL Press, Oxford.

93. Bonner, T.I., Buckley, N.J., Young, A.C. and Brann, M.R. (1987) *Science*, **237**: 527.

94. Fargin, A., Raymond, J.R., Tohse, M.J., Kobilka, B.K., Caron, M.G. and Lefkowitz, R.J. (1988) *Nature*, **335**, 358.

95. Brake, A.J., Wagenbach, M.J. and Julius, D. (1994) *Nature*, **371**, 519.

96. Valera, S., Hussy, N., Evans, R.J., Adami, N., North, R.A., Surprenant, A. and Buell, G. (1994) *Nature*, **371**, 516.

CHAPTER 15
NEURAL NETWORKS

R. Baddeley

1. INTRODUCTION

This chapter is aimed at the physiologist new to neural networks and the student wishing to experiment for the first time. The chapter will describe general principles the author has found useful, together with pointers to primary sources for greater detail on a given topic.

For a more general introduction to neural networks as applied to biology see ref. 1. For a more technical introduction to network theory see ref. 2, but be warned this is quite mathematical. A good practical guide to implementation including example code in C++ is provided by ref. 3, and refs 4 and 5 both contain examples of biological research using neural network methodology. A good textbook on general numerical computation is ref. 6.

Of the specialist journals, *Neural Computation* is consistently good as are the eight volumes of ref. 7. The World Wide Web also provides a variety of information sources. Ref. 8 provides a large depository of preprints of neural network papers, ref. 9 provides a powerful bibliography search system specializing in neural networks, and ref. 10 provides answers to many frequently asked questions.

2. RELEVANCE OF NEURAL NETWORK TECHNIQUES TO THE NEUROSCIENTIST

There are three main areas where neural network techniques can be useful to the neuroscientist. The first is for regression. In many cases the scientist has two sets of data and is interested in the relationship between them. If there is a simple linear relationship, then linear regression can be used. In neuroscience, however, the relationship between two data sets is often either highly nonlinear or some of the variables are categorical rather than real values. Here the 'biological realism' of a network is not especially important but simply its ability to be used as a statistical system. Examples of this approach include measuring the information transmission of a neuron (11, 12) and investigating binocular rivalry (13).

A second use of neural networks is the existence proof. Here the emphasis is on demonstrating that a certain principle is sufficient to explain a biological phenomenon. Examples include demonstrations that random input activity is sufficient to self-organize receptive field-like systems (14); that self-organization can result in the development of topographically organized maps (15); and that coherent firing could be used to segment a perceptual scene (16). The exact details may not be realistic but the network is intended to demonstrate some high level principle.

Lastly there is work comparing quantitative aspects of neuronal operation with neural network simulations. Here the models are constructed out of more realistic but complicated units making formal analysis more dificult. Examples include investigations of the variability in neuronal spiking (17); investigation of the potential storage capacity and operation of the hippocampus (18, 19); and investigating neuronal interactions in the visual cortex (20). The complexity of the individual units can be far greater than that encountered in the more common neural network systems and there are different constraints on how to implement such systems efficiently. Each of these three different areas has different associated techniques and problems and these will be described in the subsequent sections.

3. NEURAL NETWORKS FOR REGRESSION

If neural networks are to be used to investigate the statistical relationship between two data sets, the best understood and usually the method of choice is the famous backpropagation algorithm (21). Public domain simulators are widely available (22), and good introductory textbooks exist (3).

The most important consideration when using networks for regression is to collect sufficient data to specify the model. It would not be sensible to calculate both the mean and standard deviation of a data set based on two data points. Equally it is not sensible to estimate hundreds of network weight and threshold parameters based only on a few dozen input–output examples. Neural networks can perform reasonable interpolation ('generalization') but it is unwise to expect a network to produce a reasonable prediction when presented with input that is dissimilar to anything seen before (extrapolation). Therefore collect as much data from as many situations as possible. *Table 1* describes some other considerations when using neural networks for regression.

Table 1. Considerations when using neural networks for regression

Problems	Method
Input representation	This aspect of neural networks is very important, so experiment. Minimize the number of input units. Avoid representations where translated versions of the same pattern have the same meaning. Explore the use of statistical preprocessing such as principal components analysis or exploratory projection pursuit (23, 24)
Output representation	For binary variables use the tanh function. For variables that can assume multiple exclusive values use the Softmax function (25). Scale any ordinal data (3)
Data preparation	The more data the better. If the uncertainty in the measurements is known, include 'synthetic' noisy versions of the data samples. Adding noisy copies of the data can improve generalization
Architecture	In most cases a simple two-layer network is best. Start with too few hidden units and increase the number until the network can perform the task. Check that a single-layer network cannot perform the task

Table 1. Continued

Problems	Method
Assessment	If the outputs are categorical, use cross–entropy error; if the outputs are real valued, use sum–squared error. The error for the training set is unreliable since it fits both data and noise. For comparisons of different networks use grouped cross-validation: split the data into N groups, train with $N - 1$, then use the error for the Nth. The total cross-validation error score is the summed error across all N splits
Initialization	All weights should be set to very small random values, or consider a non-random initialization technique (26)
Presentation and search method	Given redundant input data, use on-line learning with random presentation order. With non-redundant data, consider batch learning with conjugate gradient descent (6)
Accelerating the learning	Two simple methods that can speed search are using momentum (21) (a value of about 0.9 is a reasonable first guess), and an adaptive learning rate: maintain an estimate of the recent error. If this goes down, increase the learning rate ($\mu' = 1.05\mu$ is good); if the error goes up, decrease the learning rate ($\mu' = 0.7\mu$ is good)
Efficiency issues	Nearly all computational time is taken up by computing the weighted input sum and calculating the exponential and its derivative. Optimizations considered should be the use of a linear interpolation look-up table for the exponential, and ensuring the summation is calculated efficiently

There are two situations when the backpropagation algorithm may be inappropriate. Sometimes it is not clear that there is a sensible split into predicted and predicting variables, there is much missing data, or information flow in the network is not one-directional. In this context the Boltzmann machine should be considered. In its basic form it is very slow (27), but more modern versions (28, 29) can provide significant speed improvements. A second reason not to use straight backpropagation is when the problem involves prediction across time. Here there are a number of approaches (30, 31) but these are less simple to use than straight backpropagation.

4. USING NEURAL NETWORKS TO GENERATE EXISTENCE PROOFS

Neural network methods can be used to create existence proofs or as a 'demonstration of concept'. It may be suggested that a given area of the brain is operating in a particular manner or that a particular representation is set up by a given mechanism. Showing that such a proposal can be implemented using computer simulation can add strength to such a proposal. *Table 2* describes some rules of thumb when constructing such networks.

Table 2. Considerations when implementing a 'demonstration of concept' neural network

Problem	Recommendation
Scale	If at all possible start with a simplified version of the problem and expand. Many problems such as understanding the effects of various parameters are computationally intractable in a large network. In a reduced network these problems often become trivial. Identifying the important 'principles of operation' is much easier in a smaller network
Visualization	Use as many methods to visualize the state of the network as possible. Many problems that are nearly impossible to solve based on a single summary variable can be easily understood with an appropriate picture. It is often better to avoid writing a specialized 'front end' and instead to output network state 'images' to a file in a form that is easily readable by general purpose image display software
Performance quantification	Devise as many methods as possible to quantify both the state and performance of the network. Plotting frequency distributions of various parameters is often informative. For such complex systems, the time invested devising measurements of the system state is rarely wasted
Storage	Disk space is cheap and therefore it is often worth storing the evolving state of the network. It is often only after a week long simulation that the 'interesting' variable is discovered
Parameter optimization	This is often a great problem with the systems performance depending on a large number of parameters. Resist the temptation to optimize by hand as much time can be wasted. Best is to get the computer to systematically plot performance against each individual parameter. This can give insight into which parameters are critical. Alternatively a nonlinear optimization technique can be used (6)

The main problems encountered are not usually programming problems but getting the ideas clear and finding working parameters. If this is the first attempt at model implementation, re-implement a similar previously published model where working parameters have already been found. Define beforehand what is required for the model to be classified as working since it is easy to spend much time 'improving' aspects that are irrelevant. Often, if the system is to learn, some mechanism is required to prevent all units learning the same 'features'. *Table 3* describes a number of alternative methods to avoid this problem.

5. BIOLOGICALLY REALISTIC SIMULATIONS

The most realistic simulations attempt to capture detailed characteristics of neurons. To program such a system the following are required: a spike generation model (*Table 4*); possibly a model for dendritic integration (41, 42); a synaptic interaction model (*Table 5*); possibly a learning rule (*Table 6*); and a method for solving the associated differential equations (*Table 7*).

Table 3. Methods for ensuring that different units learn different features

Method	Description
Incomplete connectivity	All units have different features since each unit is differently connected
Winner-takes-all (WTA)	The input is calculated for all 'neurons' and only the neuron with highest level of activation signals to subsequent layers
Softmax	The output of a given unit is proportional to the exponential of its input divided by a variable called the temperature (T). The constant of proportionality is set to make the sum of all outputs equal to one. It has the following virtues: the system can operate both as a winner-takes-all system ($T \to 0$) and an approximately linear system ($T \to$ inf). The system is invariant under addition to its input, the outputs sum to one, and it is related to techniques in statistics (logistic regression)
Lateral inhibition	A simple way to ensure that different units correspond to different features is to have modifiable inhibitory connections between them (32). These nets sometimes can have stability problems (33)
Gram–Schmitt orthogonalization	An effective but unbiological method. Each unit subtracts from the input the contribution in its direction. This can ensure that the features learnt are orthogonal (34, 35)

Table 4. Spike generation models commonly used to simulate neurons

Mode	Description
Integrate-and-fire	The simplest spiking neuron model. At each step the input is added to a constantly evolving internal sum. This sum is then reduced by some percentage and if this sum exceeds a threshold, the neuron fires and the sum is reset to zero. There are some relevant analytic results (36, 37) but this model lacks many characteristics of more complicated neuronal models
Morris–Lecar	A simple two-channel approximation that is computationally efficient to simulate (38). Synchronization characteristics are superior to integrate and fire neurons (39) but can become unstable
Full Hodgkin–Huxley	The most complicated and accurate neuronal model (40). The number of possible channels is large and new ones are still being found. Consider carefully before adding additional channels because many may have little functional significance for models and can greatly effect efficiency

Table 5. Synapse models

Model	Description
Instantaneous	No temporal dynamics for synaptic integration. Obviously simple to program but can hinder synchronization
Exponential filter	Simple to program: an input sum is maintained with any new input added to it. Every iteration: this sum decays by an amount dependent on the time constant of the filter used. The instantaneous rise time of this model can cause problems in some systems making the more biological alpha function preferable
Alpha filter	A non-zero rise time together with a graceful decay make this function both a good fit to the biology and computationally useful. Efficient implementations exist (43)

Table 6. Historically important 'biologically plausible' learning rules; references to other learning rules can be found in ref. 44

Rule	Description
Hebb	The earliest proposal and interpreted in many different ways. In its simplest form it states that if two neurons fire simultaneously, the synapse between them is strengthened. If the synaptic strengths into a given neuron are normalized and the system is linear, then this rule can converge to the first principal component (34) of the input. If there is a nonlinearity in the neuron then it can perform exploratory projection pursuit (45). The performance of this rule for auto-association has been heavily studied (46)
Covariance	Very similar to the Hebbian rule but takes into account the average firing rate of both neurons with the synapse being increased only if the two neurons fire together at a rate greater than chance. It has been shown that in restricted subset of Hebbian type rules, this can be optimal (47)
Stent	Based on physiological speculations (48). Has been shown to be equivalent to one of the rules proposed by Grossberg (49)
BCM (Bienenstock Cooper and Munro)	After (50). The degree of synaptic strengthening is based on a nonlinear function of the post-synaptic potential with the weights not being modified if this potential is small, reduced if intermediate, and only increased if this potential is large. Also see refs 45 and 51
ABS (Artola Brocher and Singer)	Similar to the BCM but based on physiological observations (52). It has been shown that this can implement a form of biologically plausible delta rule (53)
Anti-Hebbian	This rule attempts to make two neurons independent. The synapse is made more negative if two neurons are positively correlated (32)

Table 6. Continued

Rule	Description
Marr	Different from the others in that a synapse is considered to be binary: either there is a connection or not (54). This connection is made probabilistically if two neurons are simultaneously active
Contrastive Hebbian	Strengthens the synapse between two neurons if the level of correlation between two neurons when being trained is greater than that when the network is freely running. Under certain conditions this learning rule corresponds to gradient descent of the difference between the current 'model' inherent in the connections in the network and the probability distribution inherent in the world (27)

Table 7. Methods for solving differential equations used in modeling neurons

Method	Description
Euler	The most basic method and simplest to program and understand. Unfortunately it can behave badly if the equations are not well behaved so use it only for quick implementations or check it is working by using it with a very small step size
Runge–Kutta	The main work horse. Reasonably well behaved and not too complicated (6).
Gear	Good for badly behaved problems but possibly harder to understand and debug (6, 55)
Adaptive step size methods	Versions of the above are available that adaptively change the step size. Improves eficiency but can make book-keeping difficult (6, 55).

It is important to start small initially. Many neuronal models can be unstable and only operate sensibly when driven with current within a certain range. Therefore even when constructing a large-scale network, perform a preliminary analysis on a single isolated neuron. If learning is to be used, consider initially implementing the learning in a model without a complicated neuronal model since sorting out the neuronal and the learning aspects separately can make their final combination far simpler.

6. REFERENCES

1. Churchland, P. and Sejnowski, T.J. (1992) *The Computational Brain*, MIT Press, Cambridge, MA.

2. Hertz, J., Krogh, A. and Palmer, R.G. (1990) *Introduction to the Theory of Neural Computation*, Santa Fe Institute/Addison Wesley.

3. Masters, T. (1993) *Practical Neural Networks Recipes in C++*, Academic Press, New York.

4. Koch, C. and Segev, I. (1989) *Methods in Neuronal Modeling: From Synapses to Networks*, MIT Press, Cambridge.

5. Anderson, J.A. and Rosenfeld, E. (eds) (1988) *Neurocomputing: Foundations of Research*, MIT Press, Cambridge, MA.

6. Press, W.H., Flannery, B.P., Teukolsky, S.A. and Vetterling, W.T. (1986) *Numerical Recipes*, Cambridge University Press, Cambridge.

7. *Neural Information Processing Systems*, vols 1–7, Morgan Kaufman, San Mateo, CA, 1989–1995.

8. Neuroprose repository of paper pre-prints: ftp://archive.cis.ohiostate.edu/pub/neuroprose.

9. World Wide Web page of bibliography search tools for neural networks: http://www.emsl.pnl.gov:2080/docs/cie/neural/neural.journals.html#search.

10. A list of frequently asked questions and their answers provided as a World Wide Web page: http://wwwipd.ira.uka.de/prechelt/FAQ/neuralnet-faq.html.

11. Eskandar, E.N., Richmond, B.J., Hertz, J.A., Optican, L.M. and Kjoer, T. (1992) in *Neural Information Processing Systems*, vol. 4 (J.E. Moody *et al.*, eds), San Mateo, CA, p. 356.

12. Kjaer, T.W., Hertz, J.A. and Richmond, B.J. (1994) *J. Comput. Neurosci.*, **1**, 109.

13. Lehky, S.R. and Sejnowski, T.J. (1989) in *Methods in Neuronal Modeling* (C. Koch and I. Segev, eds), MIT Press, Cambridge, MA, p. 361.

14. Linsker, R. (1988) *Computer*, **21**(3), 105.

15. Willshaw, D. and von der Malsburg, C. (1976) *Proc. Roy. Soc. London B*, **194**, 431.

16. Eckhorn, R., Reitboeck, H.J., Arndt, M. and Dicke, P. (1989) in *Models of Brain Function* (R.M.J. Cotterill, ed.), Cambridge University Press, Cambridge.

17. Softky, W.R. and Koch, C. (1993) *J. Neurosci.*, **13**, 334.

18. Hasselmo, M., Schiell, E. and Barka, E. (1995) *J. Neurosci.*, **15**, 5249.

19. Treves, A. and Rolls, E.T. (1991) in *Network: Comput. Neural Syst.*, **70**(2), 640.

20. Wehmeier, U., Dong, D., Kock, C. and Van Essen, D. (1989) in *Methods in Neuronal Modeling* (C. Koch and I. Segev, eds), MIT Press, Cambridge, MA, p. 335.

21. Rumelhart, D.E., Hinton, G.E. and Williams, R.J. (1986) in *Parallel Distributed Processing*, vol. 1 (D.E. Rumelhart and J.L. McClelland, eds), MIT Press, Cambridge, MA, p. 318.

22. World Wide Web site offering access to a range of neural network simulators: http://www.neuronet.ph.kcl.ac.uk/neuronet/software/software.html.

23. Friedman, J.H. (1987) *Am. Statist. Assoc.*, **82**, 249.

24. Huber, P.J. (1985) *Ann. Statist.*, **13**, 435.

25. Bridle, J.S. (1990) in *Neural Information Processing*, vol. 2, D.S. Touretzky (ed.), Morgan Kaufmann, San Mateo, CA, p. 211.

26. Nguyen, D. and Widrow, B. (1990) in *International Joint Conference on Neural Networks*, vol. 3, 21.

27. Ackley, D.H., Hinton, G.E. and Sejnowski, T.J. (1985) *Cogn Sci.*, **9**, 147.

28. Kappen, H. (1995) *Neural Net.*, **8**, 537.

29. Peterson, C. and Anderson, J.R. (1987) *Complex Syst.*, **1**, 995.

30. Elman, J.L. (1990) *Cogn. Sci.*, **14**, 179.

31. Werbos, P.J. (1990) *Proc. IEEE*, **78**, 1550.

32. Foldiak, P. (1992) Technical Report CUED/F-INFENG/TR 91, University of Cambridge, Department of Engineering.

33. Leen, T.K. (1991) *Network*, 85.

34. Oja, E. (1982) *J. Mathe. Biol.*, **15**, 267.

35. Sanger, T.D. (1989) in *Advances in Neural Information Processing Systems*, vol. 1 (D.S. Touretzky, ed.), Morgan Kaufmann, San Mateo, CA, p. 11.

36. Corral, A., Perez, C.J., Diazguilera, A. and Arenas, A. (1995) *Phys. Rev. Lett.*, **74**(1), 118.

37. Treves, A. (1993) *Network*, **4**, 259.

38. Rinzel, J. and Ermentrout, G.B. (1989) in *Methods in Neuronal Modelling* (C. Koch and I. Segev, eds), MIT Press, Cambridge, MA.

39. Cairns, D.E. Baddeley, R.J. and Smith. L.S. (1993) *Neural Computation*, **5**, 260.

40. Hodgkin, A. and Huxley, A.F. (1952) *J. Physiol. London*, **117**, 500.

41. Rall, W. (1989) in *Methods in Neuronal Modeling* (C. Koch and I. Segev, eds), MIT Press, Cambridge, MA, ch. 2.

42. Segev, I., Fleshman, J.W. and Burke, R.E. (1989) in *Methods in Neuronal Modeling* (C. Koch and I. Segev, eds), MIT Press, Cambridge, MA, ch. 3.

43. Srinivasan, R. and Chiel, H.J. (1993) *Neural Comput.*, **5**, 200.

44. Hinton, G. (1989) *Artif. Intell.*, **40**, 185.

45. Fyfe, C. and Baddeley, R.J. (1995) *Biol. Cyber.*, **172**(6), 533.

46. Amit, D. (1989) *Modelling Brain Function*, Cambridge University Press, Cambridge.

47. Willshaw, D. and Dayan, P. (1990) *Neural Comput.*, **2**, 85.

48. Stent, G. (1973) *Proc. Natl. Acad. Sci.*, **70**, 997.

49. Hancock, P.J.B. and Smith, L.S. (1990) *Network*, **1**, 133.

50. Bienenstock, E.L., Cooper, L.N. and Munro, P.W. (1982) *J. Neurosci.*, **2**.

51. Intrator, N. and Cooper, L.N. (1991) *Neural Net.*, **5**, 3.

52. Artola, A., Brocher, S. and Singer, W. (1990) *Nature*, **347**, 69.

53. Hancock, P.J.B., Smith, L.S. and Phillips, W.A. (1991) *Neural Comput.*, **3**, 201.

54. Marr, D. (1971) *Phil. Trans. Roy. Soc. London B*, **262**, 23.

55. Mascagni, M.V. (1989) in *Methods in Neuronal Modeling* (C. Koch and I. Segev, eds), MIT Press, Cambridge, MA, p. 439.

CHAPTER 16
REGULATION OF GENE EXPRESSION

D.M. Flavell

1. INTRODUCTION

The nervous system consists of a heterogeneous population of cell types whose functions are performed by their mutual interactions. Experimental manipulations of the nervous system entail the use of techniques such as electrolytic lesioning of discrete nuclei and administration of selectively neurotoxic compounds. Molecular genetic techniques provide the ability to non-invasively modulate the activity of a specific cell type and so enable the study of the complex interactions that underlie the functioning of the nervous system.

2. STUDY OF NEURONAL GENE REGULATION

Study of the regulation of gene expression has traditionally involved the use of cultured cell lines which express the 'gene of interest' in order to study the effects of deletion of putative regulatory sequences on the expression of the 'gene of interest' flanking regions linked to a marker gene. Cell lines represent a homogeneous tissue and enable the rapid analysis of deletions and point mutations.

The study of neuronal gene regulation is complicated by the existence of relatively few neuronal cell lines, though immortalization by transgenesis (see section 3) or retroviral gene transfer (see section 6) has enabled the generation of neuronal cell lines. The heterogeneity of the CNS also presents problems as a cell line must express the 'gene of interest' endogeneously to represent a relatively appropriate model system. The use of transgenic animals to study neuronal gene expression is the most appropriate approach as the tissue-specificity and developmental regulation of a transgene can be determined, and results obtained from cell lines must be confirmed *in vivo*. The analysis of gene regulation is described in greater detail in ref. 1.

3. TRANSGENIC ANIMALS

Transgenic animals arise from the addition of genetic material to the genome by microinjection of DNA into fertilized eggs, or by transfection of ES cells followed by ES cell injection into blastocysts (see section 4), resulting in a random integration event into the host genome and the generation of a line of animals which transmit the transgene to their offspring. The microinjection process is outlined in *Figure 1*.

The mouse is the preferred species, though transgenic animals have also been generated in nematodes (*C. elegans*), *Drosophila*, fish, rat, rabbit, chicken, goat, pig and cow (for an overview of techniques see ref. 2). Lower organisms enable the rapid generation and analysis of mutants and share many similarities with higher organisms whilst showing less complexity, thus making analysis simpler. Mouse genetics has been studied more intensely than in any organism except humans, and spontaneous mouse mutants provided the first animal models for neurological disorders. Transgenic rats prove more difficult to generate,

Figure 1. Generation of transgenic mice.

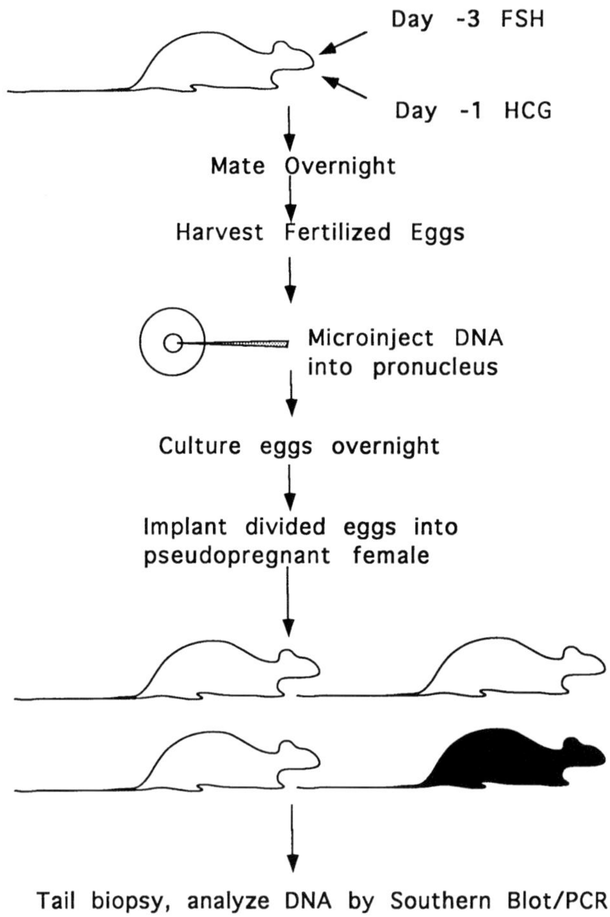

though this can be outweighed by their applicability as animal models owing to the greater understanding of their physiology and neurobiology.

3.1. Transgene expression

In order to obtain expression of a transgene in a particular cell type, it is necessary to use DNA sequences containing regulatory elements that direct transgene expression to the correct cell type at the correct stage in development. These regulatory elements include the promoter region, enhancer(s) and locus control regions (LCR) (reviewed ref. 3), and their properties are summarized in *Table 1*.

A transgene need not contain all these regulatory elements. Constructs consisting of marker genes that are promoterless or contain a weak promoter can be used in promoter and enhancer traps respectively, relying on random integration into a gene or in close proximity of an enhancer. It can be advantageous to have 'position effect' to create a range of transgene expression levels. An LCR is necessary to achieve expression levels comparable to the endogeneous gene (e.g. where a mutant protein is competing with a wild type

Table 1. DNA regulatory elements

Element	Typical features	Effect on transcription
Promoter	TATA box CCAAT box Transcription factor binding sites	Site of transcriptional initiation Low level expression Some tissue specificity
Enhancer	Transcription factor binding sites	Enhances rate of transcription Confers tissue specificity Subject to 'position effect'
Locus control region	Enhancer regions Nuclear matrix attachment regions Chromosome scaffold attachment regions	High levels of gene expression Tissue-specific Copy number dependent Integration site independent

protein). A frequent mistake is to use a short region of 5′ flanking sequence (i.e. the promoter region) when the aim is to obtain cell-specific expression. Expression in the desired cell type may be achieved, but with ectopic expression. An LCR or enhancer element can be situated anywhere around a gene (i.e. 5′, 3′ or within the exons/introns of the gene), and so a large DNA fragment should be used to maximize the chance of obtaining appropriate transgene expression. Transgenic mice with yeast artificial chromosome (YAC) transgenes of up to 250 kb have been reported (4) and YACs have been introduced into ES cells and then mice (5).

3.2. Transgene design

The choice of transgene depends on the aim of the experiment. A common choice for an initial transgenic experiment is a marker gene, as the pattern of tissue-specific and developmental stage-specific expression of the transgene should be determined before embarking on further experiments. Marker genes do not exert an effect on the host and so provide a way of identifying the cell type in which the transgene is expressed, the pattern and level of expression and whether the transgene responds to physiological stimuli. Having determined the expression pattern that a transgene produces, the investigator has numerous options available in which the function of the neuron can be experimentally manipulated. A list of types of transgenes and some of their uses is given in *Table 2*.

Table 2. Types of transgene

Type	Uses	Examples
Marker	Analysis of gene expression patterns, levels, regulation Promoter/enhancer trap Cell lineage marking	β-galactosidase (lac Z) Luciferase (luc) Chloramphenicol acetyl transferase (CAT) Green fluorescent protein (GFP)
Oncogene	Targeted oncogenesis	SV40 large T antigen
Toxin gene	Genetic ablation Inducible ablation	Diphtheria toxin Herpes simplex virus thymidine kinase (HSV-tk) Immunotoxins
Dominant mutant	Disease model	Amyloid precursor protein (Alzheimer's disease)

Table 2. Continued

Type	Uses	Examples
Wild type gene	Disease rescue	GnRH gene (hypogonadal mouse)
Bioactive gene	Phenotype disruption	Antisense expression Recombinant antibody expression Neurotransmitter expression

Examples of the uses of different transgenes in the CNS are listed in *Table 3*. Further examples can be found in refs 6 and 7. For a review of the use of transgenic animals in the neuroendocrine system, see ref. 8; and for cognition see ref. 9.

Table 3. Expression of transgenes in the nervous system

Construct	Expression (phenotype)	References
L7-lacZ	Purkinje + retinal bipolar neurons	10
SCG10-CAT	Neurons	11
GnRH-Tag	GnRH neurons (immortalization)	12
γ2-crystallin-DT	Lens cells (microphthalmia)	13
γ2-crystallin-tox176	Lens cells (clinical anophthalmia, cataracts)	14
DBH-IL2Rα	Noradrenergic neurons (inducible ablation + immunotoxin)	15
PDGF-APP V717F	Brain + other tissues (AD model)	16
L7-Human SCA1	Purkinje cells (ataxia + cell degeneration)	17
GnRH	GnRH neurons (rescue of hpg mouse)	18
MBP-antisense	Myelinating Schwann cells (shiverer)	19
hCaMKII T286D	Forebrain (altered LTP)	20
Human AChE	AChE neurons (cognitive deterioration)	21
Vgf-A Sub P mAb	vgf neurons (motor deficits)	22
Thy-1-GAP-43	Neurons (spontaneous nerve sprouting)	23

AChE = acetylcholinesterase; AD = Alzheimer's disease; APP = amyloid precursor protein; CAT = chloramphenicol acetyltransferase; DBH = dopamine β-hydroxylase; DT = diphtheria toxin A chain; GAP-43 = growth-associated protein-43; GnRH = gonadotrophin hormone releasing hormone; hCaMKII = human calcium-calmodulin kinase II; hpg = hypogonadal; IL2Rα = interleukin 2 receptor α chain; LTP = long-term potentiation; mAb = monoclonal antibody; MMTV = mouse mammary tumor virus; PDGF = platelet-derived growth factor; SCA1 = spinal cerebellar ataxia type 1 gene; Sub P = substance P; Tag = SV40 large T antigen.

3.3. Temporal control of transgene expression

Two recent reports outline successful attempts to create inducible transgene expression systems. The first report describes a hybrid tetracycline regulated transcription factor that represses transcription from a defined sequence in the presence of tetracycline. Withdrawal of tetracycline causes activation of transgene expression, though this is a slow process, dependent on tetracycline clearance (24). The second system uses the yeast transcriptional activator GAL4 DNA-binding domain linked to a mutated progesterone receptor (PR) ligand binding domain (25). The mutant PR binds the progesterone antagonist RU486, but not progesterone, and activates transcription at a concentration of RU486 that has no antiprogesterone activity. Thus by placing the chimeric transcriptional activator under a tissue-specific promoter, transgene expression can be induced via a promoter containing GAL4 DNA binding sites by a low dose of RU486.

4. HOMOLOGOUS RECOMBINATION

The development of homologous recombination has enabled the generation of mutant mice to progress from a random to a specific process. Mouse mutations have been cataloged and collected for over 30 years and many neurological mutants have been identified. Gene targeting by homologous recombination allows the analysis of the effect of either removing a gene or making a subtle modification to the gene of interest. The methodology and mechanism of the generation of knockout mice is outlined in *Figure 2*. For a more detailed description and discussion of homologous recombination, see refs 26 and 27.

Figure 2. Gene knockout by homologous recombination.

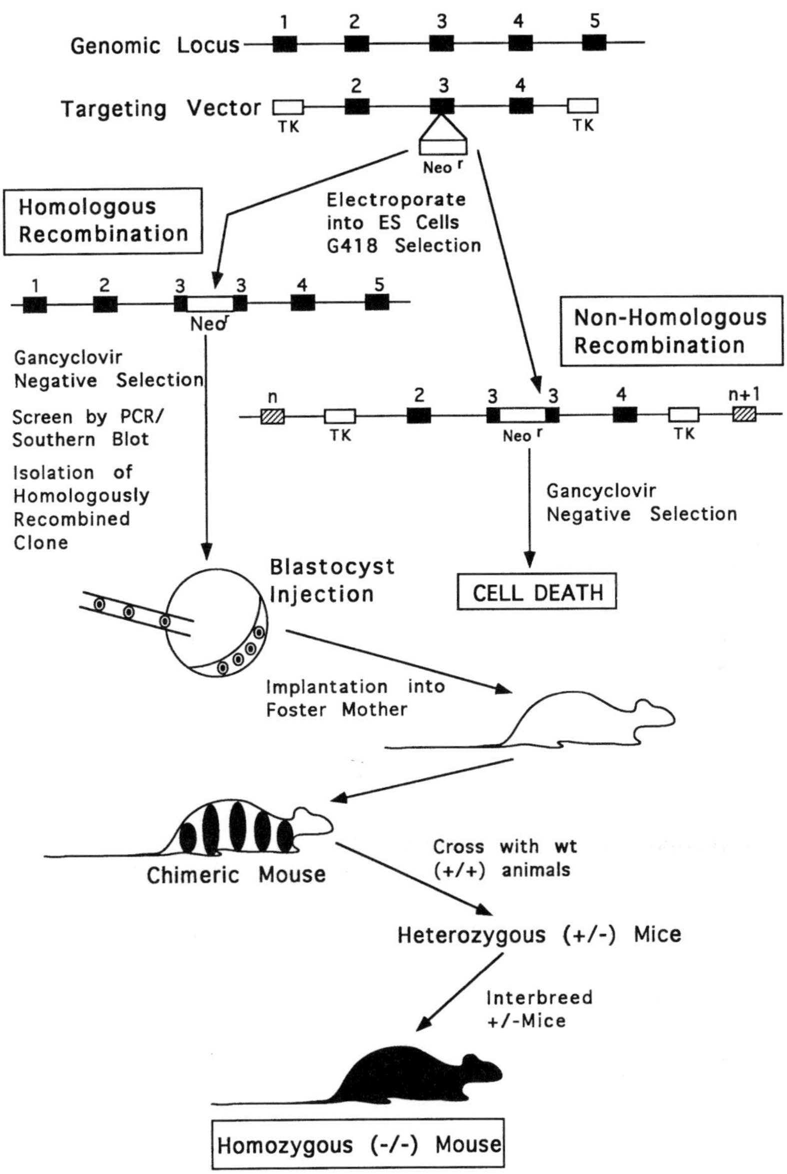

4.1 The impact of gene targeting on neurobiology

The ability to selectively introduce mutations into the mouse genome is revolutionizing the field of neurobiology and the advent of tissue-specific targeting promises even greater breakthroughs. A compendium of knockouts, comprehensive to the end of 1994, catalogues 263 null mutations (Brandon *et al.*, 1995a,b,c) of which approximately 20% affect an aspect of nervous system function, and forms the basis of a knockout database which is accessible on the Internet (via BioMedNet). Gene knockouts have proved particularly valuable in the study of developmental neurobiology (31), cognition (9, 32) and neurodegenerative disorders.

4.2. Tissue-specific and developmental stage-specific gene knockout

Some of the problems encountered in knockout experiments are summarized in *Table 4*. Knocking out a gene that is expressed in multiple tissues or is required during development can result in embryonic/neonatal lethality or pleiotropic effects. The development of tissue-specific and developmental stage-specific homologous recombination (34) should overcome these problems. The method combines transgenesis and gene targeting.

Table 4. Problems encountered in knockout experiments

Problem	Frequency	Explanation/solution	Examples
Prenatal mortality	~25%	Embryonic lethal mutation	33
		Cell-type specific targeting	34
		Inducible gene targeting	35
		Transgenic approaches	See section 3
Perinatal mortality	~10%	As for prenatal mortality	
No phenotype	~5%	Genetic redundancy	
		Closer examination	
		Cross with other knockouts	30
Pleiotropic phenotype	~15%	As for prenatal mortality	36
Different phenotype between ligand + receptor knockout	Rare	Multiple receptors + ligands	37, 38

Homologous recombination is used to introduce short DNA sequences, termed lox P sites, into DNA flanking the gene region of interest in ES cells. Mice bearing the lox P-flanked target gene can then be crossed with transgenic mice that tissue-specifically express the Cre recombinase gene from bacteriophage P1, which catalyses recombination between lox P sites, resulting in a tissue-specific recombination event that deletes the sequence between the two lox P sites. This technique has been used to delete the DNA polymerase β gene in T cells, as conventional knockout results in a lethal phenotype (34). An interferon inducible promoter has also been used to express the Cre recombinase gene, with interferon treatment resulting in deletion of the DNA polymerase β gene in liver, lymphocytes and other tissues within days (35). The FLP recombinase of the yeast 2 μm plasmid (39) has similar properties to Cre and should allow similar manipulations.

5. VIRAL GENE TRANSFER

Viral gene transfer enables transient transgenesis in a temporally and spatially controlled manner into the CNS (40). Three types of viral vector are in common use: retroviral, adenoviral and herpes simplex virus (HSV), the properties of which are summarized in *Table 5*.

Table 5. Viral vectors

Type	Advantages	Disadvantages	Uses
Retrovirus	Integration High infectivity Replication in host	Small capacity (9 kb) Insertional inactivation Fragile viral particle Low titers Infect replicating cells only	Cell lineage and migration Immortalization Tumor treatment
HSV	Large capacity (30 kb) Neurotrophic Infect quiescent cells High infectivity	Latency Genome size (152 kb) Cytotoxicity	Transient gene transfer
Adenovirus	Episomal High titers Stable viral particle Infect quiescent cells	Small capacity (8 kb) Immune response Transient treatment Retrograde transfer Cytotoxicity	Transient gene transfer

Viral gene transfer can be used directly by injection of viral particles, or indirectly by transplantation of cells that have been virally infected (or genetically modified by other means). An advantage that viral vectors have over other methods of gene transfer into the CNS is that they infect a high proportion of cells in the region to which they are delivered (41), thus enabling efficient gene transfer. The use of viral vectors in the nervous system is summarized with examples in *Table 6*.

Table 6. Uses of viral vectors in the nervous system

Use	Vector type	Examples	Reviews
Cell lineage marking	R	42	50
Transplanted cell marking	R	43	
Immortalization	R	44	51
Ex vivo gene transfer	R	45	52
Tumor treatment model	R	46	
Human brain tumor therapy	R	47	
Marker gene transfer	H	48	40
Marker gene transfer	A	41	40
Growth factor gene transfer	H	49	53

A = adenovirus; H = herpes simplex virus; R = retrovirus.

Gene therapy has been a major goal in the development of viral vectors (reviewed in ref. 54), which have been extensively used in the modeling of gene therapy for neurological disorders (reviewed in ref. 52). Two major strategies have evolved: the first is an *ex vivo* strategy involving genetic manipulation of cells *in vitro* followed by transplantation; the second is direct injection of viral particles into the target tissue. While retroviral vectors can be used for the former strategy, adult neurons are post-mitotic and thus refractory to retroviral infection, so therefore require the use of adenoviral or HSV vectors.

6. ANTISENSE OLIGONUCLEOTIDES

Antisense oligonucleotides (ASO) are short synthetic nucleic acid sequences that are complementary to the coding or sense strand of DNA and so can base-pair with mRNA, which can result in degradation of the mRNA or the inhibition of mRNA translation. The use of ASOs *in vivo* can enable a specific transient reduction of protein synthesis for a selected gene product in a localized manner, resulting in altered pharmacological, electrophysiological, behavioral and physiological responses (reviewed in refs 55–57).

ASOs have the advantage of specificity, which can not always be achieved by pharmacological agents, and so may be used to specifically downregulate a single receptor subtype; for examples see *Table 7*. Administration can be via cerebral ventricle or directly into a specific brain region, thus enabling central or localized effects to be determined; and may be a single or multiple injections, enabling acute or chronic studies, though single injections may have long lasting effects. Transgenesis presents another way of delivering an antisense RNA to a cell, one example being expression of antisense myelin basic protein which results in a shiverer phenotype (19), though there are few reports of successful transgenic antisense experiments.

Table 7. Examples of ASOs directed against specific receptor subtypes

Receptor	Specificity	Paradigm	References
Neuropeptide Y (NPY)	Y_1 *vs* Y_2	Anxiety	58
Dopamine (DA)	D_2 *vs* D_1	Rotational behavior	59
Glutamate	NMDA-R1	Neurotoxicity	60
Angiotensin (AT)	AT_1	Hypertension	61
Acetylcholine (ACh)	m1 *vs* m2		62

7. CONCLUSION

The techniques that have been outlined are of particular value in the analysis of the complex systems studied in neurobiology. From the examples that have been reviewed it can be seen that, while each technique has particular applications to which it is especially suited, none can provide a definitive approach to a problem. This is demonstrated by the use of genetic engineering to generate animal models of human neurodegenerative disorders. Dominant genetic disorders (e.g. Huntington's disease) can be modeled in transgenic animals, whereas recessive disorders must be modeled in gene knockout experiments.

Transgenesis and homologous recombination enable the addition or removal of genetic material and so enable the production of animals with a defined genetic modification and the subsequent analysis of the effects of this modification. The application of these techniques to the nervous system is reviewed in greater detail in ref. 7. Viral gene transfer allows the addition of a gene to a specific region of the nervous system, whilst antisense oligonucleotides can permit the specific inhibition of protein synthesis in a chosen brain region. In combination these techniques enable specific manipulations of the nervous system that can generate new research tools that are of enormous value in the field of neurobiology.

8. REFERENCES

1. Hames, B.D. and Higgins, S.J. (1993) *Gene Transcription: A Practical Approach*, IRL Press, Oxford.

2. Murphy, D. and Carter, D.A. (1993) *Transgenesis Techniques: Principles and Protocols*, Humana Press, New Jersey.

3. Dillon, N. and Grosveld, F. (1993) *Trends Genet.*, **9**, 134.

4. Forget, B.G. (1993) *Proc. Natl Acad. Sci. USA*, **90**, 7909.

5. Jakobovits, A. (1994) *Curr. Biol.*, **4**, 761.

6. Carter, D.A. (1993) in *Transgenesis Techniques: Principles and Protocols* (D. Murphy and D.A. Carter, eds), Humana Press, New Jersey.

7. Herrup, K. (1995) *Transgenic and ES Cell Chimeric Mice as Tools for the Study of the Nervous System*, Elsevier Press, Amsterdam.

8. Waschek, J.A. (1995) *Mol. Neurobiol.*, **10**, 205.

9. Mayford, M., Abel, T. and Kandel, E.R. (1995) *Curr. Opin. Neurobiol.*, **5**, 141.

10. Oberdick, J., Smeyne, R.J., Mann, J.R., Zackson, S. and Morgan, J.I. (1990) *Science*, **248**, 223.

11. Wuenschell, C.W., Mori, N. and Anderson, D.J. (1990) *Neuron*, **4**, 595.

12. Mellon, P.L., Windle, J.J., Goldsmith, P.C., Padula, C.A., Roberts, J.L. and Weiner, R.I. (1990) *Neuron*, **5**, 1.

13. Breitman, M.L., Clapoff, S., Rossant, J., Tsui, L.-C., Glode, L.M., Maxwell, I.H. and Bernstein, A. (1987) *Science*, **238**, 1563.

14. Breitman, M., Rombola, H., Maxwell, I., Klintworth, G. and Bernstein, A. (1990) *Mol. Cell Biol.*, **10**, 474.

15. Kobayashi, K, *et al.* (1995) *Proc. Natl Acad. Sci. USA*, **92**, 1132.

16. Games, D. *et al.* (1995) *Nature*, **373**, 523.

17. Burright, E.N., Clark, H.B., Servadio, A., Matilla, T., Feddersen, R.M., Yunis, W.S., Duvick, L.A., Zoghbi, H.Y. and Orr, H.T. (1995) *Cell*, **82**, 937.

18. Mason, A.J., Pitts, S.L., Nikolics, K., Szonyi, E., Wilcox, J.N., Seeburg, P. and Stewart, T.A. (1986) *Science*, **234**, 1372.

19. Katsuki, M., Sato, M., Kimura, M., Yokoyama, M., Kobayashi, K., Nomura, T. (1988) *Science*, **241**, 593.

20. Mayford, M., Wang, J., Kandel, E.R. and O'Dell, T.J. (1995) *Cell*, **81**, 891.

21. Beeri, R., Andres, C., LevLehman, E., Timberg, R., Huberman, T., Shani, M. and Soreq, H. (1995) *Curr. Biol.*, **5**, 1063.

22. Piccioli, P., Di Luzio, A., Amann, R., Schuligoi, R., Surani, M., Donnerer, J. and Cattaneo, A. (1995) *Neuron*, **15**, 373.

23. Aigner, L., Arber, S., Kapfhammer, T., Schneider, C., Botteri, F., Brenner, F., Brenner, H. and Caroni, P. (1995) *Cell*, **83**, 269.

24. Furth, P.A., Onge, L.S., Boger, H., Gruss, P., Gossen, M., Kistner, A., Bujard, H. and Henninghausen, L. (1994) *Proc. Natl Acad. Sci. USA*, **91**, 9302.

25. Wang, Y., O'Malley, B.W., Tsai, S.Y. and O'Malley, B.W. (1994) *Proc. Natl Acad. Sci. USA*, **91**, 8180.

26. Bronson, S.K. and Smithies, O. (1994) *J. Biol. Chem.*, **269**, 27155.

27. Soriano, P. (1995) *Ann. Rev. Neurosci.*, **18**, 1.

28. Brandon, E.P., Idzerda, R.L. and McKnight, G.S. (1995) *Curr. Biol.*, **5**, 625.

29. Brandon, E.P., Idzerda, R.L. and McKnight, G.S. (1995) *Curr. Biol.*, **5**, 758.

30. Brandon, E.P., Idzerda, R.L. and McKnight, G.S. (1995) *Curr. Biol.*, **5**, 873.

31. Joyner, A.L. and Guillemot, F. (1994) *Curr. Opin. Neurobiol*, **4**, 37.

32. Grant, S.G.N. (1994) *Curr. Opin. Neurobiol.*, **4**, 687.

33. Copp, A.J. (1995) *Trends Genet.*, **11**, 87.

34. Gu, H., Marth, J.D., Orban, P.C., Mossmann, H. and Rajewsky, K. (1994) *Science*, **265**, 103.

35. Kuhn, R., Schwenk, F., Aguet, M. and Rajewsky, K. (1995) *Science*, **269**, 1427.

36. Umemori, H., Sato, S., Yagi, T., Aizawa, S. and Yamamoto, T. (1994) *Nature*, **367**, 572.

37. DeChiara, T.M. *et al.* (1995) *Cell*, **83**, 313.

38. Masu, Y., Wolf, E., Holtmann, B., Sendtner, M., Brem, G. and Thoenen, H. (1993) *Nature*, **365**, 27.

39. Futcher, A.B. (1988) *Yeast*, **4**, 27.

40. Wolff, J.A. (1993) *Curr. Opin. Neurobiol.*, **3**, 743.

41. La Salle, G.L.G., Robert, J.J., Berrard, S., Ridoux, V., Stratford-Perricaudet, L., Perricaudet, M. and Mallet, J. (1993) *Science*, **259**, 988.

42. Williams, B.P. and Price, J. (1995) *Neuron*, **14**, 1181.

43. Groves, A.K., Barnett, S.C., Franklin, R.J.M., Crang, A.J., Mayer, M., Blakemore, W.F. and Noble, M. (1993) *Nature*, **362**, 453.

44. Renfranz, P.J., Cunningham, M.G. and McKay, R.D.G. (1991) *Cell*, **66**, 713.

45. Kawaka, M.D., Rosenberg, M.B., Yoshida, K. and Gage, F.H. (1992) *J. Neurosci.*, **12**, 2849.

46. Culver, K.W., Ram, Z., Wallbridge, S., Ishii, H., Oldfield, E.H. and Blaese, R.M. (1992) *Science*, **256**, 1550.

47. Oldfield, E.H., Ram, Z., Culver, K.W., Blaese, R.M., DeVroom, H.L. and Anderson, W.F. (1993) *Hum. Gene Ther.*, **4**, 39.

48. Andersen, J.K., Garber, D.A., Meaney, C.A., Breakefield, X.O. (1992) *Hum. Gene Ther.*, **3**, 487.

49. Federoff, H.J., Geschwind, M.D., Geller, A.L., Kessler, J.A. (1992) Expression of nerve growth factor *in vivo* from a defective herpes simplex virus I vector prevents effects of axotomy on sympathetic ganglia. *Proc. Natl. Acad. Sci.* USA, **89**, 1636–1640.

50. Price, J. (1994) *Curr. Opin. Neurobiol.*, **4**, 680.

51. Cepko, C.L. (1989) *Ann. Rev. Neurosci.*, **12**, 47.

52. Friedmann, T. (1994) *Trends Genet.*, **10**, 210.

53. Geller, A.I. (1993) *Curr. Opin. Genet. Dev.*, **3**, 81.

54. Crystal, R.G. (1995) *Science*, **270**, 404.

55. Eng, L.F. (1993) in *Disease* (S.G. Waxman, ed.), Raven Press, New York, p. 293.

56. McCarthy, M.M., Brooks, P.J., Pfaus, J., Brown, H.E., Flanagan, L.M., Schwartz-Giblin, S. and Pfaff, D.W. (1994) *Neuroprotocols*, **2**, 67.

57. Morris, M. and Lucion, A.B. (1995) *J. Neuroendocrinol.*, **7**, 493.

58. Wahlestedt, C., Pich, E.M., Koob, G.F., Yee, F. and Heilig, M. (1993) *Science*, **259**, 528.

59. Weiss, B., Zhou, L.W., Zhang, S.P. and Qin, Z.H. *Neuroscience*, **55**, 607.

60. Wahlestedt, C., Golanov, E., Yamamoto, S., Yee, F., Ericson, H., Yoo, H., Inturris, C.E. and Reis, D.J. (1993) *Nature*, **363**, 260.

61. Sakai, R.R., He, P.F., Yang, X.D., Ma, L.Y., Guo, Y.F., Reilly, J.J., Moga, C.N. and Fluharty, S.F. (1994) *J. Neurochem.*, **62**, 2053.

62. Zang, Z., Florijn, W. and Creese, I. (1994) *Biochem. Pharmacol.*, **48**, 225.

CHAPTER 17

NEURAL TRANSPLANTATION AND REGENERATION

S.B. Dunnett

Transplantation

1. INTRODUCTION

In recent years, neural transplantation has developed from being a somewhat specialist method for the study of neuronal development (more often than not in amphibia), to a powerful tool with widespread application in many aspects of basic and applied neurobiology. The major breakthroughs occurred in the 1970s with the determination of the basic techniques for achieving reliable survival of neuronal grafts in the adult mammalian brain, and in the early 1980s with the demonstration that such grafts could have a functional impact on the host animal.

Applications of neural transplantation span basic and applied neurobiology. On the one hand it has been used as a method to manipulate sources, substrates and target tissues selectively in order to elucidate the principles of axon growth and circuit formation in the brain, whether during initial development of the nervous system or under conditions of regeneration following injury. On the other hand, the fact that grafts have been seen to be functional in experimental animals has stimulated speculation about whether neural transplantation might provide a basis for radical new therapies following injury or disease in the central nervous system.

The potential of these new techniques for clinical application has received a powerful stimulus from the dramatic success of some of the first experimental trials in Parkinson's disease (1). In between the basic developmental and clinical studies, we are now seeing a wide range of applications of neural transplantation in studying the principles of structural and functional organization and reorganization of the mammalian brain (2). Moreover, whereas in most early studies the grafts were themselves the main subject of investigation, increasingly as the techniques become more widely established, transplantation is merely another routine tool (alongside lesions, drugs, behavioural analysis, *in vivo* monitoring of brain activity, post mortem histology of brain tissues, etc.) in the neurobiologist's armoury for the analysis of degeneration and regeneration in the brain.

1.1. Factors important in graft viability

Transplantation in the brain differs from transplantation of other organs and tissues of the body in that the brain exhibits a marked level of immunological privilege (3). Note that the privilege is not absolute: transplants in the brain may still be rejected if there are large differences between donor and host in both major and minor histocompatibility antigens (such as occurs with cross-species transplantation), or if immune processes are invoked by other means (for example by making additional skin grafts between the same animals). Nevertheless, several factors combine to reduce the immunological problems to be confronted in neuronal transplantation:

(i) Neural tissues express relatively low levels of histocompatibility antigens, making them less immunogenic than other tissues.

(ii) The brain itself has a poor lymphatic drainage, with the result that activated T cells do not readily trigger immunological reaction.
(iii) The blood–brain barrier is effective in excluding circulating molecules of the immune system, making it less susceptible to targeting of an immune response.

Together these factors render the brain relatively immunologically privileged, so that problems of rejection are generally not encountered when grafting neural tissues between the brains of animals of the same species. Nevertheless, in situations where an immune response is problematical, this has proved relatively easy to contain using conventional immunosuppression treatments. Daily treatment with 10 mg/kg cyclosporin A has been found to provide effective immunoprotection for brain grafts in a variety of species (3).

Two other factors have been found to be far more important in achieving reliable graft survival: the source of the donor tissue and the host environment into which the graft tissue is implanted.

Although many peripheral tissues may survive transplantation throughout life, neuronal tissues of the central nervous system have been found only to be viable when taken for transplantation during a narrow time window of embryonic or neonatal development (4, 5). The critical age is typically determined empirically, but is generally found to correspond closely to the stage of final cell division during development. Consequently, for each population of cells that are targeted for transplantation it is imperative to select an appropriate donor age (see *Table 1*).

Secondly, it is important to implant tissues into a site where they can receive adequate nutritional and metabolic support, which generally involves rapid incorporation into the host vascular circulation and cerebrospinal drainage (6, 7). When implanting chunks of donor

Table 1. Optimal ages and stages for transplanting embryonic rat tissues, based on studies following transplantation *in oculo* (5) and our own experience (4)

Region	Embryonic age[a]	Crown–rump length[b]	Carnegie stage[c]
Parietal cortex	17–19	F	F
Entorhinal cortex	15–19	21–23	21–23
Hippocampus	19–21	26–34	F
Dentate gyrus	20–22	30–36	F/N
Cerebellum	14–15	11–14	18–21
Olfactory bulb	17–19	19–25	F
Spinal cord	15–17	14–20	21–23
Caudate nucleus	14–16	11–15	18–22
Septum	14–16	11–15	18–22
Substantia nigra	14–15	10–13	17–20
Locus coeruleus	14–15	10–13	17–20
Dorsal raphe	14–15	10–13	17–20
Retina	22+	45	N

[a] Embryonic age calculated on the basis of designating the day following overnight mating as day 0.
[b] Crown–rump length is based on the sizes of freshly removed embryos, and may differ from anatomical descriptions based on formalin-fixed tissues.
[c] Carnegie stages, originally defined for human embryonic development, is based for the rat on the comparative descriptions of Butler and Juurlink (14).

tissue into the brain (so-called 'solid' grafts) it is necessary either to select one of a few target sites that have a naturally high level of vascularization or undertake additional surgical procedures to create an artificial vascular supply without which the graft will not survive (7). This factor turns out to be less important when implanting dissociated cells by injection directly into the brain (so-called 'cell suspension' grafts) (8), since these become more rapidly integrated into the microcirculation of the host neuropil, and is one (of several) reasons why the cell suspension technique is the procedure of choice for most purposes. In contrast to solid grafts, cell suspensions allow multiple graft deposits to be placed at will into deep or superficial sites within the host brain with very little additional surgical trauma. Nevertheless, solid grafts are still preferred for certain experimental purposes, such as when it is intended to implant an electrode or cannula into the grafts under visual guidance in order to monitor graft activity or to trace functional connections with the host brain.

1.2. Outline of procedures

Animal models
For many purposes the focus of experimental investigation is the repair of neurodegenerative disorders. A variety of lesion models using electrolytic, neurotoxic, ischemic and genetic manipulations have been described in earlier chapters of this volume and many have been investigated in terms of transplantation repair (2). In addition, it has been found that graft survival and outgrowth is in many situations promoted by lesions (and the consequent induction of a trophic response) of intrinsic systems of the brain (9). As a consequence, it may be preferable to make the grafts into the lesioned brain, even when the primary experimental focus is on the regenerative capacity of grafts. In yet other situations, quite arbitrary graft sites have been selected because of their ease of transplantation and suitability for monitoring graft survival *per se*; the fourth ventricle and the anterior chamber of the eye are examples (5, 10).

Experimental design
Several factors need to be borne in mind when designing transplant experiments. Most important is the need for control groups appropriate for the experimental issues under investigation. Typically the minimum requirement is for at least three groups: Normal, Lesion, and Lesion + Graft animals. If the study is on the capacity of grafts to repair lesion effects, which are themselves already well validated, it may be more appropriate to use normal animals rather than sham lesions against which to evaluate the extent of repair. Similarly, if there is an interest not only on whether a graft can have an effect, but the specificity and mechanisms of graft action, then it becomes necessary to also have control grafts. Depending on the particular circumstances (11), control grafts may involve:

(i) sham transplantation procedures,
(ii) implantation of control tissues,
(iii) removal of selected populations of cells from the experimental grafts,
(iv) implantation of dead tissues, or
(v) removing grafts after first observing their effects.

The increased number of experimental groups in a typical transplantation experiment, and the frequently increased variability of the transplantation groups, then generally requires using larger group sizes than are sufficient for simple lesion experiments, even when combined with the most powerful statistical procedures, to resolve clear and significant effects of the treatment(s).

In addition to selecting appropriate animal models and experimental designs, the techniques of neural transplantation involve several stages:

(i) identification of appropriate donor tissues,
(ii) tissue dissection,
(iii) tissue preparation,
(iv) host preparation, and
(v) host implantation.

These are considered in turn in the following sections.

2. EMBRYONIC NEURONAL TISSUES FOR TRANSPLANTATION

2.1. Staging of developing embryos

As stated in section 1.1 above, the most important factor in achieving reliable graft survival is the selection and accurate dissection of donor tissues of the appropriate donor age. Obtaining accurately staged donor material is the first substantial problem to confront the neophyte transplant surgeon.

What is the appropriate donor age?
Select the age of the donor on the following basis:

(i) Do a detailed literature search on previous transplantation experiments using the particular populations of cells of experimental interest, identifying what age previous studies have found most successful for the cells.
(ii) Consider the developmental literature; take graft tissues close to the neuronal birthdates of the particular groups of cells of experimental interest.
(iii) Always undertake a pilot study to validate a novel graft procedure, and include a sensible range of donor ages.

Note that different studies calculate the gestational age of embryos differently: some (in particular US authors) take the day after an overnight mating as embryonic day 1, whereas others (in particular Scandinavian authors) take that day as embryonic day 0. In the present context, we will follow the latter system, which has recently also been adopted in the developmental atlas of Paxinos (12). Since accurate timing of conception is anyway difficult, it is therefore safer to consider not the age of the embryo *per se*, but its stage of development or size (see *Table 1*). The stage of embryonic development has been characterized in detail for human embryos, the so-called Carnegie stages (13), and this provides direct comparison between different species. In practice, the size of the embryo measured as the flattened crown–rump length is easier to use and tables of equivalent size, embryonic age and Carnegie stage for a range of widely used experimental species (including chick, rat, mouse, marmoset and rhesus monkey) are provided in ref. 14.

If your animal breeder/supplier can provide accurately staged embryos then the simplest approach is simply to order them by specifying the relevant age for the appropriate stage of development. In practice, however, many laboratories find that their suppliers cannot provide the level of accuracy required for achieving reproducible transplants. In this case, it is easy with practice to achieve an accurate estimate of embryonic stage by lightly anesthetizing the pregnant mother (e.g. with ether, if allowed) and palpating the abdomen. Detailed guidelines for palpation and descriptions of the signs to be felt are provided in several places (4, 5), and are summarized in *Table 2*.

Having identified pregnant mothers of the correct age of gestation, the embryos are removed. This may be achieved either by sacrificing the mother by an approved method and removing

Table 2. Age, stage and crown–rump length of rat embryos (4)

Gestational age (days)[a]	Carnegie stage	Crown–rump length	Signs at palpation
4–7	3–5		Uterine horns are difficult to find and have variable thicknesses
8–9	6–9		Uterine horns have small, often closely spaced, distinct swellings
10	10–11		
11	12–13		
12	14–15	8	
13	16–17	9	Small distinct firm spheres
14	18–19	10–11	
15	20–21	12–14	
16	22–23	15–16	
17	Fetal	17–19	Elastic, somewhat ovoid enlargements
18	Fetal	21–23	Fetal structures begin to become palpable
19	Fetal	24–25	Fetal indurations appear
20–21	Fetal	30–40	Uterine horns become thick, soft and continuous
22	Neonatal	45	Day of birth

[b] Based on (4). The gestational age is defined by designating the day following overnight mating as embryonic day 0. The data on Carnegie stages for rat embryos are derived from Butler and Juurlink (14). The signs of palpation are based on the descriptions of Olson and colleagues (5).

the uterus containing the embryos or fetal donors (4), or, if the mother is to be retained (e.g., in primate studies), by Cesarean section under general anesthesia with recovery (15).

2.2. Dissection of neuronal tissues

Detailed descriptions for removing the embryonic brain and dissecting relevant areas are available elsewhere (4). Briefly, remove the brains from the donor embryos using a pair of Dumont no. 5 forceps, fine scissors and a pointed (no. 11) scalpel blade, and collect in sterile 0.6% glucose-enriched saline, or tissue culture medium.

The dissection procedures cannot be described here for all the different procedures that may be of interest. A selection of common neural tissue dissections used in a variety of grafting paradigms are described and illustrated in ref. 4, including dissections for the substantia nigra, locus coeruleus, raphe, septum, cortex, hippocampus, and striatal (ganglionic) eminence. Other procedures are described in primary reports of individual transplantation experiments. Moreover, since the first stage of tissue collection for neural transplantation is very similar to that in a variety of widely used cell culture methods, many other dissections can be found in the tissue culture literature (including many excellent methodological texts).

When dissecting tissues for neural transplantation, it is sufficient to collect the dissected pieces in a glucose-enriched saline solution. Although many authors prefer a variety of glucose-enriched balanced salt solutions and buffers, in our experience these add little to the viability of the resulting grafts.

2.3. Preparation of neuronal cell suspensions

Dissected tissues are prepared as a cell suspension for graft implantation. These have been developed from tissue culture preparations, but differ in that the final suspension is generally far denser when used for implantation. Briefly, the tissue is incubated in 0.1% trypsin at 37°C for 20–60 min, washed several times in glucose–saline with 0.04% DNase to reduce cell reaggregation, then mechanically dissociated by aspiration through a series of graded, fire-polished Pasteur pipettes. The final cell suspension is typically made up to a final concentration of 1 piece of embryonic tissue in 2–10 µl glucose–saline, and of the order of 1–5 × 10^5 cells per µl. Detailed descriptions of the standard cell preparation procedures and common variants are provided in ref. 16.

A powerful addition to enhance the reproducibility of the standard protocol is to count the actual density of cells in the injected suspension, and to estimate the viability of the cells at the time of implantation. One method that is widely used in the tissue culture literature is the trypan blue exclusion principle to identify dead cells. A more powerful method is to label living and dead cells separately by staining with the fluorescent dyes acridine orange and ethidium bromide (17). Prepare a stock solution of 0.01% acridine orange + 0.01% ethidium bromide and store at 4°C. Mix 2 µl cell suspension with 8 µl of the dyes, add the droplet to the graticule of a standard hemocytometer, and count live and dead cells (fluorescing green and red, respectively) in a standard fluorescent microscope. The raw counts are corrected for dilution and scaling to yield estimates of the density of cells in the suspension and their percentage viability.

3. OTHER SOURCES OF TISSUE FOR TRANSPLANTATION

3.1. Peripheral nervous tissues

Whereas central nervous tissue must be taken from the developing embryo if it is to be viable, neuronal tissues from peripheral glands can survive transplantation even from adult donors. This may be related in part to the greater regenerative potential of the peripheral than the central nervous system (18) and in particular to the fact that peripheral neurons undergo neurogenesis and replacement throughout life rather than only in the early periods of development.

A variety of different peripheral tissues have been used for neural transplantation into the central nervous system, the most important of which are summarized in *Table 3*.

3.2. Spinal ganglia

Spinal ganglia such as the dorsal root ganglia have been used widely in tissue culture as a prototypical tissue for the study of the effects of growth factors on cell survival and of cell adhesion molecules on neurite outgrowth. Different ganglia are associated with different neurotransmitters and are dependent upon different neurotrophic molecules for survival and the control of cell phenotype. Similarly, these ganglia were often used in early studies of graft survival in the central nervous system, identification of trophic actions, and as a source of catecholamine neurons that survived better (at least using the early techniques) than do central aminergic neurons such as the noradrenergic cells of the locus coeruleus. Different groups of spinal ganglia have provided a powerful source of information on the selective patterns of neurotransmitter specific outgrowth from grafts, in particular *in oculo* (19, 20) and in the hippocampus (21, 22), as well as a potential alternative source of dopamine-producing cells for clinical transplantation (23).

3.3. Adrenal medulla

When grown in culture in the presence of NGF and in the absence of steroids, the chromaffin cells of the adrenal medulla adopt a neuronal phenotype, with noradrenaline and some dopamine (rather than adrenaline) as their primary secretory product (24). Following the demonstration that embryonic nigral grafts can alleviate dopamine-denervation models of

Table 3. Peripheral tissues used for transplantation

Tissue	Replacement	Lesion/model system	References
As a replacement source of deficient neurochemicals			
Adrenal medulla	Dopamine	Nigrostriatal lesions	(25)
	Catecholamines/endorphins	*In oculo* grafts	(51)
		Spinal pain	(85)
Carotid ganglion	Dopamine	Nigrostriatal lesions	(23)
Superior cervical ganglion	Catecholamines	Lesioned hippocampus	(86)
		In oculo grafts	(5)
Dorsal root ganglion	Acetylcholine	Spinal cord lesions	(87)
		Olfactory bulb lesions	(88)
As a source of deficient neurotrophic molecules			
Peripheral nerve (e.g. sural nerve)	Nerve growth factor	Co-grafts with adrenal	(89, 90)
Mouse submaxillary gland	Nerve growth factor	Protection from striatal lesion	(91)
		Protection from nucleus basalis lesions	(92)
As a substrate for axon growth			
Peripheral nerve	E.g. sciatic nerve	Spinal cord lesions	(39, 93)
Purified Schwann cells	Schwann cell cultures and bridges	Nigrostriatal regeneration	(94)
		Septohippocampal lesions	(95)
		Spinal cord injury	(96)
Artificial matrices	Glial coated cellulose bridge	Callosal lesions	(97)
Tracks of substrate molecules	E.g. laminin, fibronectin	Nigrostriatal lesions	(98)
		Serotonin lesions	(99)

parkinsonism in rats (25) (and later in monkeys and man also (26)), it was considered that adrenal medulla might provide a more accessible source of dopamine-secreting cells for clinical application in Parkinson's disease. By taking autografts from the patient him/herself, both the immunological problems of tissue compatibility and the ethical problems associated with using fetal tissues would be overcome.

The first clinical trials of neural transplantation in Parkinson's disease used adrenal chromaffin autografts (27). These however were not widely successful and were associated with substantial complications of surgery so are no longer widely undertaken (28, 29). Nevertheless, adrenal grafts have opened up major novel approaches to study the mechanisms of graft function. In particular, functional recovery was demonstrated in both experimental animals and patients, even in cases where the graft itself did not survive (30, 31). This was found to be attributable to the fact that adrenal grafts can stimulate substantial sprouting of host dopamine systems spared by incomplete lesions (32), and indicated that grafts could function by exerting a trophic action, stimulating intrinsic repair processes in the host brain, rather than by a mechanism of neuronal or neurotransmitter replacement as first envisaged (33).

3.4. Retina and other ocular grafts

Regeneration of the retinotectal system in amphibians has provided one of the foundations of modern developmental biology (34). Not surprisingly, regeneration and transplantation in the visual system has provided one of the main foci for study as the techniques of neural transplantation have been developed for mammals. The studies of Lund and coworkers in particular have demonstrated the viability of retinal (as well as tectal and visual cortical) transplantation into the neonatal CNS, and a capacity for systematic and directed growth of retinal axons into and through the host brain to reach appropriate targets in the tectum and elsewhere (35). Moreover such grafts can be functional, in that the transplanted retinae can transduce light signals and the reformed retinotectal projections can transmit this information not only to drive pupillary light reflexes but also to be used by the host animal as effective visual stimuli controlling learned responses (36).

Retinal tissues survive transplantation much better if dissected from embryonic or neonatal donors. In the experiments by Lund and coworkers, the retina itself is transplanted into the brain, in sufficient proximity to make contacts with central targets in the host CNS (35, 37). Other groups have studied transplantation of retinal tissues into the posterior chamber *in oculo* in retinal degeneration models in animals, and of epithelial cells to protect the retina from degeneration in retinal pigment epithelium mutant strains of mice, in each case with some limited success (38).

3.5. Peripheral nerve and other glial grafts

Whereas the adult CNS has only limited capacity for regeneration, the PNS provides a good substrate for axonal regrowth whether derived from central or peripheral neurons. As a consequence, segments of peripheral nerve have been widely used as bridge grafts to provide a substrate for axon regrowth, not only after damage in the PNS but in the CNS also. For example, a PNS graft can bridge the transected spinal cord and allow a limited regrowth of transected central axons across the site of injury (39).

It appears that the critical cellular element in peripheral nerve grafts is the Schwann cell, and a variety of strategies have been used for their engraftment as tissue bridges. As outlined above a whole segment of peripheral nerve may be used. Alternatively, Schwann cells may be isolated or grown in culture, and then implanted into the CNS as injections of dissociated cells, or plated on substrates and implanted in layers or as 'jelly rolls' (40). The advantages of the latter approach is that the alignment of Schwann cells may be important in their guidance of directed axon growth, and such a regular parallel alignment can be achieved under appropriate culture conditions.

A second major interest for glial transplantation is remyelination in a variety of genetic and toxic models of human demyelinating disease. In these experiments, the need is for relatively pure populations of oligodendrocytes or their precursors for transplantation. Undifferentiated CNS tissues are taken from neonatal donors, grown in conditions in which the neurons do not survive, and then appropriate cell separation procedures may be applied to yield populations of oligodendrocytes and/or their precursors, with a demonstrated capacity to remyelinate areas of experimental demyelination in the rodent central nervous system (41, 42).

Other studies have used implants of astrocytes or endothelial cells, typically to study mechanisms of response to injury, and processes of regeneration in the CNS (43–45).

4. TRANSPLANTATION OF TISSUES

4.1. Dissociated cell suspensions in the adult brain

The most powerful method for neuronal tissue implantation is injection as dissociated cell suspensions. The donor tissue is prepared as outlined in section 2.3 above. The host animal is anesthetized, mounted in a stereotaxic frame, and 0.5–3 µl aliquots of graft tissue are injected stereotaxically, directly into selected sites in the host parenchyma. Injections are typically delivered via fine stainless steel needles of a 1–10 µl Hamilton syringe mounted in the stereotaxic frame. Alternatively, micrografts may be delivered using glass micropipettes attached to the microsyringe (46, 47). A detailed account of the procedure is provided in ref. 16.

The power and flexibility of the cell suspension method is indicated by several distinct factors:

(i) Stereotaxic injections involve minimal additional trauma to the host brain.
(ii) The grafts are positioned with stereotaxic accuracy into any site.
(iii) No special enriched vascular site is required unless very large volumes are to be injected.
(iv) It is easy to make multiple placements of a graft tissue, or co-grafts of different tissues into the same or different sites.

4.2. Solid grafts in the adult brain

For certain purposes it is necessary to implant grafts as solid pieces of donor tissue. This may be because the purpose of the experiments is to study the development of particular patterns of internal organization within the grafts (e.g. lamination in cortex or hippocampus), or because part of the experimental protocol involves implantation of electrodes or cannulae into the grafts under visual guidance. In other circumstances, implantation of small discrete pieces of tissue is the simplest procedure technically (e.g. placing hypothalamic tissues into the third ventricle). Whatever the purpose, in each case the essential requirement as emphasized above is to select or create a richly vascularized site in the host brain, to provide rapid incorporation of the graft into the host circulation, and to sustain the newly transplanted piece of tissue up to that time. Essentially four different procedures have been applied.

The simplest procedure is to identify naturally rich vascular surfaces in the brain. These include the choroid plexus and ependymal linings of the ventricles, and the choroidal fissure between the thalamus and hippocampus. Small pieces of embryonic neuronal tissue or peripheral ganglia will survive well following insertion into such sites. Thus the first simple method is to eject the graft from a broad gauge cannula into a ventricle adjacent to the target area, taking care not to cause undue damage or bleeding in the process (10, 48).

The second approach is to expose the surface by making an aspirative lesion. This is a typical approach for grafting onto the choroidal fissure, adjacent to the hippocampus and above the thalamus, and may be reached via either fimbria–fornix or retrosplenial approaches. The graft tissue is then placed into the cavity, oriented as appropriate, and held in place by a small piece of gel foam (7).

If a richly vascularized site is not immediately available (such as adjacent to the dorsal or lateral surfaces of the neostriatum, or the putamen in primates and man), the third approach is to create a new vascular surface surgically. A number of different approaches have been adopted, including a variety of co-grafts with richly vascularized tissues (such as muscle, omentum or iris), but the simplest approach is to use a delayed transplantation strategy. When a cavity is made through the cortex to expose the striatal surface, and then filled with gel foam, a new pial surface will reform over the following weeks. This pial surface is itself highly vascularized and provides a suitable bed for transplantation undertaken in a two-stage operation after a delay of 3–6 weeks (7).

The fourth approach is to inject small fragments of tissue directly into the host neuropil. This approach has been successfully used in a number of circumstances, including making implants into the neocortex (49). However, although this approach works well in neonates, the survival and rapid incorporation of the cells is generally not as good in adults as delayed transplantation or cell suspension methods.

4.3. Transplantation in the neonatal and embryonic brain

Implantation into the neonatal brain presents both advantages and special challenges. Such experiments are often required when manipulating parameters related to the development of structure and function. Moreover, implanting tissues into the developing brain (where axon growth and circuit formation are active and ongoing processes) typically results in the grafts becoming better incorporated into the host neuronal circuitry than occurs when the implants are made into the already mature brain. The disadvantages of implantation into the neonatal brain are that accurate stereotaxic placement of the grafts is difficult, and the implanted cells can migrate far and wide within the host rather than remaining localized at the site of injection.

As a consequence of these differences, somewhat different implantation techniques are employed in the neonatal brain. The newborn rat pup is easily anesthetized either with ether or cold. The brain can be trans-illuminated using a fiber-optic light source from below, which allows the outline of the brain and the sutures of the skull to be seen without incision in the skin. Consequently, if the grafts are loaded into a syringe needle with a cuff to stop penetration at a predetermined depth, grafts can be manually injected following natural landmarks directly into large structures (such as the neostriatum) or to lie on brain surfaces (such as on the cortical surface under the dura, or overlying the tectum). A detailed description of procedures for implanting grafts into the neonatal (and embryonic) brain is provided in ref. 50.

4.4. Intraocular grafting

The anterior chamber of the eye has proved a highly productive site for studying the biological principles of neural transplant survival, growth and connectivity. Like the brain it has a degree of immunological privilege, there is a rich vascular supply to the iris that provides a suitable bed for nourishing solid grafts, and the iris itself receives well characterized cholinergic and adrenergic innervations from discrete peripheral ganglia that can provide a source for axon ingrowth into the grafts and interact with graft derived outgrowth into the iris. For example, implantation of ganglionic and retinal grafts in the anterior eye chamber provided clear early examples of the capacity for regenerative and collateral sprouting in the nervous system (51). Because of the ease of observation through the clear cornea, the anterior eye chamber is a particularly straightforward site for implantation of neuronal grafts, and graft survival and growth can be monitored on a day-by-day basis, without the need for invasive surgery or post mortem analysis, simply by light anesthetization and direct observation.

The procedures for neuronal grafting into the anterior eye chamber have been well described (5), and there are a number of reviews of the power and flexibility of the techniques (50, 52).

4.5. Grafting in the spinal cord

Although restitution of long-distance axon connections by regeneration across the stumps of a transected spinal cord is considered by many to represent the ultimate challenge – that the lame may walk again – for neural transplantation, the procedures are essentially similar to those already considered above: injection of dissociated cell suspensions (for example to replace lost brainstem catecholaminergic innervations) and implantation of solid pieces of tissue, cultured cells or artificial matrices (for example to provide bridges for regeneration of central axons to distal targets). A recent review of the techniques, experimental paradigms, and functional status of neural transplantation in the spinal cord is provided in ref. 53.

5. PROMOTING GRAFT SURVIVAL

5.1. Aspects of the preparation techniques

The previous section identified sources for describing the standard methods for a variety of neuronal transplantation procedures. For many cases these are well worked out and provide a high degree of reliability and reproducibility. However, as in many other areas of the neurosciences, the field of neural transplantation is under rapid development, and the techniques are under continual revision and development. This becomes of particular importance in those situations which are under development for clinical application or where the present techniques are still clearly suboptimal.

The first main area of clinical application has been the implantation of embryonic mesencephalon ('nigral grafts') in Parkinson's disease. In contrast to many other similar models, dopamine neurons do not survive grafting well, and it has been estimated that only 2–10% of implanted cells survive the transplantation procedure. It has turned out to be necessary to use 5–8 donors per host striatum for a good functional response in patients (1), with the consequence that this is not likely to provide a widely available therapy using present procedures. The first and most obvious strategy has been to seek to improve the preparation techniques themselves in order to achieve more reliable graft survival and higher dopamine cell yields.

There has been a systematic investigation of the factors involved in graft preparation, including the comparison of different age of embryos, dissection methods, media and temperatures for collection handling and storage of tissues, enzymatic digestion, cell dissociation procedures, and implantation implements. Several of these factors have been found to yield modest enhancements, but all together they do not provide any substantial resolution of the relatively low yields of dopamine neurons (54). Indeed, the fact that using standard procedures for preparation of the cell suspensions the cells can survive at a level close to 100% when grown in three-dimensional culture suggests that the host environment is at least as important for good survival of grafts *in vivo* as the cell preparation procedures *per se* (55).

5.2. Trophic factors promoting graft survival

Recent attention has focused on the developmental dependence of neuronal cells on a variety of growth factors for their survival and fate. A second strategy has therefore been to address the possibility that the poor yield of certain populations of graft cells may be due to their developmental dependence on one or more trophic molecules that are no longer available at adequate levels in the adult host brain (33, 56). The issue of trophic dependence of grafted cells has been investigated in detail in two main situations, survival of the chromaffin cells of the adrenal medulla and of the dopamine cells of the substantia nigra, both in the context of improving methods for transplantation in Parkinson's disease.

The neuronal phenotype of adrenal medullary is dependent upon the availability of NGF, and supplementation of the grafts with NGF has been suggested as a potential strategy for enhancing their survival. Evidence for the viability of this approach has come from a variety of methods of delivery, including direct injection of NGF and co-grafts of peripheral nerve in both experimental studies and patients (56).

The survival of developing dopamine neurons of the substantia nigra has been shown *in vitro* to be enhanced by a variety of growth factor molecules of different classes. For example, the first studies to show the efficacy of FGF, BDNF, NT4/5 and GDNF on the survival and functional efficacy of nigral dopamine neurons employed direct injections of the growth factors via microcannulae implanted into the vicinity of the grafts in the striatum (57–59). However, this method is not particularly efficient since daily injections are required, and an even distribution of spread and temporal stability of the growth factor molecules at physiological levels is difficult to obtain by such phasic injection procedures. Chronic delivery of NGF into the ventricles has been achieved using Alzet minipumps. However, due to poor penetration of many growth factor molecules, intraventricular administration is not ideal for accessing grafts in deep brain sites, and long-term delivery directly into the host parenchyma proximal to the grafts has proved problematic because of tissue necrosis at the site of injection. Perhaps the most promising approach has been the recent development of cells engineered to secrete a variety of growth factor molecules, for implantation as co-grafts, and recent studies are now providing the first indications of the potency of this approach in the nigrostriatal lesion model (60–62).

5.3. Inhibiting cell death in grafts

The mismatch between the good viability of dopamine cells in three-dimensional cultures and their poor survival following implantation *in vivo* (55), suggests that an exclusive focus on graft factors *per se* may be misdirected, and attention should also be directed at the possibility that the adult host environment is itself toxic to the grafts. Recent studies have shown that treatment with antioxidants (such as lazaroids), and enhancing mechanisms of free radical scavenging (such as increased superoxide dismutase activity using transgenic mice as hosts), can enhance the viability of dopamine neurons not only *in vitro* but also in nigral grafts (63, 64). Although this approach is still in its infancy, it is extremely likely that the analysis and manipulation of the host environment will become a major focus in developing graft methodology in the next few years.

5.4. Cryopreservation and hibernation

A final alternative that may prove effective is the exploration of techniques to pool tissue collected over a period of time. Although nigral cells can both be cryopreserved and grown in culture prior to transplantation (65, 66), both of these methods result in significantly lower yields of dopamine neurons than when the same amount of embryonic donor tissue is implanted fresh. More success may be achieved by cool storage of cells in a 'hibernation' medium that is low in calcium and slows down tissue metabolism (67). This cool storage method can sustain tissue viability, at least for up to 7 days storage prior to transplantation, which is often sufficient to collate tissue from multiple donors, and undertake a variety of preparation and safety screens prior to implantation (68).

6. ALTERNATIVES TO EMBRYONIC NEURONAL TISSUE TRANSPLANTATION

In the light of the limitations surrounding use of freshly derived embryonic donor tissues, in particular when considering use of human fetal tissues for clinical applications, a number of alternative sources of neuronal and other tissues for transplantation are under active investigation. The advantages of these alternatives lie not only in the realms of practicality, but also in the possibility of the improved experimental designs that would arise from identifying flexible, reliable and reproducible supplies.

6.1. Expanding precursor cell populations for transplantation

The first strategy attracting widespread attention involves the search for methods to achieve sustained expansion of stem and/or precursor cells in culture. Whereas once differentiated, neurons do not subsequently divide so that their numbers are fixed, it has proved possible to isolate precursor cells from proliferative regions of the embryonic brain, the numbers of which can then proliferate when maintained *in vitro* in the presence of high concentrations of epidermal or fibroblast growth factors (69, 70). These cells can be expanded exponentially over several generations, followed by transplantation into the brain (71, 72). The present limitations of this technique hinge particularly on the fact that the expanded cells give rise to predominantly glial rather than neuronal cells. However, as the factors that regulate differentiation in normal development become better understood, we can expect to see rapid advances in the possibilities of controlling the phenotype of cells for transplantation by directing their differentiation down alternative phenotypic pathways *in vitro* prior to implantation.

6.2. Engineering cells for transplantation

A second strategy is to recruit the powerful techniques of genetic engineering in order to manipulate non-neuronal cells for transplantation (73). Here it may be possible to harvest cells on the basis of other advantages, such as skin cells that will be non-immunogenic by virtue of collection from the graft recipients themselves, which will grow readily *in vitro*, and can be expanded and engineered prior to implantation back into the brain of the host animal or patient. This strategy is likely to prove particularly effective for transplantation of cells that secrete individual molecules for replacement, such as growth factors to confer specific patterns of neuroprotection (60, 74) or transmitter-related enzymes to restore pharmacological receptor activation in diffusely organized neurotransmitter systems (75, 76). Many reviews are now appearing on the power and potential of engineering strategies (77–80).

A particular aspect of engineering cells for transplantation involves the insertion of genes, such as the Sv40 large-T antigen, that confer temperature-sensitive immortalization of the cells. They can then be grown as a proliferating cell line *in vitro* at 33°C, but undergo differentiation when maintained at 37–39°C, either *in vitro* or at the body temperature which the cells encounter following transplantation *in vivo*. This line of investigation has been pursued in particular in the attempt to establish immortal cell lines derived from the embryonic nervous system, whilst avoiding the lethal proliferative characteristics of cell lines derived from neuronal or glial tumors. The plasticity of such cells is a topic of active investigation, in particular their apparent ability under certain circumstances to adopt a neuronal phenotype characteristic of the vicinity into which they are implanted (81, 82).

The techniques of genetic engineering of cells for transplantation is at an early stage of development, and while restricted to the site-specific but diffuse delivery of identified neuroactive molecules there will remain many conditions not responsive to such therapy. Moreover, there remain some formidable technical problems, such as the fact that at present, in most model systems, gene inserts that are stably expressed *in vitro* shut down within a short time following implantation *in vivo*. Nevertheless, the techniques themselves are advancing rapidly, not least in the insertion of multiple genes that can transfer combinations of properties, such as immortalization in combination with expression of particular gene products (83).

Of perhaps greater impact will be the development of novel promoters that enable specification of which cells will express particular gene products at what times and under what circumstances.

7. ACKNOWLEDGMENTS

The author's studies on neural transplantation are funded by grants from the Medical Research Council, the Wellcome Trust and the Hereditary Disease Foundation.

8. REFERENCES

1. Lindvall, O. (1994) in *Functional Neural Transplantation* (S.B. Dunnett and A. Björklund, eds), Raven Press, New York, p. 103.

2. Dunnett S.B. and Björklund, A. (1994) *Functional Neural Transplantation*, Raven Press, New York.

3. Lund, R.D. and Bannerjee, R. (1992) in *Neural Transplantation: A Practical Approach* (S.B. Dunnett and A. Björklund, eds). IRL Press, Oxford, p. 161.

4. Dunnett, S.B. and Björklund, A. (1992) in *Neural Transplantation: A Practical Approach* (S.B. Dunnett and A. Björklund, eds), IRL Press, Oxford, p. 1.

5. Olson, L., Seiger, Å. and Strömberg, I. (1983) *Adv. Cell. Neurobiol.*, **4**, 407.

6. Stenevi, U., Björklund, A. and Svendgaard, N.-A. (1976) *Brain Res.*, **114**, 1.

7. Stenevi, U., Kromer, L.F., Gage, F.H. and Björklund, A. (1985) in *Neural Grafting in the Mammalian CNS* (A. Björklund and U. Stenevi, eds). Elsevier, Amsterdam, p. 41.

8. Schmidt, R.H., Björklund, A. and Stenevi, U. (1981) *Brain Res.*, **218**, 347.

9. Björklund, A. and Stenevi, U. (1981) *Brain Res.*, **229**, 403.

10. Rosenstein, J.M. and Brightman, M.W. (1983) *Science*, **221**, 879.

11. Dunnett, S.B. and Björklund, A. (1994) in *Functional Neural Transplantation* (S.B. Dunnett and A. Björklund, eds), Raven Press, New York, p. 531.

12. Paxinos, G., Ashwell, K.W.S. and Törk, I. (1994) *Atlas of the Developing Rat Nervous System*, Academic Press, San Diego.

13. O'Rahilly, R. and Müller, F. (1987) *Developmental Stages in Human Embryos*, Carnegie Institute, Washington.

14. Butler, H. and Juurlink, B.H.J. (1987) *An Atlas for Staging Mammalian and Chick Embryos*, CRC Press, Boca Raton.

15. Annett, L.E. and Ridley, R.M. (1992) in *Neural Transplantation: A Practical Approach* (Dunnett, S.B. and Björklund, A., eds), IRL Press, Oxford, p. 123.

16. Björklund, A. and Dunnett, S.B. (1992) in *Neural Transplantation: A Practical Approach* (S.B. Dunnett and A. Björklund, eds), IRL Press, Oxford, p. 57.

17. Brundin, P., Isacson, O. and Björklund, A. (1985) *Brain Res.*, **331**, 251.

18. Fawcett, J.W. (1992) in *Neurodegeneration* (A.J. Hunter and M. Clark, eds). Academic Press, New York, p. 81.

19. Olson, L. and Seiger, Å. (1983) in *Autonomic Ganglia* (L.G. Elfvin, ed.). Wiley, New York, p. 507.

20. Stieg, P., Strömberg, I. and Olson, L. (1991) *Exp. Brain Res.*, **85**, 55.

21. Björklund, A. and Stenevi, U. (1977) *Brain Res.*, **138**, 259.

22. Cenci, M.A., Nilsson, O.G., Kalén, P. and Björklund, A. (1993) *Exp. Neurol.*, **122**, 73.

23. Bing, G., Notter, M.F.D., Hansen, J.T. and Gash, D.M. (1988) *Brain Res. Bull.*, **20**, 399.

24. Tischler, A.S., Perlman, R.L., Nunnemacher, G., Morse, G.M., DeLellis, R., Wolfe, H.J. and Sheard, B.E. (1982) *Cell Tiss. Res.*, **225**, 525.

25. Freed, W.J., Morihisa, J.M., Spoor, E., Hoffer, B.J., Olson, L., Seiger, Å. and Wyatt, R.J. (1981) *Nature*, **292**, 351.

26. Plunkett, R.J., Bankiewicz, K.S., Miletich, R.S., Schwartz, J. and Oldfield, E.H. (1989) *J. Neurosurg.*, **70**, 918.

27. Madrazo, I., Drucker-Colín, R., Díaz, V., Martínez-Mata, J., Torres, C. and Becerril, J.J. (1987) *New Engl. J. Med.*, **316**, 831.

28. Goetz, C.G. *et al.* (1989) *New Engl. J. Med.*, **320**, 337.

29. Quinn, N.P. (1990) *Prog. Brain Res.*, **82**, 619.

30. Date, I., Felten, S.Y. and Felten, D.L. (1991) *Brain Res.*, **545**, 191.

31. Kordower, J.H., Cochran, E., Penn, R.D. and Goetz, C.G. (1991) *Ann. Neurol.*, **29**, 405.

32. Bohn, M.C., Cupit, L., Marciano, F. and Gash, D.M. (1987) *Science*, **237**, 913.

33. Dunnett, S.B. and Mayer, E. (1992) in *Neurodegeneration* (A.J. Hunter and M. Clarke, eds), Academic Press, New York, p. 183.

34. Lund, R.D. (1978) *Development and Plasticity of the Brain: An Introduction*, Oxford University Press, Oxford.

35. McLoon, S.C. and McLoon, L.K. (1984) in *Neural Transplants: Development and Function* (J.R. Sladek and D.M. Gash, eds), Plenum Press, New York, p. 99.

36. Coffey, P.J., Lund, R.D. and Rawlins, J.N.P. (1990) *Prog. Brain Res.*, **82**, 269.

37. Lund, R.D., Rao, K., Hankin, M.H., Kunz, H.W. and Gill, T.J. (1987) *Ann. NY Acad. Sci.*, **495**, 227.

38. Del Cerro, M. (1991) in *Progress in Retinal Research* (N. Osborne and J. Chader, eds). Pergamon, Oxford, p. 229.

39. David, S. and Aguayo, A.J. (1981) *Science*, **214**, 931.

40. Paino, C.L. and Bunge, M.B. (1991) *Exp. Neurol.*, **114**, 254.

41. Blakemore, W.F. and Crang, A.J. (1988) *Dev. Neurosci.*, **10**, 1.

42. Jiao, S.S., Gurevich, V. and Wolff, J.A. (1993) *Nature*, **362**, 450.

43. Franklin, R.J.M. and Blakemore, W.F. (1995) *Microscop. Res. Tech.*, **32**, 295.

44. Andersson, C., Tytell, M. and Brunso-Bechtold, J. (1993) *Int. J. Dev. Neurosci.*, **11**, 555.

45. Wunderlich, G., Stichel, C.C., Schroeder, W.O. and Müller H.W. (1994) *Glia.*, **10**, 49.

46. Davies, S.J.A., Field, P.M. and Raisman, G. (1993) *Eur. J. Neurosci.*, **5**, 95.

47. Nikkhah, G., Olsson, M., Eberhard, J., Bentlage, C., Cunningham, M.G. and Björklund, A. (1994) *Neuroscience*, **63**, 57.

48. Freed, W.J., Perlow, M.J., Karoum, F., Seiger, Å., Olson, L., Hoffer, B.J. and Wyatt, R.J. (1980) *Ann. Neurol.*, **8**, 510.

49. Smith, L.M. and Ebner, F.F. (1986) in *Neural Transplantation and Regeneration* (G.D. Das and R.B. Wallace, eds), Springer Verlag, New York, p. 81.

50. Olson, L., Björklund, H., Freedman, R., Goldowitz, D., Hoffer, B.J., Palmer, M.R., Seiger, Å., Taylor, D. and Woodward, D.J. (1983) in *Developing and Regenerating Vertebrate Nervous Systems*, Alan R. Liss, New York, p. 121.

51. Olson, L. and Malmfors, T. (1970) *Acta Physiol. Scand. suppl.*, **348**, 1.

52. Olson, L., Björklund, H. and Hoffer, B.J. (1984) in *Neural Transplants: Development and Function* (J.R. Sladek and D.M. Gash, eds), Plenum Press, New York, p. 125.

53. Bregman, B.S. (1994) in *Functional Neural Transplantation* (S.B. Dunnett and A. Björklund, eds), Raven Press, New York, p. 489.

54. Barker, R.A., Fricker, R.A., Abrous, D.N., Fawcett, J.W. and Dunnett, S.B. (1995) *Cell Transplant.*, **4**, 173.

55. Fawcett, J.W., Barker, R.A. and Dunnett, S.B. (1995) *Exp. Brain Res.*, **106**, 275.

56. Gash, D.M., Mresjanac, M., Junn, F. and Zhang, Z. (1994) in *Functional Neural Transplantation* (S.B. Dunnett and A. Björklund, eds), Raven Press, New York, p. 139.

57. Sauer, H., Fischer, W., Nikkhah, G., Wiegand, S.J., Brundin, P., Lindsay, R.M. and Björklund, A. (1993) *Brain Res.*, **626**, 37.

58. Mayer, E., Fawcett, J.W. and Dunnett, S.B. (1993) *Neuroscience*, **56**, 389.

59. Haque, N.S.K., Hlavin, M.-L., Fawcett, J.W. and Dunnett, S.B. (1995) *Brain Res.*, **712**, 45.

60. Frim, D.M., Uhler, T.A., Short, M.P., Ezzedine, Z.D., Klagsbrun, M., Breakefield, X.O. and Isacson, O. (1993) *NeuroReport*, **4**, 367.

61. Isacson, O., Frim, D.M., Galpern, W.R., Tatter, S.B., Breakefield, X.O. and Schumacher, J.M. (1995) *Rest. Neurol. Neurosci.*, **8**, 59.

62. Takayama, H., Ray, J., Raymon, H.K., Baird, A., Hogg, J., Fisher, L.J. and Gage, F.H. (1995) *Nature Med.*, **1**, 53.

63. Nakao, N., Frodl, E.M., Duan, W.-M., Widner, H. and Brundin, P. (1994) *Proc. Natl Acad. Sci. USA*, **91**, 12408.

64. Nakao, N., Frodl, E.M., Widner, H., Carlson, E., Eggerding, F.A., Epstein, C.J. and Brundin, P. (1995) *Nature Med.*, **1**, 226.

65. Gage, F.H., Brundin, P., Isacson, O. and Björklund, A. (1985) *Neurosci. Lett.*, **60**, 133.

66. Redmond, D.E., Naftolin, F., Collier, T.J., Leranth, C., Robbins, R.J., Sladek, C.D., Roth, R.H. and Sladek, J.R. (1988) *Science*, **242**, 768.

67. Sauer, H. and Brundin, P. (1991) *Rest. Neurol. Neurosci.*, **2**, 123.

68. Brundin, P. (1992) in *Neural Transplantation: A Practical Approach* (S.B. Dunnett and A. Björklund, eds), IRL Press, Oxford, p. 139.

69. Reynolds, B.A. and Weiss, S. (1992) *Science*, **255**, 1707.

70. Ray, J., Peterson, D.A., Schinstine, M. and Gage, F.H. (1993) *Proc. Natl Acad. Sci. USA*, **90**, 3602.

71. Sabaté, O., Horellou, P., Vigne, E., Colin, P., Perricaudet, M., Buc-Caron, M.-H. and Mallet, J. (1995) *Nature Genet.*, **9**, 256.

72. Svendsen, C.N., Clarke, D.J., Rosser, A.E. and Dunnett, S.B. (1995) *Exp. Neurol.*, **137**, 376.

73. Gage, F.H., Wolff, J.A., Rosenberg, M.B., Xu, L., Yee, J.K., Shults, C. and Friedmann, T. (1987) *Neuroscience*, **23**, 795.

74. Rosenberg, M.B., Friedmann, T., Robertson, R.C., Tuszynski, M., Wolff, J.A., Breakefield, X.O. and Gage, F.H. (1988) *Science*, **242**, 1575.

75. Wolff, J.A. *et al.* (1989) *Proc. Natl Acad. Sci. USA*, **86**, 9011.

76. Horellou, P., Vigne, E., Castel, M.-N., Barnéoud, P., Perricaudet, M., Delaère, P. and Mallet, J. (1995) *Rest. Neurol. Neurosci.*, **8**, 63.

77. Shimohama, S., Fisher, L.J. and Gage, F.H. (1993) *Adv. Neurol.*, **60**, 744.

78. Suhr, S.T. and Gage, F.H. (1993) *Arch. Neurol.*, **50**, 1252.

79. Breakefield, X.O. (1987) *Mol. Neurobiol.*, **1**, 339.

80. Friedmann, T. (1994) *Trends Genet.*, **10**, 210.

81. Onifer, S.M., Whittemore, S.R. and Holets, V.R. (1993) *Exp. Neurol.*, **122**, 130.

82. Snyder, E.Y. (1994) *Curr. Opin. Neurobiol.*, **4**, 742.

83. Martínez-Serrano, A., Lundberg, C., Horellou, P., Fischer, W., Bentlage, C., Campbell, K., McKay, R.D.G., Mallet, J. and Björklund, A. (1995) *J. Neurosci.*, **15**, 5668.

84. Lavoie, M.P., Palmatier, M., Gentile, F.T., Kaplan, F.A., Fiore, D.M., Hazlett, T.F., Bell, W.J. and Flanagan, T.R. (1993) *Cell Transplant.*, **2**, 163.

85. Sagen, J., Pappas, G.D. and Perlow, M.J. (1986) *Brain Res.*, **384**, 189.

86. Björklund, A., Stenevi, U. and Svendgaard, N.-A. (1976) *Nature*, **262**, 787.

87. Kuhlengel, K.R., Bunge, M.B. and Bunge, R.P. (1990) *J. Comp. Neurol.*, **293**, 63.

88. Sekerková, G., Malatová, Z., Orendácová, J. and Zigová, T. (1993) *Rest. Neurol. Neurosci.*, **6**, 1.

89. Kordower, J.H., Cochran, E., Penn, R.D. and Goetz, C.G. (1990) *J. Neurosurg.*, **73**, 418.

90. Watts, R.L., Bakay, R.A.E., Herring, C.J., Sweeney, K.M., Colbassani, H.J., Mandir, A., Byrd, L.D. and Iuvone, P.M. (1990) *Prog. Brain Res.*, **82**, 581.

91. Fischer, W., Wictorin, K., Isacson, O. and Björklund, A. (1988) *Prog. Brain Res.*, **78**, 409.

92. Springer, J.E., Collier, T.J., Sladek, J.R. and Loy, R. (1988) *J. Neurosci. Res.*, **19**, 291.

93. Maeda, T., Mackinnon, S.E., Best, T.J., Evans, P.J., Hunter, D.A. and Midha, R.T.R. (1993) *Brain Res.*, **618**, 196.

94. Brecknell, J.E., Haque, N.S.K., Du, J.-S., Muir, E., Hlavin, M.-L., Fawcett, J.W. and Dunnett, S.B. (1996) *Neuroscience*, **71**, 913.

95. Kromer, L.F. and Cornbrooks, C.J. (1985) *Proc. Natl Acad. Sci. USA*, **82**, 6330.

96. Guénard, V., Kleitman, N., Morrissey, T.K., Bunge, R.P. and Aebischer, P. (1992) *J. Neurosci.*, **12**, 3310.

97. Silver, J. and Ogawa, M.Y. (1983) *Science*, **220**, 1067.

98. Zhou, F.C. (1990) *Prog. Brain Res.*, **82**, 441.

99. Zhou, F.C. and Azmitia, E.C. (1988) *Prog. Brain Res.*, **78**, 413.

CHAPTER 18
SINGLE-CELL RECORDING IN AWAKE, BEHAVING RATS

J.L. Kubie, R.U. Muller and E.S. Hawley

1. INTRODUCTION

The purpose of this chapter is to orient the investigator to methods for extracellular recordings from individual neurons in freely moving rodents. The chapter will emphasize strategic decisions in methodology, rather than 'cookbook' details. We will focus on methods used in our lab over the past ten years. In a few areas we will give detailed construction procedures (headstage and cable construction). These are areas in which we have made considerable investment of time and effort; and similar descriptions are not available elsewhere.

In many places we describe a particular solution to a problem. These should be viewed as examples rather than ideal choices. The methods are extended to include techniques for recording the animal's head position, head direction and other aspects of its behavior in parallel with the electrical activity. Although these behavioral methods were developed with hippocampal neurons in mind, they have broad applicability in studying neuron–behavior relationships in other regions of the brain.

The sections of this chapter correspond to the blocks in the schematic in *Figure 1*.

Figure 1. Block diagram of a recording system.

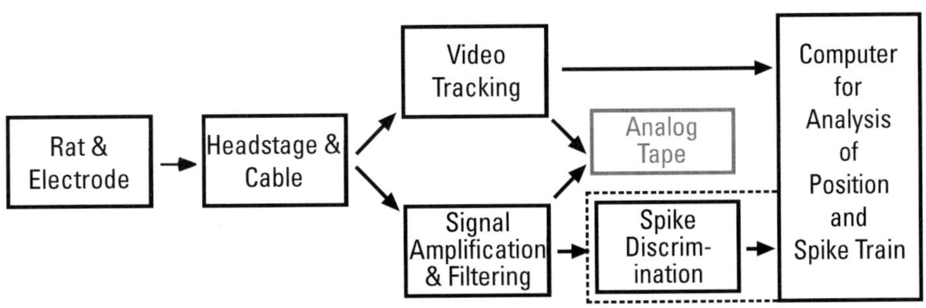

2. ELECTRODES, THE HEADSTAGE AND CABLE

2.1. Electrodes

Although there is a never-ending controversy about what makes the best electrodes for single-cell recording, it now is clear that good-quality recordings can be obtained with a variety of methods. In the 1940s and 1950s, when single-cell recording was established, it was assumed that fine-tipped electrodes were essential and small capillaries were used. The fine tip was thought to be necessary for positioning the electrode near the membrane of one, and only one, neuron. Later, fine-tipped, etched, metal electrodes were introduced (1), and shortly thereafter

Strumwasser introduced larger metal electrodes (2). Now it is clear that large electrodes, made from cross-cut wires, ranging from 17 to 67 microns in diameter, do work (3).

Fine-tipped electrodes record large-amplitude, extracellular, single-cell potentials, but they also produce large baseline noise owing to their high tip impedance. In our experience, the signal-to-noise ratio from small-tipped electrodes is about the same as with cross-cut wires. An important advantage of large electrodes is their ability to reject artifacts from large, distant voltage sources, such as muscle contraction. With large electrodes it is common practice to record 'differentially' using a quiet nearby electrode as the reference for active electrodes. Use of a nearby electrode, matched in impedance, eliminates large signals common to both the reference and active electrodes. Such common-mode rejection is difficult to implement with fine-tipped electrodes.

Our lab routinely uses 25 micron Nichrome (nickel–chrome alloy) electrodes whose tips are exposed by cutting with scissors. We have also used 17 micron Nichrome wire but find that the smaller wire is more difficult to handle and does not produce better cell yield. Nichrome is stiffer than most other electrode metals, so it bends less when advanced through the brain. When iron-containing Nichrome electrodes are used, they can be used to mark the final electrode position with a Prussian blue spot (4). These marks are made by passing a small amount of current through the recording wire, which deposits a small quantity of iron in the region of the electrode tip. The iron spot is made blue by soaking the brain in potassium ferrocyanide after formalin perfusion.

Our basic electrode designs are modifications of the driveable bundle system of Kubie (5). This system is stable and permits selection from 10 or more recording electrodes. It can readily be adapted to stereotrode or tetrode configurations (6–8). The principal disadvantage is that the bundle must be advanced as a unit, and electrodes cannot be moved independently. Recently, a modification of this design has been successfully miniaturized for use with mice (9) and modified for simultaneous intracerebral microdialysis (10, 11).

2.2. The headstage

The headstage contains a set of small impedance-lowering amplifiers that are carried on the rat's head. These are essential for reducing output impedance to prevent the small extracellular signals from neurons (typically 200 μV at $Z = 150$ kΩ with wire electrodes) from being obscured by huge artifacts from cable movements.

Early headstages were made with discrete field-effect transistors (FETs) (12, 13). Even though modern FETs have become more robust, they are still quite sensitive to damage by misconnection and electrostatic discharge. In our hands, FET headstages last less than a year, most failing in dry weather. FET failure often results in passing substantial amounts of current through the electrode, thus lesioning the tissue under study.

Currently, operational amplifiers (op-amps) with FET inputs are available in very small packages. The Texas Instruments TLC2274ACD is a quad, CMOS, FET op-amp in an SOIC package (4 mm × 8.5 mm × 1.5 mm, plus pins) which has electrical specifications ideally suited for headstage construction: low noise (9 nV/$\sqrt{\text{Hz}}$ at $f = 1$ kHz), very high input impedance (10^{12} Ω), low input bias current (typically 1 pA), low output impedance (130 Ω), and protection from electrostatic discharge (≤2000 V).

Design and construction of our op-amp headstage

Three TLC2274 chips (12 op-amps; 2 are not used) are glued upside down on an Augat 8058-1G34 plug, modified by adding a center pin, Augat #LST-1AG4-1 (drill a #54 hole in a jeweler's lathe), and wired under 7× magnification in a non-inverting unity gain configuration using 32-gauge polyurethane insulated magnet wire. The result is a 10-channel, virtually indestructible headstage (none has failed), with a gain of exactly 1, of very small size, and inconsequential noise. Though the TLC2274 will operate on a single-sided supply, we power it with a split supply (±5 V) so that there can be a real ground rather than the virtual ground needed with AC signals. Thus there is no signal voltage offset. In case of momentary power interruption or noise, 10 μF tantalum capacitors (Mallory #TAC-106K010P02) are wired across the power lines in the headstage.

Construction continues with wiring the op-amp outputs to an Augat 8058-1G51 12-pin connector, modified by adding a center pin. An extra Augat 8058-1G34 is added to the bottom of the headstage to protect the exposed pins, and the whole headstage is potted in Hygenic dental acrylic in a cylindrical mold made from RTV (Dow-Corning 3110 silicone rubber).

2.3. The recording cable

The recording cable plugs into the headstage at the bottom and a slip-ring commutator at the top. It carries physiological signals to the commutator and power to the op-amps and tracking LEDs. We currently use a headstage with 10 op-amps, and need a signal wire for each. The cable conductors are fine-gauge insulated stainless-steel wires held together by threading them through braided fishing line. Additional conductors are needed for the tracking LEDs, for powering the op-amps and for animal ground.

Specifics of cable construction

The headstage end of the cable has an Augat 8058-1G51 12-pin connector, modified by adding a center pin, and the commutator-end has a subD-25 male connector. In addition to the 8058-1G51 connector, accessory wires are led out of that plug to power tracking LEDs. Other specific arrangements can be made, for instance where concurrent stimulation and recording is occurring or theta rhythm recordings are being made, with a supplemental headstage. The slip-ring commutator is obtained from Crist Instruments (Damascus, MD).

The running cable is made with between 15 and 23 conductors in a casing. Cooner AS632 38-gauge stainless steel wire (15/50-gauge, Teflon insulated; Cooner Wire, Chatsworth, CA) is wrapped around two pins 3.2 mm apart. The wires are picked up at the continuous end with a loop of 10 lb test nylon monofilament fishing line which is then threaded through the eye of a blunt carpet needle and folded back. The needle is threaded through the center of a length of black 45 lb test braided Nylon 'Ice Fisherman' tilt line or 130 lb test braided Dacron 'Troll-A-Line' (US Line Co., Westfield, MA), the size dependent upon the number of wires. The bundle of wire is pulled through. The function of the casing is to enclose the wires, not to take strain. Thus in final form it should be fairly loose on the wires. If it is too tight it will bind the wires and they will fail prematurely.

The wire bundle is trimmed flush and 3 mm of insulation is stripped from each wire with a coarsely ground #5 forceps; grinding leaves burs which grab and tear the Teflon. Each stripped wire is tinned by submerging it just below the surface of Oatey #5 zinc/ammonium chloride paste flux, and touching it, under the flux, with a 650°F tinned soldering iron. Do not use an acid flux. A strain relief spring (#302 stainless steel, i.d. = 0.091 inches (spec 1204) or 0.118 inches (spec 2414), length 1½ inches, stiffness = 8.6 (Mechanical Spring Corp., Brooklyn, NY) is slipped over the wire/casing bundle. The connector is prepared by adding a center pin (see above) and drilling a #72 hole for each accessory wire (such as for LED power) in a concentric pattern. A 4 cm U-shaped piece of 22 a.w.g. tinned copper wire (Alpha

#298) is pushed through two adjacent accessory wire holes from the pin side of the connector. Under 7× magnification, the cable wires are then wrapped around each of the accessory wires or pushed into the Augat sockets and soldered into place. For strength after potting (see below), a 1–2 mm loop of 30 a.w.g. wire is soldered into each socket at the same time. The connector is cleaned by sonication in lacquer thinner.

The connector is made into a plug by potting it with Hygenic dental acrylic. To do that, the accessory wires are bent to the outside at a 60° angle from vertical. The frayed casing is trimmed, lightly melted with a soldering iron, and then carefully pushed down on the wires, making them bunch up, and the strain relief spring is pushed down on the casing. All the wires are checked to make sure none is touching. A cylindrical mold is made from silicone RTV or wax. The connector is placed upright in it, and acrylic is poured to a level sufficient to embed both the cable casing and the strain relief spring by a few mm. The excess acrylic and the U ends of the accessory wires are carefully trimmed off with a dental burr in a Dremel Moto-Tool.

The accessory wires are bent down parallel to the acrylic and trimmed to be flush with the face of the pin side of the connector. LED power is carried by #32-7/40 wires (Alpha #1850), each terminated with a male connector (Cannon 'Centi-loc' #031-9540-000). These are soldered to the accessory wires. Both ends of the wires are insulated with heat-shrink tubing.

The other end of the cable is treated similarly, with the additional concern of conforming to a wiring diagram. Wires are identified with an ohmmeter and soldered into the appropriate cup of the subD-25 connector. After cleaning, the connector, wires, casing, and strain relief spring are potted into a subD-25 hood.

In order to suspend the cable an attachment eye is added. A 10 cm piece of small gauge PVC tubing is slit lengthwise and slipped over the cable, a 10 cm piece of 18-gauge stiff wire (steel or nichrome) is bent into an eye at one end and a hook on the other, the wire is aligned over the middle of the PVC tubing with the eye up, a 4 cm piece of 5/16-inch heat-shrink tubing is collapsed in place, and the ends of the PVC tubing are secured with 24-gauge hook-up wire. This attachment can be loosened and moved up and down the cable. Note that the heat-shrink tubing must be put on the cable before the second connector; in practice several pieces are put on, the extras to be held in reserve.

Several modifications of cable construction are under study. It is anticipated that the Augat connectors will be eventually replaced with Mill-Max series 852 and 853 dual-row, 0.05-inch spacing connectors. This denser connector will obviate any need for accessory wires, and allow the construction of electrode assemblies with greater numbers of wires (an experimental 17-channel headstage has been constructed; a 20-channel one, next on the list, would fully utilize five quad op-amp packages). The Mill-Max male connector pins are two-ended, each end capable of serving as a plug (even though one is intended to be a PCB through-hole connector). Each connection will be tripart, involving a female termination of cable, headstage, and electrode, in each case with a male connector in between. Thus the most fragile part of the connection, the pins, will be easily replaceable. In addition, the cabling wire may be replaced with 36 a.w.g. (7/42-gauge) CS-95 silver-plated high-strength copper alloy, 2 mil PFA insulation, Temp-Flex Cable, Inc. #P36H07S-2.0-1 wire. This is about one-sixtieth the resistance of the stainless steel wire, less expensive, and seems to be about as strong. Though a larger wire, because of the thin insulation its o.d. is smaller (0.0097 inches versus 0.012 inches for the AS632). It is, however, only available in custom lots.

The cable above the commutator is terminated in a patch panel that contains no electronics. The purpose of the patch panel is to select among electrodes for oscilloscope display and data acquisition. The output cable of the patch panel connects to standard AC amplifiers.

3. AMPLIFICATION, FILTERING AND STORAGE

3.1. Amplification and filtering

We routinely amplify unit signals 10 thousand times, to a final signal level of 1–5 V, before they are displayed on an oscilloscope or digitized for computer analysis. Unit signals are passively filtered between 300 Hz and 10 kHz, which are standard settings on most amplifiers. Highpass active filtering, with a cutoff of 500 Hz, is sometimes required to eliminate an unstable baseline.

The passive 10 kHz lowpass filter found on most amplifiers sometimes is not sufficient to eliminate high-frequency artifacts. These artifacts must be removed before digitizing by a computer. If a digitized signal includes frequencies that are greater than half the computer sampling rate, aliasing will occur. Action potentials will appear to have impossible components, and there will be great variance in their apparent waveform. Therefore, all signals must be carefully filtered for power in frequencies greater than half the sampling frequency. If there is a nearby source of high frequencies, passive lowpass filtering at 10 kHz diminishes, but does not eliminate, the noise. This noise may be caused by nearby elevator motors and video monitors. Lowpass filtering with active filters helps, but eliminating the noise source or careful shielding of the recording chamber is more effective.

3.2. Audio or video tape for data storage

Before the microcomputer revolution it was common to record electrophysiological signals on tape prior to computer analysis. The rationale was that tape faithfully stored spike waveforms, while early computer-based systems registered only the timing of spikes. Now that most computer acquisition systems, such as DataWave, can record spike waveforms, this rationale has disappeared. Now we do not routinely store experimental data on audio or video tape. Nevertheless, there are still reasons for using magnetic tape as a temporary data store.

Standard audio tape, which can record signals between 50 Hz and 10 kHz, can do an excellent job of recording spike waveforms, and an inexpensive audio tape recorder can do an adequate job. Video tape is useful because it can store a simultaneous record of behavior and raw (or discriminated) unit activity. Videotape systems that support stereo can record two unit channels. Standard audio cannot, however, store signals lower than 50 Hz, and, therefore, cannot store EEG. For EEG or other slow signals, FM encoding or digital encoding of the signal is essential. Inexpensive machines are available that work in conjunction with standard videotape recorders for storage of up to eight channels of slow physiological data. Four simultaneous channels can be recorded at settings appropriate for units.

There are several disadvantages with taping. First, taping can add an extra, time-consuming step when the data must be replayed in real-time for computer data acquisition. Second, for our purposes, tracking the rat from video tape, especially two-spot tracking, is prone to serious error due to loss of color and image quality.

As an aside, it is likely that analog video tape will disappear in the near future, and be replaced by digital video storage. Digitized video offers dramatic opportunities for scientific analysis of behavior and physiology. Fenton and Muller (14) have developed a prototype system for collecting digital video in parallel with physiological signals, and synchronizing the two data streams. There are a number of advantages of their system. First, digitized video permits the use of computer-based pattern detection. Second, cross correlations between physiology and behavior are relatively easy to perform. For example, behavior can be automatically selected for occurring during certain physiological patterns or physiological signals can be automatically selected for occurring during certain behaviors. Third, automatic analysis is much easier. The current barrier to digital video is data storage. It takes roughly a gigabyte of space to store 30 minutes of low-resolution video. Nevertheless, as storage media become cheaper and faster and video compression technologies improve, storage of large amounts of digitized video will become standard.

4. RAT TRACKING

Our group has used two methods of rat tracking: a one-spot (black-and-white) method for tracking a rat's location (15); and a two-spot (color) method for tracking location and head direction (16). Both methods are covered in greater detail in publications (15, 16). Since the two-spot system is derived from the one-spot system the one-spot system will be described first. The general strategy for spot detection is described below.

4.1. One-spot tracking

For one-spot tracking a pair of red light-emitting diodes (LEDs) mounted on the rat's headstage are tracked. The details of our implementation follow.

Tracking LEDs must have a wide viewing angle, because the animal's position and angle can change dramatically relative to the tracking camera. I.I. Stanley currently provides surface-mount technology LEDs in both high-efficiency red (#BR1101W, $\lambda = 660$) and pure green (#BG1101W, $\lambda = 555$). These LEDs have a viewing angle of about 130°. For one-spot tracking, two red LEDs are soldered to wires mounted on a small piece of Perfboard. The spacing is such that at least one LED can be viewed no matter which way the headstage is tilted. The wires are soldered to a female connector (Cannon 'Centi-loc' #030-9542-001) which mates with the connector on a cable accessory wire. Because of the stainless steel cable wire resistance, the power supply voltage drop is actually lower if the LEDs are wired in series rather than parallel. The Perfboard is fixed to the front of the headstage with dental acrylic, making sure to fix and protect the LEDs, and the connectors are fixed to the rear.

The luminance of the LED is adjusted for optimal detection. The output of a standard video camera, mounted directly overhead, is fed to the tracking hardware. Sixty times per second the camera sends a video 'field' to the tracker. The field is a raster of 255 horizontal lines. Each line increments a Y-axis counter. When properly adjusted, the LED produces the highest voltage signal in each field. A voltage threshold device in the tracker is set so that only the signal from the LED exceeds threshold. The first signal to exceed threshold in a frame is marked. The number of the line it occurs on is used to determine the LED's vertical position (Y) within the field. A fast counter (driven by a phase-locked-loop) is reset at the beginning of each horizontal sweep. The number in this counter is stored when threshold is first crossed, and is used to determine the horizontal position (X) of the LEDs. Sixty times per second, two numbers, the X and Y coordinates of the LEDs, are transferred to the computer.

4.2. Two-spot tracking

Two-spot tracking is essentially a color-coded version of the one-spot system. The details of our implementation follow.

Red and green LEDs are arranged on the midline of the head, 2.5 cm in front of, and 3.5 cm behind the headstage as measured in the horizontal plane. The position of the rear green LED is over the shoulders of the animal, about 2.5 cm lower than the headstage, while the front red

LED is at the same level as the headstage. The LEDs themselves are horizontal. The fixture holding the LEDs is constructed of soft PVC tubing just the size of the outside of the headstage. Eighteen-gauge enameled copper wire is wrapped in a groove around the outside of the tubing. Separate pairs of wires are twisted fore and aft and an LED mounted on a tiny piece of Perfboard is soldered to the end of each pair of twisted wires. The Cannon connectors are soldered to the wires more centrally. The fixture is protected and made more rigid by adding dental acrylic at points of stress and at the LEDs and connectors. The whole thing is colored black with a permanent felt-tip marker. This fixture allows accurate alignment of the LEDs down the midline of the head even if the electrode assembly is slightly out of line.

The overhead camera must be color, and must have separable red, green and blue output channels. The red and blue outputs are each fed to a spot tracker, one for tracking the red LED, the other for the green LED. Prior to each session it is important to set up the trackers, with the aid of an oscilloscope, to ensure that they are tracking the desired spots reliably and there is no cross talk. The two-spot tracker sends four data points per video field: red-x, red-y, green-x and green-y.

4.3. Comments on tracking techniques and parameters

Other methods for one- and two-spot tracking are in use. Some are based on computer capture of an image of the lights and online extraction of the position data (e.g. 17, 18). We feel the hardware LED tracking procedure is a little more difficult to set up than software methods, but it is very stable and reliable. Errors are easy to detect prior to recording sessions and rarely contaminate data. The 60 Hz sampling rate we use is limited by the television field rate and, ideally, could be higher. Note that 60 Hz position information imposes virtually no load on disk space. For two-spot tracking, where head angle information is sought, 60 Hz sampling is critical because a rat's head orientation can change much more rapidly than its location. We currently use a spatial resolution of 256-by-256. This corresponds reasonably well with the resolution of NTSC video signals. If the image of our 76 cm recording cylinder fills the screen, 256-by-256 resolution produces pixels that are 4 mm squares on the floor. For positional analysis, we collapse these small pixels into larger ones 8:1 in each dimension, so that the floor of the cylinder contains 600 pixels.

5. SPIKE DISCRIMINATION

Action potential propagation is caused by local-circuit currents that flow inward at the active membrane and outward at resting and repolarized membrane. The local circuits are completed by current flow inside the cell. The signal detected in extracellular recordings is the voltage drop caused by the current flow through the resistance of the extracellular fluid. In practice, the magnitude of the signals that are used ranges between 150 and 350 microvolts, although it is often possible to use signals as small as 125–150 µV and it is not uncommon to find signals larger than 400 µV. We assume that we are recording neurons whose cell bodies are in the vicinity of the electrode tip. It is possible that axons or dendrites are occasionally recorded, but the general finding that unit recording is easiest from cell layers argues strongly that recordings are made from nearby cell bodies. We also assume, from passive extracellular electrical properties, that larger spikes come from neurons whose cell bodies are very near the electrode tip, while those with smaller spikes are further away.

Since the recording noise is only about 40 µV, why aren't waveforms less than about 100 µV very useful? The reason is that they all look the same. The magnitude of action potential currents depends in large part on the distance between the electrode and the cell. It is always difficult to tell if a small recurring waveform is produced by one or several distant cells. The problem would remain even if there were no recording noise at all. In a similar vein, we may ask why there is an upper limit for the signal amplitude. The reason again has to do with distance from the cell to the electrode; if the signal is big, the distance is small and the cell is likely to be damaged by the electrode after a short interval.

Methods of spike discrimination may be classified as either hardware or software based. Hardware methods are beginning to disappear, but are useful for comparison, and continue to have certain strengths and advantages. The hardware spike discriminator is typified by Bak's DIS-2 model which is used in conjunction with a storage oscilloscope (*Figure 2*). A signal containing potential spikes is connected to the input. Whenever voltage exceeds a 'trigger' level, two timing devices are set into action. These permit the user to set two independent time-and-amplitude gates with upper and lower limits. Thus, any waveform that exceeds the trigger and passes through the limits of the two gates will be accepted and an acceptance (TTL) pulse will be generated. The acceptance pulses are sent to a computer interface. (At early stages of our work, the computer interface was a simple accumulator that added spikes fired each 1/60th of a second – the video tracking rate – and sent this number to the computer (15). This system is completely obsolete, suffering from poor temporal resolving power compared with the time-stamp systems that will be described below.)

The fundamental weakness of hardware discrimination is that it is very difficult to track or correct errors in discrimination. Once a discrimination criterion is set, it is generally impossible to change or correct discrimination in the middle of an experiment. A second major drawback is difficulty in recording (and monitoring) more than one cell at a time. In our experience, we felt comfortable recording a maximum of two cells simultaneously.

Software methods of spike discrimination are fundamentally similar to the hardware methods in that both compare spike waveform. The advantages of software methods are that the spikes are stored in the computer so that many criteria can be applied for spike classification and judgments about classification can be modified offline.

Figure 2. Operation of a typical window discriminator. Dashed vertical lines indicate moveable time points. t_0 is the time of threshold offset; t_1 is the time of first window and t_2 the time of the second window.

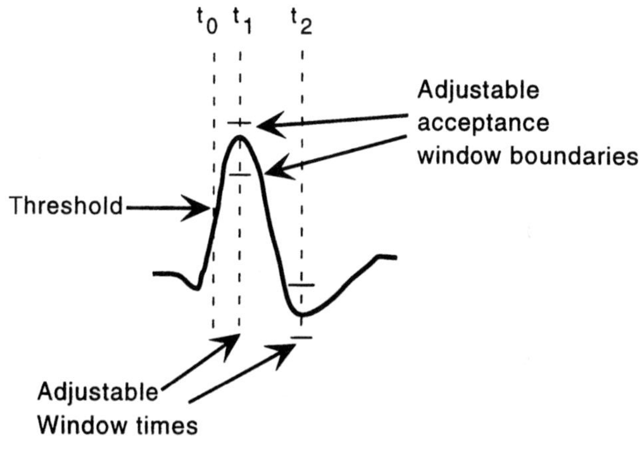

The most important step in computer-based spike discrimination is capturing each spike. Data are continuously fed to a computer input port and whenever voltage exceeds a user-selectable threshold, the spike is 'saved'. For each saved spike, equally-spaced points on the waveform are digitized and saved to disk as a single record. Typically, the data points include several pre-trigger values. For most systems, such as the DataWave (Longmont, CO) system we use, the sampling rate is user-selectable. Typical sampling rates range from 15 kHz to 50 kHz. Thirty-two data points sampled at 33 kHz yield a sampling burst of 1 msec. (An alternative data-collection strategy would be to digitize continuously, save all data to disk, and compress the file later by clipping out spikes.)

Sampling is a critical issue in software spike recognition. If the digitization rate is too slow, the task of finding the voltage and time of a spike component, such as peak time, is more difficult. Consider, for instance, *Figure 3*. Copies of identical waveforms are imagined to be digitally sampled asynchronously; the samples are indicated by dots. DataWave Systems would choose the closed dot as the peak of the spike. It is clear that the true peak value and the peak samples are different, both in time and amplitude, owing to digitization error. Digitization error can be reduced by increasing the sampling rate. The briefer the waveform, the higher the sampling rate must be to accurately capture the peak and other features. Sampling 400 μsec waveforms at 40 kHz produces a range of peak values that is about 5% of the total spike amplitude. It is possible to better estimate the voltage and time of the peak by fitting a smooth curve, such as cubic spline, between the sampled points (19). Alternatively, the waveform can be reconstructed with Fourier techniques.

Clearly, one answer is to sample as fast as possible, but there are two limitations to fast sampling in the DataWave system. First, since there are always 32 data points per spike, if sampling is too fast (100 kHz), the end of the spike may be missed. Second, analog-to-digital boards operate at a sampling rate that must be divided across channels. If a very high sampling rate per channel is selected, the total number of channels that can be recorded is less.

Other factors that must be set for data acquisition are the number of channels, and gain per channel. Neither of these are usually critical issues. We usually sample output from any

Figure 3. Representation of how three identical waveforms can be digitized differently. For A, B and C, the waveforms and sampling rates are identical; the only differences are in the time points of intersection of sampling and the waveform. The 'peak' sample is indicated by a filled circle. Both the time and amplitude of these peaks vary. Time variance is generally greater than amplitude variance.

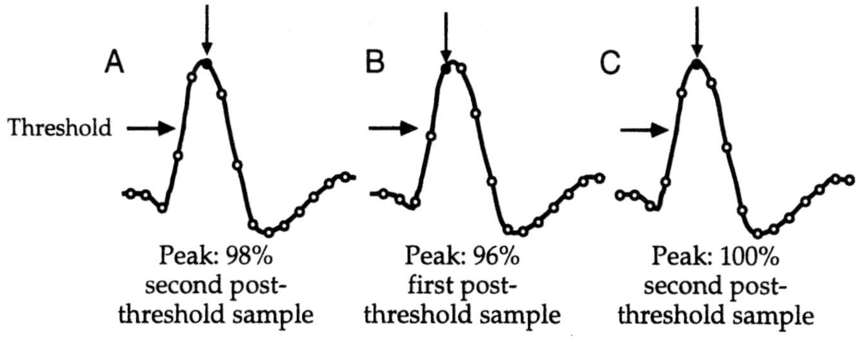

Single-cell Recording

electrodes which potentially discriminates between cells. The gain setting is not a problem, if the resolution of the A-to-D board is 12-bits; this produces steps of 0.25 mV for a 1 V signal. One should be careful that no waveform 'saturates' by reaching the highest possible recording value, because saturated waveforms are impossible to discriminate. Finally, the threshold for spike acquisition should be carefully set. If the threshold is too high, low-amplitude spikes are not stored. This is a particular problem in the hippocampus where the later spikes in a complex burst are small. In contrast, setting the threshold too low causes acquisition of noise and smaller waveforms that cannot be discriminated. A low threshold leads to three problems. First, it creates data files that are large and unwieldy. Second, the process of offline spike discrimination is made more difficult by many small, undiscriminated spikes. Third, low-amplitude triggers occur with increasing likelihood and occupy the sampling capacity of the system, occluding or distorting waveforms from interesting units.

The goal of spike discrimination ('spike sorting') is to classify every spike emitted from a single neuron into a single category. There are two types of classification errors: classifying waveforms from other neurons (or artifact) as spikes from a target neuron (analogous to type 1 error in statistics), and rejecting waveforms that, in fact, were caused by action potentials from the target neuron (analogous to type 2 error). Type 1 errors are caused by being too liberal, and type 2 are caused by being too conservative. Unfortunately, except for very clean signals, it is generally impossible to completely eliminate either of these two types of error. This is in part due to the large variance of waveforms from single cells, especially hippocampal pyramids. Several methods can be used to check for the adequacy of discrimination.

(i) On the average, properties of units should not vary with spike amplitude. For instance, if one routinely finds that small-amplitude candidate units fire faster than large-amplitude candidates, the hypothesis that several small cells make up the small unit must be entertained.
(ii) Properties of a candidate unit should not vary with behavioral variables or states. For instance, if a candidate place cell is discriminated and found to have two firing fields, the average discriminated waveform from one firing field should be identical to the average waveform obtained in the other firing field. If the average waveforms are different, it seems likely that the two fields come from two different cells.
(iii) A clear 'dead time', corresponding to a refractory period, should occur after discriminated spikes (20).
(iv) It is generally assumed that waveform properties coming from a single neuron ought to have a mean and normally-distributed scatter, producing a scatter plot, with a dense cluster around the mean. Therefore, for a well discriminated cell, clustering should be evident on all waveform parameters. For parameters that are critically used in discrimination (e.g., spike amplitude) there ought to be a clear gap separating the cluster of accepted waveforms from the clusters of rejected waveforms.

When does one decide to risk accepting bad spikes in the interest of ensuring that all spikes from the target neuron are accepted? Alternatively, when should one risk rejecting a few target-neuron spikes in the interest of rejecting all spikes from other cells? The answer varies with the goal of the study. In certain studies, where absolute firing rate or interspike interval are critical variables, only extremely well-discriminated units are acceptable. For other studies, compromises can be made. This is frequently necessary when it is important to discriminate multiple neurons from a single electrode. In such work, conservative boundaries are applied in spike sorting, reducing type 1 errors. There are the less-common cases where type 2 errors (errors due to liberal boundaries) are preferable to type 1 errors. This is the case when one wants to know the firing rate of a fast-firing cell whose waveforms overlap those of a slower firing cell. Another example comes from hippocampal place cells, where it is generally found that estimates of in-field firing rates are most accurate when waveform parameter boundaries are liberal.

This chapter does not deal with computational methods of spike discrimination. This is an area of active development and a number of different strategies are emerging (21). These strategies have in common the process of extracting parameters (attributes) of each waveform, and looking for groupings and differences across parameter sets. The most direct strategy is to extract parameters that are equivalent to those analyzed by hardware discriminators – e.g., several voltage-at-time parameters. A variety of other parameters are possible, and many have been tried. Among the possible parameters are principal components, ascending slope, descending slope, point of curvature, spike width, etc. Choosing sensitive parameters remains one of the most important factors in effective discrimination.

Until recently, most computer spike sorting was done 'by hand' with the investigator choosing parameters and separation points that produced clear-cut separations. Automatic clustering algorithms are emerging that may prove more time-efficient, effective and objective (22–24). A few cautions are in order. First, until automatic clustering techniques have become more accepted and standardized, any automatic decisions should be carefully checked. Any automatic method should have a feedback option for displaying differences between accepted and rejected waveforms and these options should be rigorously used. For instance, an oscilloscope-like display of discriminated and undiscriminated waveforms would suffice. Applying these tools, a unit should meet common-sense as well as automatic criteria for isolation. When automatic methods are applied to tens of simultaneously-recorded neurons, potential problems are compounded. Every discriminated unit should present unequivocal evidence of isolation.

6. ANALYSIS OF TIME-SERIES DATA

Meaningful analysis of discriminated spikes depends on each spike being reliably and accurately 'time-stamped'. In a similar fashion, behavioral events and other physiological parameters must be time-stamped. The resolution of the time-stamp must be appropriate for the speed that the data can change state; thus 10 μsec resolution is generally needed for spikes, whereas 20 msec or so (video frame rates) is adequate for behavioral analyses. The time stamps must also be synchronized. This is, of course, no problem if a clock in a single computer time stamps all events. It is common, however, for different types of data to be collected by different systems. For instance, when if one computer is used to collect spikes and a second to record behavior, synchronization is necessary. Fenton and Muller (14) used a flashing diode as a 'clapboard' to synchronize spike data acquired on one computer with digitized video of behavior acquired on a second computer.

7. MULTIPLE CELL RECORDING

As an aside, we briefly consider the advantages of simultaneous recording from several neurons at one time. For some purposes, recording one cell at a time is sufficient. Imagine a class of neurons in the retina, lateral geniculate or primary visual cortex. It is often of interest to understand how cells in such a class respond to specific stimuli. Under these circumstances, it is reasonable to assume that the population of cells does the same thing each time a certain stimulus is presented, so that as much would be learned by recording from cells one after the other as from simultaneous recordings. For such experiments, the main advantage of simultaneous recordings would therefore be to increase the rate of acquiring data.

In other circumstances, however, simultaneous recordings can answer questions that are in principle unapproachable if only a single cell is recorded. For example, one proposed solution to the problem of binding the component features of a single object into a stable percept involves synchronized firing of cells in different places in the visual system (25, 26). Evidence for discharge that is tightly coupled in time requires simultaneous recordings from at least two units. Other cases that require simultaneous recordings arise in place cell or head direction cell research, where it is desired to correlate moment-to-moment discharge with the position or head direction of a freely moving rat. Here, the state of the animal is very difficult to specify, so that it is hard to know if non-simultaneous recordings were made under 'similar enough' conditions. The problem is removed if the cells are recorded at the same time. In a similar vein, it is important to know if place cell or head direction cell discharge is modified on an independent, cell-by-cell basis when conditions are changed, or if the discharge of all the cells is modified in such a way that the relationships among cell properties are fixed. In the second case, one would conclude that the system acted as a whole, and not as a set of independently tuned units. Questions about the consistency of the population representation are best answered with simultaneous recordings.

Simultaneous recordings from multiple cells can be accomplished either by multiple recording electrodes or by having several discriminable units on a single electrode. Recording from many electrodes simultaneously clearly increases yield and permits analyses of simultaneous units. For each electrode, additional op-amps, cable channels, commutator channels, preamplifiers and A-to-D channels to the computer are needed. In the hippocampus, where neurons are tightly packed, it is common to have several waveforms on a single wire. Using single-electrode discrimination techniques, we are able to discriminate up to four

Figure 4. Two methods of spike sorting: on the left, single electrode discrimination of multiple waveforms; on the right, a two-probe stereotrode. The waveform of each unit (*a*, *b* or *c*) is different on probes A and B.

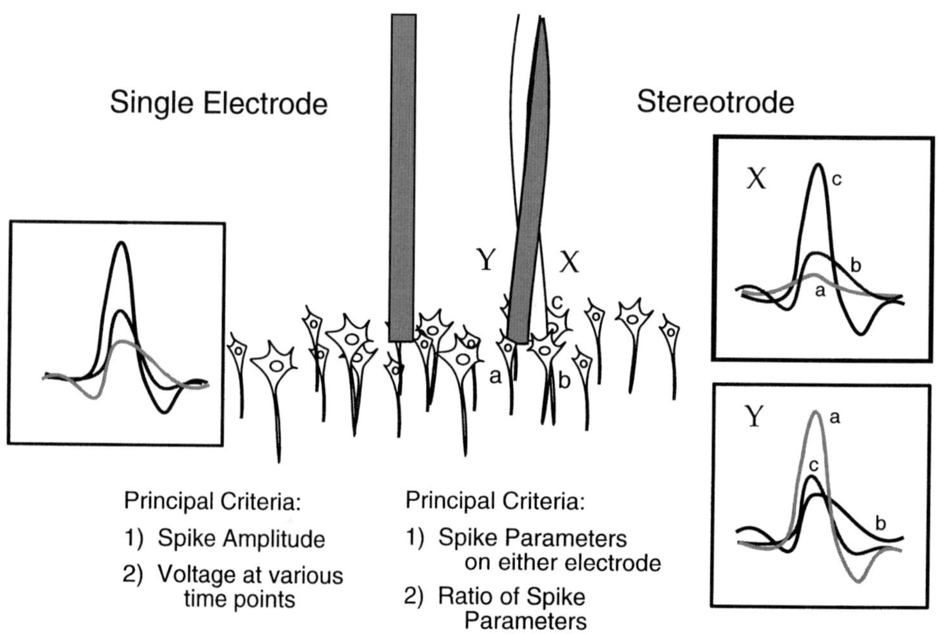

neurons from a single electrode; more frequently, discrimination of one or two units is possible. The alternative to single electrodes are twisted pairs of wires called 'stereotrodes' (6) or twisted groups of four wires called 'tetrodes' (7, 8, 18). The idea is that waveforms from two (or more) cells that are too similar on a single wire to allow separation will appear different on one or more other wires (*Figure 4*). There are many reasons why waveforms from a single cell might differ from wire to wire, the simplest being proximity of each wire tip to the cell. Regardless of the reason, however, it is clear that the different 'view' of a set of cells provided by two or four neighboring wires allows spike train separations not possible with a single wire. The most extensive application of simultaneous recording methods we know of comes from Wilson and McNaughton (17) who, using multiple tetrodes, were able to record in excess of 100 neurons at a time.

8. REFERENCES

1. Hubel, D.H. (1957) *Science*, **125**, 549.
2. Strumwasser, F. (1958) *Science*, **127**, 469.
3. Loeb, G.E. (1995) *J. Neurosci. Meth.*, **63**, 175.
4. Marshall, P. (1940) *Stain Techn.*, **15**, 133.
5. Kubie, J.L. (1984) *Physiol. Behav.*, **32**, 115.
6. McNaughton, B.L., O'Keefe, J. and Barnes, C.A. (1983) *J. Neurosci. Meth.*, **8**, 391.
7. O'Keefe, J. and Recce, M.L. (1993) *Hippocampus*, **3**, 317.
8. Cray, L.M., Maldonado, P.E., Wilson, M. and McNaughton, B. (1995) *J. Neurosci. Meth.*, **63**, 43.
9. Rotenberg, A., Mayford, M., Hawkins, R.D., Kandel, E.R. and Muller, R.U. (1996) *Soc. Neurosci. Abstr.*, **22**, 912.
10. Ludvig, N., Potter, P.E. and Fox, S.E. (1994) *J. Neurosci. Meth.*, **55**, 31.
11. Ludvig, N., Potter, P.E. and Fox, S.E. (1994) *Neurosci. Proto.*, **94**, module 4 (section 070), 1–17.
12. O'Keefe, J. and Dostrovsky, J. (1971) *Brain Res.*, **34**, 171.
13. Ranck, J.B. Jr (1973) *Exper. Neurol.*, **41**, 461.
14. Fenton, A.A. and Muller, R.U. (1996) *J. Neurosci. Meth.*, in press.
15. Muller, R.U., Kubie, J.L. and Ranck, J.B. Jr (1987) *J. Neurosci.*, **7**, 1935.
16. Taube, J.S., Muller, R.U. and Ranck, J.B. Jr (1990) *J. Neurosci.*, **10**, 420.
17. Wilson, M. and McNaughton, B.L. (1993) *Science*, **261**, 1055.
18. Gothard, K.M., Skaggs, W.E., Moore, K.M. and McNaughton, B.L. (1996) *J. Neurosci.*, **16**, 823.
19. Kubie, J.L., Stead, M., Fenton, A.A. and Muller, R.U. (1996) *Soc. Neurosci. Abstr.*, **22**, 1870.
20. Kubie, J.L., Muller, R.U. and Bostock, E. (1990) *J. Neurosci.*, **10**, 1110.
21. Zardosti, M. and Wheeler, B. (1992) in *Proceedings of Computational Neuroscience Symposium* (M.A. Penna, ed.), p. 229.
22. Kreiter, A.K., Aertsen, A.M. and Gerstein, G.L. (1969) *J. Neurosci. Meth.*, **30**, 59.
23. Jansen, R.F. and Ter Maat, A. (1992) *J. Neurosci. Meth.*, **41**, 123.
24. Ohberg, F., Johansson, H., Bergenheim, M., Pedersen, J. and Djupsijobacka, M. (1996) *J. Neurosci. Meth.*, **64**, 181.
25. Gray, C.M. and Singer, W. (1989) *Proc. Natl Acad. Sci. USA*, **86**, 1698.
26. Gray, C.M. and Singer, W. (1995) *Ann. Rev. Neurosci.*, **18**, 555.

INDEX

Index